Human Chromosome Atlas

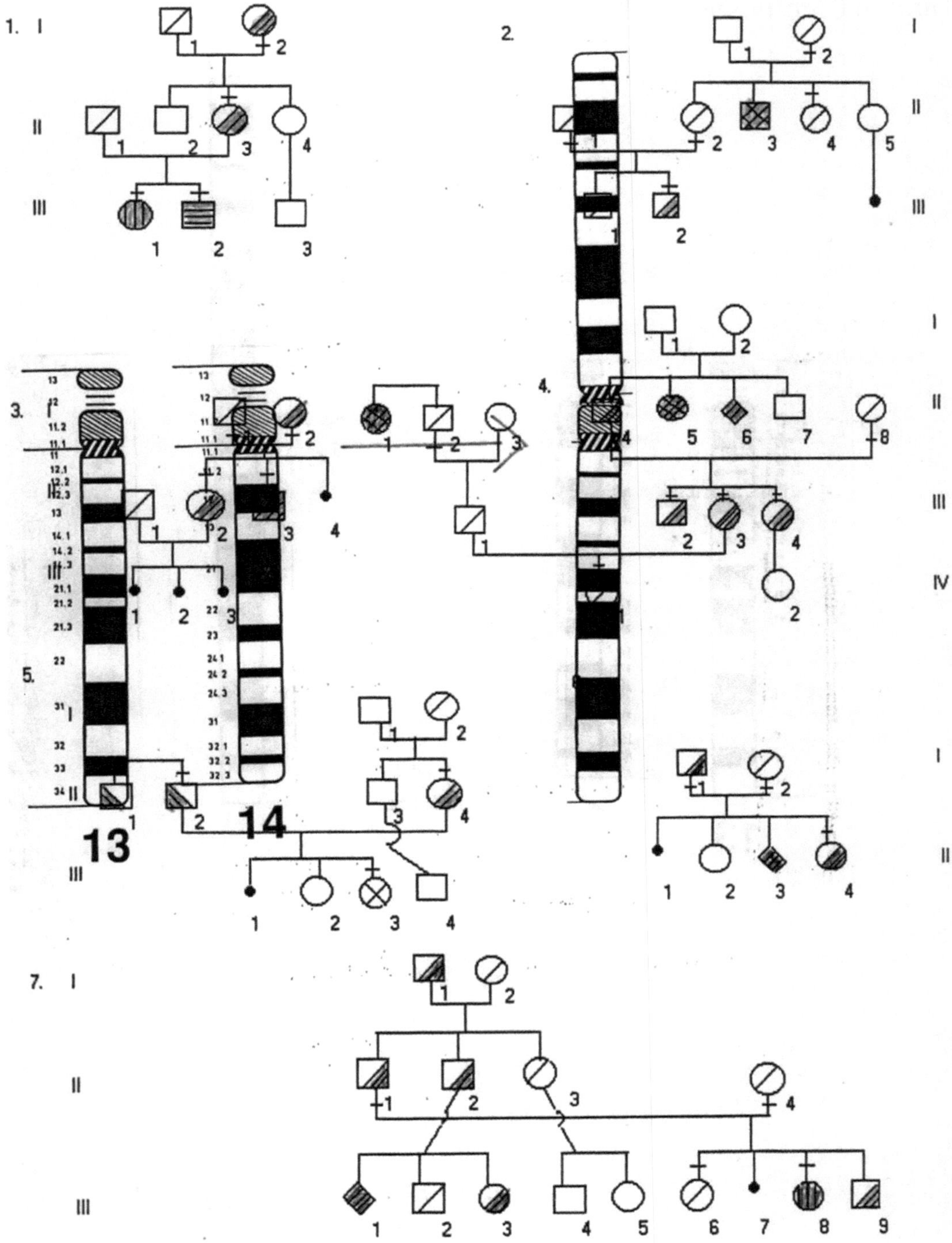

1.
2.
3.
4.
5.
7.
I
II
III
IV
13
14

Claudia Behrend • Javad Karimzad Hagh
Parvin Mehdipour • Heinz Schott
Gesa Schwanitz

Human Chromosome Atlas

Introduction to Diagnostics of Structural Aberrations

Second Edition

Technical Assistance: Ute Schlütter

Springer

Claudia Behrend
Department of Human Genetics
MVZ Düsseldorf Zentrum
Düsseldorf, Germany

Javad Karimzad Hagh
Department of Human Genetics
MVZ Düsseldorf Zentrum
Düsseldorf, Germany

Parvin Mehdipour
Department of Medical Genetics
Tehran University of Medical Sciences
Tehran, Iran

Heinz Schott
Institute of Medical History
Universitätsklinikum Bonn
Bonn, Germany

Gesa Schwanitz
Institute of Human Genetics
Universitätsklinikum Bonn
Bonn, Germany

ISBN 978-3-031-10590-6 ISBN 978-3-031-10588-3 (eBook)
https://doi.org/10.1007/978-3-031-10588-3

This Springer imprint is published by the registered company Springer Nature Switzerland AG
The registered company address is: Gewerbestrasse 11, 6330 Cham, Switzerland

Preface

Cytogenetic is a significant subject of research in the field of Genetics, especially in Human Genetics. Over the past 66 years, it has developed explosively, and during this time has been divided into various research and application areas with the most relevant of them being theoretical Cytogenetics as a basic science, followed by Tumour Cytogenetics, investigations on Mutagenesis, Evolution, and Population Cytogenetics. In addition to and in parallel to these fields of research, cytogeneticists established a close cooperation with related areas of research such as Molecular Genetics, Clinical Genetics, Cell Biology, Biology of Reproduction, Oncology, and Pathology.

Progress in the field has always been in conjunction with progression in investigation methods which generally lead to new focus areas of research and to improved capabilities of diagnosis in the investigation of carriers of chromosome aberrations. From the start, cytogenetics excelled in intensive international cooperation, which included joint decisions in questions of chromosome nomenclature (ISCN), specific symposia, workshops, seminars, and events for continuing education on national and international levels. These activities led to close personal contact between scientists and research groups, as well as to counselling of young scientists thus serving as an impetus for many future-oriented projects.

It is desirable that this inspiring atmosphere will accompany cytogenetic scientists in the future as it did in the past.

The Atlas presented here is composed of three major topics:

First, to introduce general social specifics concerning the acceptance of handicapped persons, second, to initiate those ethic principals which have to be taken into account, and third to present the fundamental principles that guarantee the desirable quality in diagnostics and counselling. This is underlined and supported by the results of genetic counselling in cases of chromosome aberrations, specifics in the course of life of carriers of chromosome syndromes, the development of support groups, and the main pathways in the course of diagnostic investigations.

The presentation of 88 single cases of structural chromosome aberrations is divided into intra-, interchromosomal abnormalities and in mutations in polymorphic regions.

The present atlas of structural chromosomal aberrations is intended to be a diagnostic assistance in applied cytogenetics in order to analyse the human karyotype.

It illustrates the spectrum of chromosome abnormalities through a combination of karyotypes and ideograms, it emphasizes the expressiveness of different banding techniques, and it expresses the karyotype formula and morphological features of each case presented. Furthermore, it reveals potential problems in the detecting of aberrations and lists necessary additional investigations and particularities that have to be taken into account when counselling carriers of a chromosome aberration or their relatives.

It is our intention to present a broad spectrum of chromosome abnormalities. We are not suggesting that we can give a complete collection of all aberration types possible, but we are attempting to give an impression of the spectrum of findings that an institute of applied cytogenetics and clinical genetics can offer.

Most likely, many users of this atlas experienced other additional important cases of aberrations or developed new strategies while investigating problematic cases. We would appreciate them to share their knowledge with us.

To all colleagues working in the field of cytogenetics: Let's aim for a successful future and satisfying research projects. Hopefully, the chromosome atlas presented here will give a helping hand.

Claudia Behrend
Düsseldorf, Germany

Javad Karimzad Hagh
Heinsberg, Germany

Parvin Mehdipour
Tehran, Iran

Heinz Schott
Bonn, Germany

Gesa Schwanitz
Bonn, Germany

Acknowledgement

Work for the second edition of the Chromosome Atlas warranted extensive research due to the added subject topics.

It was supported by two colleagues—Herdit Schueler, Institute of Human Genetics, University Aachen, and Lieselotte Kalz, Institute of Human Genetics Bonn. They were of great assistance with their specialised suggestions in the wording of the new chapters.

We feel obliged to Werner Jung and Ute Schlütter for their assistance of the case presentations.

Tanja Weyandt and Dr. Eliana Acosta—Associate Editors of Biomedicine and Life Sciences, Springer Verlag Heidelberg—were a valued contact persons concerning all the problems that occurred in structuring the second edition of the Atlas, for formal issues and for the desirable progress of the work. Without their help, the process would not have run so smoothly.

Contents

I Introduction

1 **Social Attitude Towards Disabled People: Historical and Cultural Aspects** 3
1.1 Monstrosities: Consequence of Sin Versus Play of Nature 4
1.2 "Cripples": The Power of Imagination, Crippled Mind, and Show Business 6
1.3 Changing Social Attitude 9

2 **The Role of Ethics Committees** 11

3 **Quality Control in the Field of Diagnostics** 17

4 **Types of Chromosome Mutations** 21
4.1 Intrachromosomal Rearrangements 22
4.1.1 Inversions 22
4.1.2 Deletions 23
4.1.3 Duplications 24
4.1.4 Ring Chromosomes 25
4.1.5 Isochromosomes 26
4.2 Interchromosomal Rearrangements 26
4.2.1 Translocations 27
4.2.2 Insertions 30
4.2.3 Complex Chromosome Rearrangements (CCR) 31
4.2.4 Marker Chromosomes (Chromosome Derivates) 32

5 **Heteromorphismes: Mutations in Non-Coding-DNA Regions** 33
5.1 Euchromatic Variants 34
5.2 Satellites 34
5.3 Nucleolus Organising Regions (NOR-Region) 35
5.4 Pericentromeric Heterochromatin 35
5.5 Centromere 36
5.6 Heterochromatic Blocks 36

II Single Case Presentations of Intrachromosomal Rearrangements

6 **Inversions** 41
6.1 Pericentric Inversions 42
6.2 Paracentric Inversions 94

6.3 Summary ... 163

7 **Deletions** ... 165
7.1 Deletions ... 166
7.2 Summary ... 227

8 **Duplications** ... 229
8.1 Duplications ... 230
8.2 Summary ... 240

9 **Ring Chromosomes** ... 241
9.1 Ring Chromosomes ... 242
9.2 Summary ... 251

10 **Isochromosomes** ... 253
10.1 Isochromosomes ... 254
10.2 Summary ... 269

III Single Case Presentations of Interchromosomal Rearrangements

11 **Translocations** ... 273
11.1 Reciprocal Translocations ... 274
11.2 Summary ... 303
11.3 Robertsonian Translocations ... 304
11.4 Summary ... 308

12 **Insertions** ... 309
12.1 Insertions ... 310
12.2 Summary ... 314

13 **Complex Chromosome Rearrangements (CCR)** ... 315
13.1 Complex Chromosome Rearrangements (CCR) ... 316
13.2 Summary ... 326

14 **Marker Chromosomes** ... 327
14.1 Marker Chromosomes ... 328
14.2 Summary ... 338

15 **Heteromorphismes: Mutations in Non-Coding-DNA Regions** ... 339
15.1 Euchromatic Variants ... 340
15.2 Satellites ... 341
15.3 Nucleolus Organising Regions (NOR-Regions) ... 343
15.4 Pericentromeric Heterochromatin ... 344
15.5 Centromeres ... 345
15.6 Heterochromatic Blocks ... 346

IV Guidelines

16 **Genetic Counselling Procedures** 349
16.1 **Overview** 350
16.2 **Case Reports** 352
16.2.1 Carrier of Unbalanced Translocation 16/21 de Novo 352
16.2.2 Familial Translocation Y/22 353
16.2.3 Familial Translocation 4/22 354
16.2.4 Homologeous Translocation 2/2 de Novo 356
16.2.5 Mosaic Translocation Trisomy 21 356

17 **Life Courses** 359
17.1 **Overview** 360
17.2 **Case Reports** 360
17.2.1 Carrier of Interstitial Deletion 9q de Novo 360
17.2.2 Carrier of Unbalanced Translocation 14/21 362
17.2.3 Carrier of a Duplication 3q Derived from a Familial Translocation 363
17.2.4 Carrier of an Interstitial Duplication 3q de Novo 364
17.2.5 Carrier of a Terminal Deletion 18q de Novo 366

18 **Support Groups** 369

19 **Diagnostic Procedure** 375

20 **Outlook** 381

Appendix 386

Bibliography 387

Introduction

Contents

Chapter 1 **Social Attitude Towards Disabled People: Historical and Cultural Aspects** – 3

Chapter 2 **The Role of Ethics Committees** – 11

Chapter 3 **Quality Control in the Field of Diagnostics** – 17

Chapter 4 **Types of Chromosome Mutations** – 21

Chapter 5 **Heteromorphismes: Mutations in Non-Coding-DNA Regions** – 33

Social Attitude Towards Disabled People: Historical and Cultural Aspects

Contents

1.1 Monstrosities: Consequence of Sin Versus Play of Nature – 4

1.2 “Cripples”: The Power of Imagination, Crippled Mind, and Show Business – 6

1.3 Changing Social Attitude – 9

C. Behrend et al., *Human Chromosome Atlas*, https://doi.org/10.1007/978-3-031-10588-3_1

Discrimination and repulsion of disabled people can be observed throughout human history. The tendency to exclude them from "normal" society is a common attitude. Even to kill them may be an option to be realised under certain circumstances. In particular, that has become apparent in Nazi Germany during World War II with its (even then illegal) "euthanasia" programme. It displayed a sophisticated system of murdering ten thousands of patients of mental hospitals in special killing centres on the eve of the holocaust. The motivation for such a criminal atrocity shows characteristic traits of such an extermination strategy. There are three relevant dimensions: (1) The economic argument stresses the financial burdens caused by the disabled seen as unproductive eaters weakening the normal society (this idea became first an ideological argument in the face of the misery of World War I); (2) the biological reluctance assumes that disability caused by hereditary diseases would produce a general degeneration of the "vital race" according to the doctrine of eugenics which was widely accepted in the academic world at the end of the nineteenth century; and (3) for social psychological reasons, the disabled were supposed to be "inferior" and a threat for fellow men because of their typical mind of evil resentment (see below: *Krüppelseele*, crippled mind).

On the other hand, there has been always a benevolent attitude towards the disabled, too. There are impressive examples of support and shelter for those people traditionally common in all cultural areas and religions: asylums, charitable orders, auxiliary institutions, sheltered workshops etc. In the second half of the twentieth century, the idea of medical, social, and economical rehabilitation emerged and became an important issue of social medicine and health policy. The reintegration of the disabled is a social political objective under the slogan "inclusion". So, many public and private institutions target this mission (see below "conclusion"). Nevertheless, there remain a lot of problems for disabled people in their everyday life. They originate often from a more or less unconscious rejection of their otherness according to their appearance, behaviour, and insufficient performance. In this respect, it is useful to consider some cultural historical patterns affecting the respective social attitude towards the disabled. This chapter focusses on the treatment of the humans with congenital deformities and physical and/or mental anomalies, for example, people with Down syndrome.

1.1 Monstrosities: Consequence of Sin Versus Play of Nature

Newborns with congenital deformities were traditionally called "monsters" (from Latin monstrum = warning sign). This concept originated from pre-scientific superstition. But it is often inherent today regarding the horror and defence of "normal" people meeting disabled persons. The emergence of monsters seemed to display future calamities according to the art or science of ancient divination performed by erudite priests and other experts (e.g. augurs in the Roman Empire), along with the interpretation of dreams, astrology, hepatoscopy, or bird flight as methods of prophecy. The clay tablets with cuneiform inscriptions are the earliest written documents showing such a divination belonging to the library of King Assurbanipal (seventh century B.C.) of the Assyrian Empire in Nineveh. There are precise interpretations of malformations of the newborn. For example, "If a woman gives birth to a child who's right ear is small, the house in which it is born will be destroyed". Or: "If the child has no feet, the water streets of the country will be interrupted and the house destroyed".

So, monsters were perceived as signs of impending doom. Even in the renaissance, this attitude flourished again. Although monsters could also refer to lucky circumstances, their meaning as bad omens prevailed among medics in particular. There is a famous example for the interpretation of a monstrosity as a divine sign in a satirical form: "*Der Bapstesel zu Rom*" (The pope donkey of Rome), a flyer edited by Martin Luther and Philipp Melanchthon in 1523.

The idea that the deities punish the humans for committed iniquities with misery and disease originates from antiquity, too. In particular, it should explain the outbreak of an epidemic. Later on, the Christian tradition intensified this theory of sin claiming that the "original sin" would also cause misfits. So, among others, Paracelsus (i.e. Theophrastus Bombast von Hohenheim, 1493/94–1541) interpreted malformations as signs of the evil: "In the same way as hangman marks his children with evil signs, the evil ascendants mark their children with unnatural blasphemous signs, so that one should abstain from such people like from hangmen". Insofar, the physical stigmas also indicated moral deficiencies and were equally signatures of the evil. It was common in early modern medicine to characterise monsters as evidence of punishment for committed fornication, debauchery, and excessiveness. In general, the explanation of diseases as a consequence of sin is essential for medical doctrines of disease throughout modern times. So, mental disorder was deduced from the sinful life of the respective patient still in early nineteenth century by some psychiatrists (shrinks). In a secularised manner, it was (and is) also subversively present in psychosomatics, psychotherapy, and psychoanalysis nowadays in regard to the theory of an "unconscious guilt" producing harmful symptoms. Analogically, also physical medicine provides the theory of bad, i.e. unhealthy behaviour as a sinful one resulting in disorder and disease. In this view, for example, lung cancer appears to be a punishment for chain smoking, a misconduct which is penalised by the abused body. It should be stressed that disease as well as disablement are thought to be causally linked with a previous reprehensible activity.

In the renaissance—within the context of natural magic, astrology, and the belief in witchcraft—the assumption spread that misfits and "cretins" would directly descend from the devil. For Paracelsus and his contemporaries, it was obvious that misfits would be procreated by the conjunction of devils and witches. In the sixteenth century, it was seriously discussed, whether it was allowed or forbidden to kill the monsters. Paracelsus denied killing them. They should remain as creatures of God: "So, I just let them alive like animals". In contrast, "The Hammer of Witches" (*Hexenhammer*), the most important textbook of the inquisition of the Roman Catholic church, approved the extinction of the monsters. In the course of the witch-hunt, even small children were tortured and killed.

A special product of devil's work was the so-called changeling. It was assumed that shortly after the birth the devil or other evil demons would exchange the newborn with a devil child. Such "changelings" (*Wechselbälge*) would show all signs of the evil: extremely voracious, lazy, and misshapen. The most cruel mistreatment of the changeling should enforce the return of the original child. And even its killing was allowed according to Luther, who said in one of his table talks in 1541 that such changelings would only be a piece of flesh (*massa carnis*) because there would be no soul in it which could be done by the devil. So, even young children were treated brutally like presumptive witches and possessed by the devil. Only in the eighteenth century, they were finally seen as medical objects and a challenge for scientific research, for example, changelings were identified sometimes as children suffering from rickets. Now, disability was no longer interpreted as a punishment for sins or negligence of the parents but as a natural disorder or consequence of the "play of nature" (*ludibrium naturae*).

After the French Revolution, for the first time, there was a systematic practical approach to treat the disabled medically. So, lunatics, blind people, or deaf-mutes in particular were cared in special institutions. They were taken seriously as severely disabled who had to be supported according to human dignity. From today's perspective, that guiding principle was contrasted by a brutal reality, for example, in lunatic asylums in the nineteenth and twentieth centuries.

Nevertheless, one can state a general shift from repulsion and exclusion of the disabled to an attempt to adapt them to normal life by physical and psychological treatments.

1.2 "Cripples": The Power of Imagination, Crippled Mind, and Show Business

Besides the religious or demonological explanation of congenital malformations, there was also a sort of psychosomatic theory. The "accidental look" (German: *Versehen*) of a pregnant woman could produce an introjection of an object of the environment having an immediate impact on the foetus inside her body. This seemed to be due to the power of "imagination" (Latin: *imaginatio*), a central concept of early modern medical anthropology. Paracelsus among others used this concept to explain congenital malformations. It has a long tradition in popular culture. The terms "*Hasenscharte*" (unilateral cleft lip and palate) and "*Wolfsrachen*" (bilateral cleft lip and palate) point at this tradition. A woman catching sight of a hare or frightened by a wolf could involuntarily coin her unborn child by respective introjected images. Therefore, it seemed to be dangerous for a pregnant woman to look at cripples or otherwise disabled persons (e.g. lepers) which was supposed to cause eventually monstrosities. In a way, the stigma of the evil seemed to produce itself another stigma of the evil. It seemed to be contagious. So, for centuries, pregnant women were recommended to avoid encountering disabled persons. For example, in 1708 the city council of Nuremberg banned deformed and mutilated people to stay at the marketplaces in order to avoid the disastrous consequences when pregnant woman looked at them. The prevention to be infected by the evil was just to ban the respective figures from the public sphere. At the same time, the imagination doctrine provided an apologetic explanation. Not a sinful behaviour would generate deformation of the child, but the evil influence emanating from crippled beggars, mutilated troopers, or certain animals. So, the doctrine of imagination (bad look of a pregnant woman in particular) did not replace the theory of the disablement as a consequence of sin, but rather implied it. It projected it into an evil impact from outside analogously the "evil eye" producing harm and disorders in somebody else.

Monstrosities like changelings were supposed to exist without a soul, being just a "piece of flesh". But if a soul was attributed to them, what was likely to be the case in milder forms of malformation or disability, the soul of the respective individuals seemed to have also a malformed or disabled soul. One supposed that a physical deformity would hint at a mental one due to the traditional signature theory founding also the popular doctrine of physiognomics: Physical signs of natural things would indicate their hidden inner qualities. So, the hunchback seemed to have a wicked "hunchback soul" (*Buckelseele*). Even in the early twentieth century, this theory was officially discussed in sociomedical literature. The influential German pioneer of the special education of disabled young people (the so-called *Krüppelpädagogik*) Hans Würtz (1875–1958) propagated a theory based on the "individual psychology" of Alfred Adler. In this perspective, the hunchback or other disabled individuals would suffer from an "inferiority complex" and therefore strive to overcome it by a maximum performance or "over-compensation". So, crippled persons seemed to be particularly sophisticated, calculated, and devious, but also witty and sensitive. Of course, such an inferiority theory is refused by modern medical psychology. One does not accept that the disabled should have generally an abnormal mental attitude.

The bad creep, the child-eating monster, the furious giant are well-known figures of popular culture identifying the misshapen body with the respective character. Insofar, the disabled are not only stigmatised, victims of powerful evil influences, but also themselves origins of mental malevolence. Moreover, there is a common assumption that the deformed person in the role of a victim tends to take revenge on normal people. There are novels illustrating this mechanism, for example, "Lady Chatterley's Lover" by D. H. Lawrence (1928). Herein, the paraplegic Sir Clifford is described becoming more and more a wicked cripple. Hunt (1966) called this phenomenon "Chatterley syndrome".

In modern times, cripples played an illustrious role in public. They performed as figures in shows on markets and in circus tents. It is remarkable that just during the Victorian Era, when Darwinism and social Darwinism boomed, the public interest in oddities and abnormalities increased. Some misshaped figures were assumed as a "missing link" between the animal realm and the human world. So, the American entertainer P. T. Barnum (1810–1891) invented the "side show". It was performed in a separate tent. "He included all possible 'missing links': mermaids, lion men, direct descendants of Aztecs, men of a Bornean tribe, etc. Barnum had talent to judge rightly the preference and mental capacity of his auditory. His "missing links" and freaks (snake men, hunger artists, hair and beard men, Siamese twins, giants, and midgets, etc.) attracted then a mass public. (J. Kunze and I. Nippert: Genetik und Kunst, Berlin 1986, p. 13) portraits of persons with genetically caused anomalies from different cultural epochs.

But this curiosity at the end of the nineteenth century originated not only from pure pleasure to demonstrate abnormalities. Rather, it reflected also a state of anxiety because of a general doom and gloom in the face of the suspected damages caused by pathogenic factors of civilisation. A new dimension of disease emerged: A sort of inner deformity and mutilation of humans by social vices and psychological abnormalities resulting in neurasthenia, neurosis, hysteria, psychopathy. Nietzsche called this general increasing weakening of vitality—long before Freud's concept of neurosis—the "modern disease". Now, the social normality itself was taken for pathogenic. Nevertheless, the fearful vision of a general decadence, degeneration, and biological downfall of modern society became predominant in the scientific and public discourse. Especially in Germany, the social disturbances after World War I and its disastrous consequences fostered obviously harmful behaviour with bad consequences: alcoholism, sexual excesses, venereal diseases, subversive political activities were assumed to extinguish the healthy, fit, reproductive-capable and—worthy members of the nation (*Volksgenossen*). This ideology within the context of race biology became then a ruling guideline in Nazi-Germany. Its practical consequences were fatal as mentioned above.

1.3 Changing Social Attitude

In spite of all efforts to educate the public and to integrate the disabled into normal life, the old "superstitions" are more or less still noticeable. The stigma of a severe disablement stimulates subconsciously the impulse to repel the evil. So, it may be useful to know about some cultural historical sources of such an impulse. Although the critique of poor supply of the disabled is often justified, one has to state that the contemporary society (at least in European countries) cares more about the disabled than societies in former times. According to the general accepted guidelines of the official institutions, the attitude towards disabled people is characterised by the idea of equal human rights and humane dealing with diseased or disabled members of society. It is remarkable that the differentiation between physical and mental disablement regarding the stigmatisation of the respective group of persons has—at least theoretically—disappeared. So, severe chronical psychiatric disorders like schizophrenia or deep depression are judged as medical disorders equal to physical ones entitled to best medical care.

On the other hand, the modern culture of the consumer world, the wellness movement, and the idea of enhancement hide the images of disease and disablement and flee from them. This attitude comes to light in regard to prenatal diagnostics looking for malformations or other disorders of the foetus. In the case of a pathological finding, abortion may be an alternative to childbirth—constituting a tough ethical problem for the mother, respectively, the parents apart from other players. This problem is worsened by the novel method of Non-Invasive Prenatal Diagnosis (NIPD), a genetic test screening for birth defects and hereditary diseases. It can be easily carried out using a simple blood sample of the pregnant mother. So, it is a much more convenient method to diagnose, for example, growing children with Down syndrome than it was before, when invasive and not quite undangerous procedures were necessary. There is a certain tendency to exclude the birth of disabled children. At the same time, there are many initiatives and institutions, public and private, which care for them with highly sophisticated medical and social means. This has to be seen within the context of the emerging concept of medical as well social rehabilitation after World War II. The leading idea was the integration of disabled and handicapped people into ordinary life. They should not be excluded from the working and cultural world and make their own lives as far as possible.

The contemporary concept of inclusion originates from this background of rehabilitation. Today, it is a global goal: The "Convention on the Rights of Persons with Disabilities" (CRPD), an international human rights treaty of the United Nations, became effective in 2008 intending to protect the rights and dignity of persons with disabilities. This created officially quite a new social attitude towards disabled persons. Never before more attention has been drawn to them. But constitution and reality are different things as elsewhere in society. However, the social attitude towards the disabled has become more sensible endeavouring a humane and productive cooperation with them. Progressive approaches should be mentioned like the project "Ohrenkuss ... da rein, da raus" (i.e. ear kiss … here in, there out), a cultural magazine produced by persons with and without Down syndrome. It is published twice a year in Bonn (Germany) since 1998 by the Downtown-Werkstatt für Kultur und Wissenschaft (i.e. Downtown studio for culture and science). This project shows paradigmatically how deeply the social attitude towards disabled persons can be modified and encourage a creative interaction between disabled and non-disabled people.

The Role of Ethics Committees

C. Behrend et al., *Human Chromosome Atlas*, https://doi.org/10.1007/978-3-031-10588-3_2

The *German ethics commission* was founded in the year of 2013, in general accordance with the rules of the European commission. It was established by the government in order to monitor methods of genetic investigations in reference to existing laws. It was preceded by the establishment of the Gene-Diagnostics-Commission in the year of 2010, which set up the guidelines for genetic investigations on humans, relying on the current status of approved methods.

Genetic diagnostics, screening, analyses, pre-implantation diagnostics, germ cell selection, heterological egg, and semen donation are examples for current problem areas that have a high significance in patient-oriented diagnostics due to rapidly evolving capabilities.

In recent decades, genetics experienced crucial impulses in basic and applied research with the help of these new techniques.

Subsequently, debates occurred in science and in the public about potential abuse of those investigation methods or the manipulation of findings.

The hereby given challenges led to the establishment of a counsel for bioethics by the department of education and research, including specialists in life sciences, medicine, ethics, law, social studies, theology, and politics.

The new developments in life sciences were documented and their potential application in medicine apprehended, with the understanding that focus areas might change quickly or had to be extended.

In analysing ongoing applications of new discoveries though, it is not always possible to assess potentially adverse side-effects.

Furthermore, the imposed rules can only apply for Germany, since in other European countries as well as globally, moral principles and cultural specifications can make a difference, to some extent.

In addition to the German Ethics Counsel in Berlin, universities founded ethics commissions and academic chairs of their own. Beyond the universities, ethics commissions and similar organisations were established by provincial governments, churches, political parties, and unions. These groups thus launched the objective to expend internationally—primarily in Europe, in order to define and document universal moral principles. Within this framework, joint centres for documentation and communication were established.

Numerous journals with the subject of bioethics were launched and publications such as the dictionary for bioethics were the bases for decisions with areas of application, with reference to specialties such as genetic engineering, neurology, tumour diagnostics and therapy, stem cell research and therapy, and prenatal diagnostics.

An internationally mandatory view to bioethics is represented in the declaration of the Helsinki World Medical Association. Furthermore, the Council of Europe and UNESCO represent the basics for ethnic issues in medicine.

To sum up, at present the topics in deciding bioethic questions of eminent actuality are as follows:

- Genetic investigations of children or adolescents without the ability of consent
- Humans with dementia
- Diagnostics and therapy in disorders of sex development (e.g. intersexuality)
- Fertility preservation with mental retardation
- Carrier testing
- Obligation to inform within a family (see ▶ Sect. 17.2.1)
- Decisions to immediate consequences of preventive tests
- Evaluative response to the various capabilities of prenatal diagnostics

Ethical principles have always been a relevant issue in medicine and have defined the conduct of those affected. In doing so, the evaluation of the applied techniques as well as their consideration for a treatment approach play a part.

Nowadays, medicine is divided into basic research and patient-oriented domains. Both areas intersect frequently causing the allegations of ethical principles to be complicated and to be debated controversially in many cases.

Legislative power provides the foundation for ethical conduct. Legislation is responsible for the safety of life and the prevention of abuse. It is based on the principles of self-responsibility, on the verified benefits of a therapy for the patient, on the protection of personality, and the preservation of justice.

The rules of bioethics are of trans-cultural importance. Culture-specifically though, they might be interpreted differently.

Leading ethicists like Professor Kueng of the Tuebingen University have played a substantive part in contributing with their publications to the overlapping emergence of principles of ethical conduct in interreligious and intercultural dialogue since 1993.

These statements were of extraordinary importance in the field of ethics in medicine since everyone may be affected by medical-ethic decisions as a proband participating in a research project or as a patient in a potential therapy.

As a consequence, the publications about fundamental ethics in medicine prompted a range of meetings and led to a response of the "Deutscher Ethik Rat", among others.

The various approaches resulted in the "Declaration Toward a Global Ethic" in the year 1993 in Chicago and 2 years later in the foundation "Weltethos", founded by Professor Kueng. This project is aimed at raising awareness of basic joint values in all countries.

Information to ethical principles and their competent supervisory authorities should not be limited to German or even European data. For comparison, the applicable ethical principles of a different cultural sphere are being depicted.

Since two of the authors of the Chromosome Atlas are Iranians, the valid ethical guidelines for the assessment and approval of genetic investigations in humans for this country will be documented.

To some extent, they are based on moral values that are deeply affected by different religious and cultural principles, at variance with the conditions in Germany and Europe.

Academic exchange and social contacts of participating researchers at joint scientific projects lead to regular information about the respective current requirements for the cooperation between both countries.

The *Iran ethics commission* has a long tradition of medical legislation regarding the protection of patients.

It was recorded in writing as early as in pre-Islamic times already, among others in publications of the Sassanids and the Zoroastrians. Following the establishment of the Islam as the official religion, bioethic principles were an important issue in medicine.

Based on this tradition, already in the years 1953 and 1963, respectively, the Tehran university issued publications concerning the significance of ethical standards. Additional publications by the "Ministry of Health and Medical Education" followed in the years 1990 and 1991. In the year 1993, the "Center for Medical Ethic Studies and Research" was established, and the "First International Congress of Medical Ethics" was held in Tehran in that same year. Meetings and workshops followed on a regular basis.

Consequently, the "International Congress of Bioethics" was conducted in collaboration with UNESCO in Tehran. The organisers were the "Ministry of Science Research and Technology" and the "National Research Center of Genetic Engineering and Biotechnology" of Iran.

This congress was used as a forum for interchange of ideas and information between Iranian and international researchers, the second congress followed in the year 2008 already.

Main topics for discussion were:

- Insufficient continuing training of licensed physicians concerning ethical standards in the care of patients
- The lack of specialists for medical ethics
- The need for guidelines to culturally adapt bioethics while training medical students
- The lack of popular scientific publications to inform the population

Subsequently, the Iran established a library for publications in the field of bioethics and published a journal on the subject (Iranian Journal of Medical Ethics and History of Medicine).

Legislation and standards were drawn up to blend Islamic and ethical principles in a society defined by its religion.

In the area of human genetics, the key aspects were as follows:

- Therapeutical abortion
- Egg cell and semen donation for infertile couples
- Assisted reproductive technologies
- Pre-implantation diagnostics
- Cytogenetic diagnostics
- Investigations of gametes and embryos
- Examination of patients unable to consent

Nowadays, the "Medical Ethics Research Center" takes responsibility for nationwide discussions on the university level. In this context, the "National Committee" has to define the Islamic, legal, and ethical principles for biomedical research in order to:

- ensure the protection of test subjects
- appoint monitors for university and private institutes as well as for industrial projectsThe placement by the "National Committee" integrated the guidelines of the "Declaration of Helsinki", and the view of the "Head of the National Research Council".

Today, the standards for each physician include—with due regard to religious regulations and specific cultural principles—an "informed consent", which is a statement in writing, for the patient or proband. It insures accountability of the project leader for any negative consequences of the study or therapy.

The composition of the ethical committee in the Iran presents a broad range of professionals: physicians, statisticians, lawyers, ethnic specialists, workers' representatives, and the spiritual leaders.

The forthcoming analysis of the situation of ethical committees was supported by the WHO.

The comparative account of the development of medical ethics in Germany and the Iran emphasises the relevance of international cooperation in this field and, at the same time, may give suggestions for future cooperation.

Quality Control in the Field of Diagnostics

C. Behrend et al., *Human Chromosome Atlas*, https://doi.org/10.1007/978-3-031-10588-3_3

Human genetics is a special field including basic research as well as patient-related studies that are frequently strongly related, thus stimulating each other. Results in one key aspect may accentuate a new focus in the other. Therefore, quality control of results is crucial.

In the following, because of the topic given, only patient-related areas of human cytogenetics will be referred to.

The society for human genetics and the professional organisation of German human geneticists have established mandatory guidelines in order to guarantee the quality of information in the area of applied human genetics.

Before achieving an occupation in the area of human genetics, an advanced qualification process has to be undergone, lasting 5 years, provided, a patient-related position is being the objective.

In general, the trainees have a degree in medicine or biology. They will then work on gaining the qualification of a specialist in human genetics.

In the course of this education, knowledge in clinical, cytogenetic, biochemical, and molecular genetic diagnostics of genetically related disorders will be conveyed, plus, in addition, the counselling of persons affected and their families.

They will be introduced into the basic principles of genetic processes, the formation, and heredity of mutations, plus the procedures of epidemiological surveys.

Of particular relevance, in the course of the advanced training, is the transfer of fundamental data concerning ethical, psychological, and legal knowledge in diagnostics and counselling.

The emphasis can lie either on molecular genetics or on cytogenetics.

The topics within cytogenetics include the complete study of all methods in the area of pre- and postnatal cell cultures, the required steps for preparation, and the subsequent techniques of cytogenetic, molecular cytogenetic, array-based and sequencing-based nature.

A defined amount of independent investigations, including diagnostic results, is required, covering all the areas above.

Further education includes regular participation in symposia as well as local seminars and workshops.

The continuing education may occur at university or non-university institutions with accreditation. Instrumental for this purpose is a specific catalogue listing the goals.

Alongside this further education in Germany, a European centre for accreditation was installed. Within the context of an exam, the acquired knowledge is being tested, thus establishing a professional standard within all of Europe.

Having completed this continuing training, the human geneticist has several options for an additional qualification:

The German "Akademie Humangenetik" regularly offers seminars with topics like quality management or psychosocial aspects of human genetics, but also to special subjects like "Next Generation Sequencing Diagnostics" or "Variant Interpretation".

The "Deutsche Gesellschaft für Humangenetik" and the "Berufsverband deutscher Humangenetiker" are representing an association of German human geneticists.

For more than 20 years, the so-called Ringversuche are being performed for the purpose of external quality control, thus supporting the quality of the labs' results.

In departments with focus on investigations in cytogenetics or molecular genetics, the laboratory head will pursue the accreditation of his institution. For this purpose, documentation has to be submitted to a professional commission in order to prove competence. The panel will then perform an inspection on location and possibly recommend changes in the work implementation. After fulfilment of the necessary requirements, the facility will be accredited. This on-location inspection will be repeated at regular intervals.

This documented evidence of quality as the result of the accreditation normally will be included in their logo.

Following are the official terms and concepts for the various steps of the appropriate quality control in Germany:

- *Weiterbildung (Diploma in Medical Genetics).*
- *Weiterbildungsordnung der Ärztekammer.*
- *Berufsverband deutscher Humangenetiker (*► www.bvdh.de*).*
- *Akademie Humangenetik (*► www.akademie-humangenetik.de; info@akademie-humangenetik.de*).*
- *Leitlinien (*► https://www.awmf.org/leitlinien.html*).*
- *Ringversuche: Richtlinie der Bundesärztekammer zur Qualitätssicherung laboratoriumsmedizinischer Untersuchungen (RiLi – BÄK).*
- *Akkreditierung:* ► https://www.dakks.de/content/was-ist-akkeditierung

► Chapter 2 presented a comparative portrayal of the ethical principles that have to be considered regarding genetic diagnostics in humans in *Germany* as well as in *Iran*.

In like manner, the following account will depict education and training for human genetics and quality assurance for cytogenetics in Germany versus Iran.

In Iran as well as in Germany, human genetics is a specialised field that gained very much in importance over the last decades and thus became an integral part of studies in medicine and life science. In each case, there is a lecture covering the basics. In addition, both also offer seminars and workshops. For cytogenetics, these include cell cultures, the microscoping of specimens with a normal or pathological karyotype as well as the practice of the various staining techniques and hybridisations.

Training for the bachelor, masters, and doctoral degrees are being provided.

Frequently, in Iran the training for a doctoral degree involves a stay abroad in order to study and apply newly established investigation methods. This is usually accomplished in cooperation of the doctoral advisor with a foreign colleague and is being sponsored by a scholarship.

Types of Chromosome Mutations

Contents

4.1 Intrachromosomal Rearrangements – 22
4.1.1 Inversions – 22
4.1.2 Deletions – 23
4.1.3 Duplications – 24
4.1.4 Ring Chromosomes – 25
4.1.5 Isochromosomes – 26

4.2 Interchromosomal Rearrangements – 26
4.2.1 Translocations – 27
4.2.2 Insertions – 30
4.2.3 Complex Chromosome Rearrangements (CCR) – 31
4.2.4 Marker Chromosomes (Chromosome Derivates) – 32

C. Behrend et al., *Human Chromosome Atlas*, https://doi.org/10.1007/978-3-031-10588-3_4

Structural chromosome aberrations or chromosome mutations represent apart from aneuploidies the most frequent pathologic findings in applied chromosome diagnostics. But they also occur as combinations of both types as in unstable rearrangements and in chaotic karyotypes of early stages of embryonic development.

Depending on the indication for chromosome analysis—child with developmental retardation, spontaneous abortions, sterility—the majority of aberrant karyotypes will be unbalanced or balanced. Furthermore, it could be shown that the frequency of de novo and familial aberrations is depending on the type of the aberration. Chromosome mutations are divided into intrachromosomal and interchromosomal aberrations. In addition, complex chromosome rearrangements (CCR) represent a special group of abnormalities, where intrachromosomal and interchromosomal aberrations can be combined in one genome. Their investigation affords the combination of chromosome analyses, microarray and FISH, and their characterisation is still under development.

In the following intrachromosomal aberrations are shown first, followed by the interchromosomal aberrations.

4.1 Intrachromosomal Rearrangements

The following types of aberrations are included in this group:

Deletions, duplications, inversions (para- and pericentric), ring chromosomes, isochromosomes.

4.1.1 Inversions

This chromosome atlas is based on pathologic chromosome findings, collected at the institute in Düsseldorf. As the main focus of this institute is on the investigation of infertile couples before ICSI therapy, the majority of pathologic findings are balanced rearrangements in the group of inversions (60%). As they extremely well represent the spectrum of intrachromosomal rearrangements, the methodological requirements, and the order of genetic counselling of the carrier of an inversion, this group will be represented in detail in the following chapter. Particularly, it will be pointed out, in which cases molecular-cytogenetic and molecular genetic investigations (FISH), microarray are additionally required. However, it should be understood that, if a balanced chromosome rearrangement has to be expected in the patient, chromosome analyses remain the most relevant investigation.

4.1.1.1 Pericentric Inversions

This type of aberration is characterised by the position of the centromere within the inverted region. If the two breakpoints in the short (p) and long (q) arm show different distances to the centromere, the centromere index will be changed. The recombination frequency in meiosis will depend on the chromosome involved and the length of the inverted region. Very small inversions can be inherited over a number of generations exclusively in the balanced condition. Caused by an abnormal course of meiosis carrier of balanced inversions have an increased risk of the development of uniparental disomy (UPD) in the offspring. Inversions of the X-chromosome often lead, in contrast to inversions of the autosomes, to the development of gene-function irregularities.

4.1.1.2 Paracentric Inversions

This type of inversion is characterised by the position of both breakpoints in the same chromosome arm (p or q), the centromere index has not changed, and the diagnosis is only possible by the altered banding pattern in the inverted region. A high structure resolution of the chromosomes is required (at least 550 bands per genome) for a reliable diagnosis. That is why paracentric inversions were only rarely diagnosed in the first decades of cytogenetic investigations. Nowadays, it is well known that paracentric inversions represent a frequent type of aberrations.

When analysing this type of abnormalities, it always has to be kept in mind that they must be differentiated from insertions by FISH, as the latter group is characterised by a high genetic risk whereas that of paracentric inversions can usually be considered as low (about 1‰).

Paracentric inversions of the X-chromosome are rare aberrations, special breakpoints in the long arm can lead to gonadal dysfunction.

4.1.2 Deletions

Deletions are also defined as partial monosomy. In a wider sense, ring chromosomes can be attributed to this group. But in the presentation given here, they were described in a chapter of their own as their morphologic peculiarities and the often-observed secondary changes make it practical. The size of the deleted regions shows high variability from a length of single base pairs (detectable by molecular genetic methods), several kb up to 5 Mb (analysable by specific FISH probes in cases suspicious for micro-deletion, or by molecular techniques) and finally deletions of 5 Mb to more than 20 Mb, diagnosed by chromosome analyses. Only the latter group will be described here in detail.

This type of aberration is further subdivided in terminal deletions, which are caused by only one chromosome break and interstitial deletions which are caused by two breaks, followed by the reunion of the breakpoints.

Deletions occur in short (p) as well as in the long (q) arm of a chromosome and they have been observed for all human chromosomes. Large deletions are rare. The interchromosomal distribution of deletions shows differential frequencies, of which the most frequent ones represent well-defined chromosome syndromes (i.e. 4p-, 5p-, 18p-, 18q-). Deletions have already been diagnosed and described in detail in the prebanding area (i.e. Cri-du-chat syndrome, de Grouchy syndrome I and II). Recent investigations have shown that terminal deletions usually have a telomere, which stabilises the end of the chromosome. Terminal deletions are about seven times as frequent as interstitial ones. Only part of deletions has originated in the carrier de novo, others are the unbalanced inherited form of a parental translocation. Regarding the deletions 4p and 5p the percentage of inherited aberrations amounts to 8–15%. Therefore, when a deletion is analysed in a patient this always leads to the indication for chromosome investigation in the parents. In the detection of translocations by additional analyses, it is notable that the second chromosome is not randomly distributed in the genome, but there are preferred combinations.

More than 80% of deletions have been originated de novo in the carrier, the majority of them in the paternal meiosis. In single cases, if the retardation of the carrier is only marginal, the deletion can be inherited to an offspring.

4.1.3 Duplications

Chromosome duplications are also defined as partial trisomies. The duplicated segments originate from replication errors within one chromatid, by an unequal exchange between sister chromatids or by translocation between homologue chromosomes. Often duplications are combined with deletions. The phenotypic changes of a patient with segmental duplication are usually less serious than a deletion of the same region. In some cases, duplications and deletions of same chromosome region are characterised as "type" and "counter-type" (i.e. 4p, 9p, 22q11.2).

Duplications are subdivided in tandem arrangements of the aberrant region and in inversion duplications. In single cases, even triplications of chromosome segments have been described, and in these patients the gene-dosis-effect was significantly elevated.

Depending on the mode of development, the origin of duplications can be uniparental or biparental. They can be the result of a meiotic or postzygotic error; in the latter case, the carrier will show the aberration as a mosaic. Special forms of duplication are the supernumerary marker chromosomes and isochromosomes. The great majority of duplications has originated de novo in the carrier. When estimating the recurrence risk in the course of genetic counselling, the possibility of a gonadal mosaic in one of the parents has to be taken into account. Single cases of familial duplications have been reported, but in these cases the carrier parent showed only minor phenotypic changes. Microarray investigations during the last years lead to a significant increase of well-defined microduplication syndromes.

4.1.4 Ring Chromosomes

Three different types of ring formation have been observed:
1. The fusion of two breakpoints in one or two euchromatic regions and loss of the acentric fragments in the short and/or the long arm (p, q).
2. The fusion of the two telomeres without loss of euchromatic segments. But it leads to mitotic abnormalities and frequent losses of the ring.
3. A fission in the centromere is followed by a U-type exchange.

The frequency of ring chromosomes is estimated to range from 1:28 to 1:62,000, with chromosomes 13 and 18 being most frequently involved. Familial cases are, as expected, extremely rare (about 1%) and they are only observed in ring chromosomes 21 and 22. The majority of rings are of paternal origin. The phenotype of the patients is very variable caused by the different breakpoints leading to a different size and thus gene content of the deleted regions. Caused by irregularities in the S-phase of the cell cycle partial duplications can originate.

Ring chromosomes are often difficult to analyse:
1. Very small rings are not easy to distinguish from derivative chromosomes.
2. If the ring is balanced or the deleted region is only submicroscopic in size (less than 5 Mb), the indication for chromosome investigation may not be given.
3. If the ring is restricted to one germ layer or even one tissue, the cell system analysed may not contain the ring.
4. In cases of low-grade mosaicism, it can be difficult to find mitoses with the ring chromosome.

Often a karyotype with a ring presents as "dynamic mosaic". In these cases, especially if the ring has two centromeres, it is unstable in the course of the cell cycle and new cell lines develop continuously. Therefore, after prenatal diagnosis of an unstable ring it will be impossible to predict the phenotype of the child. Ring chromosomes are observed in cases with a normal chromosome number and as additional structures. The microscopic analysis of rings is not always easy, as they have a three-dimensional structure and therefore they appear different in view depending which angle you look it from.

In the investigation group represented here, patients with ring chromosomes were diagnosed only in a few cases as the majority of them was analysed because of fertility problems and was phenotypically without peculiarities (investigations before ICSI-therapy).

4.1.5 Isochromosomes

After a transverse separation of a centromere, i.e. chromosome fission, both arms (p and q) can be duplicated. The result will be two isochromosomes, one of the short and one of the long arm. The karyotype of the proband can be balanced, in case the second normal homologue chromosome is lost by the so-called aneuploidy rescue. Carrier of this particular rearrangement cannot have healthy offsprings.

The more frequent type of isochromosome formation starts by an unequal centromere fission resulting in one arm containing the whole centromere and the second being acentric. The acentric arm gets lost and the result is the development of an isochromosome in a karyotype with normal chromosome number but deletion of one and duplication of the other chromosome arm. A frequent example of this type of aberration is the isochromosome Xq. Besides this type of aberration isochromosomes can also be observed as additional derivates. In these cases, the karyotype is tetrasomic for the aberrant chromosome arm, as in isochromosome 12p, the Pallister-Killian syndrome. Furthermore, by inversion duplication a dicentric isochromosome can develop, which is often unstable in the course of mitoses caused by an incomplete inactivation of the second centromere which in the course of the somatic cell divisions frequently gets lost and leads to monosomy of the chromosome involved. This type of secondary aberration is more often observed in gonosomes than in autosomes, leading to mosaic formation with a X0 cell line. These observations lead to the conclusion, that in all cases of dicentric isochromosomes the possibility of a mosaic formation has to be considered and thus the number of mitoses analysed has to be at least 30–50.

The majority of isochromosomes develops during meiosis. But they can also result from postzygotic cell divisions and then present as mosaic. If such a mosaic is restricted to the gonads of a healthy proband, the aberration can appear as an aberration in all cells in the offspring.

4.2 Interchromosomal Rearrangements

Interchromosomal caused changes of chromosome structure are a heterogenous group of aberrations. The exchanges of segments afford at least two breakpoints, but it can be also up to seven. The size of the rearranged chromosome regions differs between 5 and more than 50 Mb. Interchromosomal rearrangements can be caused de novo in the carrier or they are of familial origin. In one individuum, different types of aberration can be combined. The interchromosomal aberrations are categorised in four main groups of which the translocations are further subdivided in two subtypes. The main groups are translocations with the two subgroups reciprocal and Robertsonian translocations, complex chromosome rearrangements (CCR), insertions, and part of marker chromosomes.

4.2.1 Translocations

The majority of translocations concern rearrangements between heterologous chromosomes, only a few occur between homologous, and they represent a very variform type of aberrations. Most of the cases of heterologous exchanges take place between two chromosomes. The breakpoints can be localised both in heterochromatic regions, both in the euchromatin or in the combination of hetero- and euchromatic segments. In cases, when one of the translocation products is significantly reduced in size in comparison to the original chromosome and therefore has left only small regions in common with that one, the course of meiosis is often pathologic and leads to mistakes in its distribution and as a consequence to an increased genetic risk.

The majority of translocation chromosomes are stable in their structure, but instable rearrangements do exist, and they are defined as "jumping translocations". Translocations present with a varying genetic risk of the carrier, reaching from less than 1% up to 100% in cases of homologous translocations. At the same time, the group of carriers of translocations of the homologs gives us the possibility to define the frequency of de novo mutations in this type of aberrations.

4.2.1.1 Reciprocal Translocations

Exchanges of segments between chromosomes are a frequent type of structural chromosome aberration, and they are among others characterised by specific "hot spots" of the breakpoints. Additional investigations by FISH methods and microarray increase the amount of reciprocal translocations significantly and the localisation of the breakpoints becomes more precise. The distribution pattern of the aberrations reveals significant differences when the investigation group consists of probands from the general population or by selection of carriers with phenotypic abnormalities.

A special group of aberrations are exchanges between an autosome and a gonosome, as in these cases additional risk factors exist in carriers of the balanced rearrangement caused by pairing abnormalities in meiosis and an abnormal X-inactivation pattern in females.

More than 95% of reciprocal translocations occur between two heterologous chromosomes. The genetic risk of a carrier of the balanced translocation depends on the size and the gene content of the segments exchanged. In the majority of cases, the unbalanced karyotype reveals the combination of a partial duplication and deletion. Additionally, unbalances caused by an unequal crossing over in the breakpoint regions during meiosis are observed. Furthermore, there exists a risk factor caused by malsegregation of the smaller translocation chromosome during meiosis. A main risk factor stated in the course of genetic counselling of a carrier and/or his relatives is the probability for birth of a handicapped child, but additionally the increased risk of spontaneous abortion and still birth as well-reduced fertility of the translocation carrier have to be taken into account. Depending on the type of translocation, these three risk factors can be reduced to one, either if all forms of unbalanced karyotype are lethal (i.e. whole arm translocations), then only the fertility is reduced in the carrier. The second possibility is realised when the unbalanced karyotype permits only the first steps of ontogenetic development, then the conceptus dies off and leads to a spontaneous abortion, but handicapped children are never born.

When the carrier of an unbalanced translocation is viable, usually also an increased risk of spontaneous abortion is given for the balanced parent and besides his fertility is reduced.

The risk for the birth of children with unbalanced karyotype and for an increased rate of abortions is statistically double in females compared to men (20:10%). This is caused by a strong selection against unbalanced karyotypes in spermatogenesis. Depending on the different types of translocation, it varies from less than 1% to almost 50%. Therefore, it is most important in cases of familial translocations to perform additional investigations in the relatives, to get more precise information on the specific translocation in the family coming for genetic counselling as to the number of healthy children, pathologic pregnancies, and the inability to have children of the further translocation carriers in the family.

Epidemiologic investigations showed that reciprocal translocations have a frequency of about 1:500 in the general population. They occur de novo and familial, and in the latter group it has been possible to trace back single translocations over a period of more than 300 years.

4.2.1.2 Robertsonian Translocations or Centric Fusions

Robertsonian translocations are also classified as whole arm translocations. This group comprises the most frequent structural chromosome aberrations with an incidence of 1 in 1000 newborns. Involved in this type of rearrangement are the acrocentric chromosomes, i.e. 13, 14, 15, 21, 22. The most frequently observed breakpoint is p11.2. The chromosome fusion takes place between homologous and heterologous chromosomes.

The original assumption that fusions of the nucleoli or tight associations of NOR-regions play a role in the origin of Robertsonian translocations could not be confirmed by more extended investigations. In heterologous fusions, the two translocation chromosomes are usually derived from one parent, and about 90% of them develop during maternal meiosis I. The breakpoints in the combinations 13/14 and 14/21 are practically always the same caused by inverted homologous sequences in p11.2. All other combinations show a significant variation as to their breakpoints. In meiosis I, trivalents are formed by pairing of two normal homologous and the translocation chromosome. By an abnormal course during the further phases of meiosis up to seven different types of gametes can originate. The two most frequent heterologous fusions are 13/14 and 14/21 with 30% each. All fusions of homologous chromosomes occur in a frequency between 0.5 and 3%. Zygotic and postzygotic fusions seem to be rare events. They can be delineated when biparental in origin.

The frequency of fertilisation is not reduced in Robertsonian translocations, but a selection against conceptus with unbalanced karyotype occurs in the postzygotic period.

Uniparental disomy (UPD) can develop by trisomy or monosomy rescue, but its frequency is estimated to be less than 0.5%. If chromosomes 14 or 15 are involved, UPD leads to characteristic malformation syndromes, such as Prader Willi or Angelman syndrome. Male carriers of Robertsonian translocations have a risk of infertility which is about seven-fold increased compared to the general population. The karyotype of the conceptus after PID was pathologic in 90% of the cases and even chaotic in 60%. One cause of these abnormalities is suspected in an abnormal course of meiosis I: the short arms (p) of the acrocentrics remain unpaired during the formation of a triradial and in the next phase they pair with the X/Y bivalent which then leads to a fragmentation of the nucleus in the following late prophase of meiosis I. But this pathologic course is not obligatory and a large intrafamilial variation of this effect has been observed.

The carrier of a balanced Robertsonian translocation usually has 45 chromosomes because the heterochromatic short arms without a centromere get lost, while the carrier of the unbalanced karyotype has 46 chromosomes in the majority of cases, in rare cases it can be 47 chromosomes. If the balanced carrier is offspring of two related parents, each of them carrying the same chromosome fusion, he can have the translocation twice and his karyotype comprises 44 chromosomes. In cases when both products of a Robertsonian translocation receive a centromere, the heterochromatic short arm regions occur supernumerary in the karyotype. These heterochromatic derivates of the short arms can insert euchromatic segments by unequal crossing over in the meiosis I and are then inherited to the offspring as chromosome aberrations leading to phenotypic abnormalities in the carrier.

4.2.2 Insertions

Insertions are defined as complex chromosome rearrangements, which afford three breaks for their formation, two of them to cut the segment from the donor and one for insertion in the recipient chromosome. The frequency of insertions in life born children is 1:5000. Regarding the cases of *intrachromosomal* rearrangement, the dislocation of the cut-out segment occurs within the donor chromosome. In cases of paracentric insertion, the centromeric index remains unchanged, but in pericentric insertions the segment is transferred to the other chromosome arm and then the centromere index changes.

After *interchromosomal* insertion, the centromere index of donor and recipient changes. The final position of the fragment compared to its alignment in the donor chromosome and compared to the localisation of the centromere will be either direct or inverted (180°). The inserted segments are either built from euchromatin or heterochromatin and in the latter case a special group exists only of the NOR-region. The size of the inserted regions reaches from 5 to more than 10 Mb in cases which are analysable by microscope, but the majority of them are cryptic, i.e. less than 5 Mb and they often contain transposons.

During the meiosis, when the homologue chromosomes are pairing, an intrachromosomal, insertion often forms loops which can lead unequal recombination products with partial deletions and duplications. The theoretical risk for the formation of an unbalanced karyotype is about 50%, but depending on the size and the gene content of the segment the risk for the birth of a handicapped child is small. When the insertion segments are large, they can form double loops and in the unbalanced karyotype of the offspring deletions can be combined with duplications. Interchromosomal insertions often develop during meiosis and give rise to four different types of gametes, two of them showing an unbalanced karyotype.

A special feature during the development of insertions is the formation of fragile sites in the chromosome, which lead to an increased tumour risk if the region contains tumour suppressor genes.

Insertions reveal an unequal intra- and interchromosomal frequency as to their distribution in the karyotype, frequent loci being 2q, 5q, 7q, and 13q. The risk for the birth of a child with unbalanced karyotype, non-regarding size, and gene content if one of the parents is balanced carrier amounts to 36% for the mother and 32% for the father. This risk amounts to 50% in certain types of insertion and thus it represents the highest genetic risk of chromosome rearrangements, apart from fusions of the homologous chromosomes.

Recent investigations lead to the observation that insertions can cause imprinting diseases.

4.2.3 Complex Chromosome Rearrangements (CCR)

Complex chromosome rearrangements can be restricted to only one chromosome, but also three and more chromosomes can be involved. They are characterised by more than two breakpoints in one or more chromosomes. The types of rearrangement which lead to a CCR are translocations as the main group, besides inversions and insertions, and they can be combined with deletions and duplication when the karyotype is unbalanced.

A special type of CCRs is the combination of different independent rearrangements in one karyotype, as for example a reciprocal and a Robertsonian translocation or a translocation and an inversion.

CCRs can occur as balanced rearrangements. But the majority of them, when they are de novo aberrations in the carrier, are unbalanced and cause phenotypic abnormalities in the carrier such as mental retardation, which is diagnosed in 10–15% of them. When the CCR appears balanced additional molecular genetic investigations often reveal the presence of cryptic aberrations or gene disruptions. When estimating the genetic risk of the carrier presenting a CCR, it is necessary to take into account the chromosomes involved and the number and localisation of the breakpoints, as these factors influence the course of meiosis by the formation of different types of multi-valents and therefore the frequency of unbalanced gametes. For example, in a proband with a three chromosome rearrangement interphase FISH revealed 20 different karyotypes in the carrier, of which 18 were unbalanced. Additionally, to the dominating structural aberrations meiotic non-disjunction can occur and lead to an aneuploidisation of the gametes.

It has been assumed that the development of a CCR can be caused by chromosome breaks which become unstable after an incomplete repair. Up till now more than 250 cases of CCR are known from the literature. Familial cases show that the majority of them are of maternal origin, as in the male carrier of a balanced CCR the abnormality often leads to a reduced fertility or even to sterility. In the majority of familial cases, the number of breakpoints is less than in de novo cases, that means four and less.

The chromosomes involved in CCR formation show an unequal distribution in their frequency. The large chromosomes are more often involved than the small ones and chromosomes 3, 7, 1, 5, 8, and 6 are most frequently diagnosed in this type of aberration. Intrachromosomal analyses of the breakpoints revealed hot spots of aberration such as 4q28.3,

18q21.3, 18q11.2, 21q11.2, and 22q11.2. The position of the breakpoints is located in the euchromatin in 85%, in the centromere in 10%, and in 5% in the telomeric region. When analysing the position of the breakpoints in the chromosome 65% of them are located intermediate, 20% distal and 15% proximal.

4.2.4 Marker Chromosomes (Chromosome Derivates)

Marker chromosomes are a special group of chromosome derivates, which cannot be classified according the ISCN nomenclature, as they don't show normal banding pattern over a larger segment which enables their identification. They are diagnosed either in addition to the normal chromosomes—supernumerary marker chromosomes (SNMC), or as structurally abnormal chromosomes (ESAC)—or they are observed in a karyotype, with normal chromosome number. In cases of additional marker chromosomes, only one is present in the majority of cases, but there are also cases published with two different or two times the same marker in the pathologic karyotype. Their frequency is about 1:1000 in the normal population and 3:1000 in probands with mental retardation.

In the majority of cases, marker chromosomes are less in size than chromosome 21—the smallest human chromosome—and they contain increased amounts of heterochromatin. According to their structure, they are divided into three main groups: marker with satellites on both sides, on one side, and without satellites. Further characteristic features are the number of centromeres—mono- or dicentric—and the composition of only heterochromatic segments or the combination of euchromatic and heterochromatic regions. The majority of SMC are derivates of the acrocentrics (about 70%) of these nearly 40% are derived from chromosome 15. Maker chromosome occurs de novo and familial. A peculiarity of the latter group is the frequent formation of mosaics (about 50%). In familial cases with a heterochromatic marker in a healthy carrier, an increased genetic risk is given for his children by the possibility of the occurrence of an unequal crossing over in meiosis with the normal original chromosome leading to the insertion of euchromatin into the marker which then causes malformations and/or retardation in the progeny. Marker chromosomes can be unstable in the course of postzygotic mitoses and thus decrease in frequency pre- and postnatal, and these changes show a different speed of reduction in different cell systems such as lymphocytes or fibroblasts.

Heteromorphismes: Mutations in Non-Coding-DNA Regions

Contents

5.1 Euchromatic Variants – 34

5.2 Satellites – 34

5.3 Nucleolus Organising Regions (NOR-Region) – 35

5.4 Pericentromeric Heterochromatin – 35

5.5 Centromere – 36

5.6 Heterochromatic Blocks – 36

C. Behrend et al., *Human Chromosome Atlas*, https://doi.org/10.1007/978-3-031-10588-3_5

For a long time, the constitutive heterochromatin has been regarded as neglectable as to its significance of mutations in chromosome structure and function. Recent investigations made it obvious that it is relevant for the course of sex differentiation and mitoses. The different types of heterochromatin were characterised in detail which led to the definition of subtypes and then the consequences of mutations in these regions were analysed in detail.

In the following, five different types of heterochromatin and one of the genetic non-relevant euchromatin are presented in detail.

5.1 Euchromatic Variants

In this type of structural variants, the recognition in most cases depends on size differences of one band, usually AT-rich, between the two homologous chromosomes of one pair. It originates either by duplication or by partial deletion of the region. If the genes located in these bands encode for quantitative features, the changes can be compensated by other genes being connected in a network of comparable functions.

Up till now, 17 different euchromatic variants have been characterised. The most frequent are 4p16, 8p23.1, 9p12, q12, 9q13, 15q11.2, and 16p11.2. Further frequent variants have been analysed in the last years in the subtelomeric regions of 1q, 9p, and Xp. Their number can be expected to increase with increasing investigations by microarray. If the healthy parents and their handicapped child present with the same euchromatic variant, this can be caused by a number of facts:

- The gene presents with variable penetrance or reduced expression,
- In meiosis of the carrier parent unequal crossing over lead to partial duplication or deletion of relevant neighbouring genes,
- Modification of genes plays a role,
- By insertion of parts of the variable band into its sister chromatid a tolerable gene content is exceeded.

Therefore, genetic counselling is always difficult in prenatal cases. The type of aberration are deletions and duplications with a size of 3–30 Mb.

5.2 Satellites

They appear as stalked appendices of the short arms (p) of the acrocentric chromosomes. They consist of satellite DNA with a high number of repetitions. They are strongly contracted and therefore stain intensively after Giemsa-staining. Mutations in these regions affect their number, size, and staining characters. The types of mutations are deletions (up to 4 of the 10 regions), duplications, and translocations with non-acrocentric chromosomes. The most frequent translocation Y/15 is diagnosed in 1:3500 persons of the general population.

Other rearrangements are jumping translocations and translocations of euchromatic segments on to the region p13. If the translocation of euchromatic regions is excluded, satellite mutations are without genetic relevance.

5.3 Nucleolus Organising Regions (NOR-Region)

The nucleolus is a cell structure which can be observed in the nucleus during interphase and which is dissolved in the prophase of mitosis and newly synthesised after termination of mitosis. It is the locus of chromosomal rRNA synthesis and is localised in the nucleolus organizing regions (NOR) in the short arms (p12) of the acrocentric chromosomes. During transcription, it is decondensed. It is built up by different gene families, only a part of them are active in transcription. Mutations in these regions are deletions, duplications, amplifications, and translocations. Translocations are mainly terminal, but interstitial insertions are also diagnosed. The chromosomes most frequently involved are 1, 2, and 4. Mosaic cases of postzygotic origin have been observed.

A secondary change to genetic relevant mutations can take place by an unequal crossing over in meiosis. If the insertion of the NOR-region leads to disruption of a tumour suppressor gene, it can induce tumour genesis.

Genetic counselling has to be offered in familial cases and in the prenatal period then the indication for microarray analysis is given.

5.4 Pericentromeric Heterochromatin

All human chromosomes are built up from different types of heterochromatin distal from the centromere (p11.1, and q11.1 and p11.2 in the acrocentrics). The size of these regions shows interchromosomal differences and can even vary in the single chromosome (i.e. 15q11.1). The band p11.2 in acrocentrics is characterised by its variation in staining intensity (QFQ and DA/DAPI). The pericentromeric regions consist of satellite DNA, β in p11.1 and q11.1 and of the fractions I, II, and III in p11.2. Mutations in these regions are very frequent as they have no negative effect on the carrier. Structural mutations are deletions, duplications, inversions, and translocations. The different types of mutations show a different intra- and interchromosomal frequency. Pericentromeric mutations in chromosome

18 can be of diagnostic relevance. Especially deletions can lead to a misinterpretation of interphase-FISH. Therefore, some diagnostic centres for prenatal aneuploidy screening changed the 18cen probe for a euchromatic probe in 18q.

5.5 Centromere

The centromere is a complex structure of highly repetitive DNA. Its position in the chromosome presents as primary constriction of meta-, submeta-, and acrocentric chromosomes. In interphase, the centromeres remain more condensed than the rest of the chromosome. Special staining is performed by CBG and DA/DAPI banding, the active centromere in dicentric chromosomes can be characterised by Cd-staining. Centromeres have a complex structure of central elements (α-satellite DNA), distal followed by repetitive regions and laterally associated of the protein complex of the kinetochore. In meristematic cells, the combination of centromere and kinetochore is necessary for the correct separation of the chromatids in anaphase of the mitosis, up to this phase the centromere saves the chromatids from premature separation. A high number of mutations in the centromere regions have been observed. Besides deletions, duplications, amplifications, and translocations, splitting of the centromere in inversions, fusions, fissions, insertions, and centromere-telomere fusions are diagnosed, as well as cases of premature separation of the centromere. Instability of the centromere structure can lead to the development of a multi-radial structure or to premature separation (i.e. Roberts syndrome). Centromere mutations are of genetic relevance; therefore, genetic counselling should be offered to the carrier.

5.6 Heterochromatic Blocks

Large heterochromatic regions or blocks are a characteristic feature in the autosomes 1, 9, 16 with position proximal in 1q12, 9q12, 16q11.2, and distal in the long arm of the Y-chromosome (Yq12). Less extended blocks visualised by QFQ-banding are 3q11.2 and 4q11.2. The heterochromatic blocks contain about 10% of the human genome. During the cell cycle, they are late replicating. The regions in 1q12 and 9q12 are built in two parts which is the reason for the frequency of partial inversions. In interphase, the heterochromatin blocks tend to develop fusions. They are preferential positions for breaks and rearrangements. It is assumed that they have a regulatory function in the cell. Special staining methods are CBG, QFQ, and DAPI. The heterochromatic regions consist of highly repetitive DNA of the classes Sat I–IV. Mutations in the regions are deletions, duplications, amplifications, and partial or complete inversions. These rearrangements are frequent in chromosomes 1, 3, 4, and 9 but rare in 16. Interchromosomal rearrangements are translocations and insertions.

They are analysed to differ between mono- and dizygotic twins, to characterise specific ethnic groups, to differ between donor and recipient after bone marrow transplantation, in the analysis of complete moles and in cases of chimerism.

Up to our present knowledge, mutations in heterochromatic regions are usually without genetic relevance. After homogenous staining (Giemsa, Orcein), the chromatids show an increased contraction of the heterochromatic blocks. The size of these regions can amount up to 35% of the total length of the chromosome with this type of heteromorphism.

Proximal of these blocks in the autosomes, they are followed by the heterochromatic band q11.1, distal by the euchromatic GC-rich band in q13. In the majority of cases, pericentric inversions concerning these regions are stable when inherited. An exception is observed in chromosome 9. In single cases when the breakpoint in the long arm is within the band q13, this can lead to an unequal crossing over in meiosis and thus to partial euchromatic duplications or deletions in the progeny resulting in a characteristic phenotype. In these special cases, diagnostic investigations afford microarray analyses.

Single Case Presentations of Intrachromosomal Rearrangements

A total of 88 chromosome aberrations is presented in detail. For each type of abnormality, the cases are summarised and the relevant facts are discussed. In the single case, the karyotype is described and illustrated in detail, the indication for chromosome analysis is given, the morphology of the aberrant chromosome(s) is noted including the necessary resolution for receiving a reliable diagnosis.

It is explicitly stated for each case, if special facts have to be taken into account when the patient and/or his family are recommended genetic counselling. If additional investigations are required, this is noted in detail.

As an example, in a total of **10** cases frequent variants of the chromosomes—we chose pericentric inversions of the heterochromatin blocks **1**, **3**, **4**, **9**—with changes of their position from the q- to the p-arm were given. This is to give the reader an impression on the altered phenotype of the chromosome and on the investigation method, which has to be chosen to analyse them (CBG- or QFQ-banding).

Contents

Chapter 6 **Inversions** – 41

Chapter 7 **Deletions** – 163

Chapter 8 **Duplications** – 225

Chapter 9 **Ring Chromosomes** – 235

Chapter 10 **Isochromosomes** – 245

Inversions

Contents

6.1 Pericentric Inversions – 42

6.2 Paracentric Inversions – 94

6.3 Summary – 163

C. Behrend et al., *Human Chromosome Atlas*, https://doi.org/10.1007/978-3-031-10588-3_6

The indication for chromosome analysis was infertility in all cases; therefore, this item is not mentioned in the detailed case presentations.

6.1 Pericentric Inversions

1. 46,XY,inv(1)(p12q12),inv(1)(p13.1q21)
2. 46,XY,inv(2)(p11.2q13)
3. 46,XY,inv(3)(p12.2q26.32)
4. 46,XX,inv(4)(p12q21.1)
5. 46,XY,inv(5)(p15.1q33.2)
6. 46,XY,inv(5)(p13.2q13.2)
7. 46,XY,inv(7)(p10q11.23)
8. 46,XX,inv(7)(p22q34)
9. 46,XX,inv(8)(p11.2q21.12)
10. 46,XX,inv(10)(p11.2q21.2)
11. 46,XY,inv(11)(p13q21),inv(9)(p12q12)
12. 46,XX,inv(16)(p13.2q11.1)
13. 46,XY,inv(16)(p12q12.2)
14. 46,XX,inv(20)(p13q13.2)
15. 46,X,inv(X)(p22.1q26)
16. 46,X,inv(X)(p22.13q24),inv(3)(p11.2q12)
17. 46,X,inv(Y)(p11.2q11.23)

▪▪ Case 1: Karyotype: 46,XY,inv(1)(p12q12),inv(1)(p13.1q21)

The proband shows pericentric inversions in both homologues chromosomes.

▪ Morphology

1. inv(1)(p12q12),

   The breakpoints of this proximal inversion are both located in heterochromatic regions. In the short arm, the breakpoint is in p12 of the pericentromeric heterochromatin and in q12 in the heterochromatic block which is thus translocated into the short arm. This inversion is known as a polymorphism without genetic relevance. The region encloses 8–15% of the total length of chromosome 1, its frequency in the European population about 1‰ (see: ► Sect. 15.6).
2. inv(1)(p13.1q21).

   Presented is a very small proximal pericentric inversion. The inverted segment comprises about 7% of the total length of chromosome 1. As the heterochromatic block q12 has a high interindividual variation in length, an inversion with differing heteromorphisms can result in a different proportion of the rearranged region. One breakpoint is localised in the GC-rich band p13.1, the second one in the proximal part of the heterochromatic block q12. Here a euchromatic polymorphic band is located which is a preferential breakpoint in pure heterochromatic inversions, as it is presented in the second chromosome 1 of the carrier.

Beside GTG- and QFQ-banding CBG staining is required for the presentation of the inversion. A resolution of at least 550 bands per genome is necessary for the diagnosis of the rearrangement.

▪ Cave

Compared to inversions with similar size but both breakpoints in euchromatic segments the genetic risk is reduced in the present case where one breakpoint is located in a heterochromatic region (◘ Fig. 6.1).

6

a

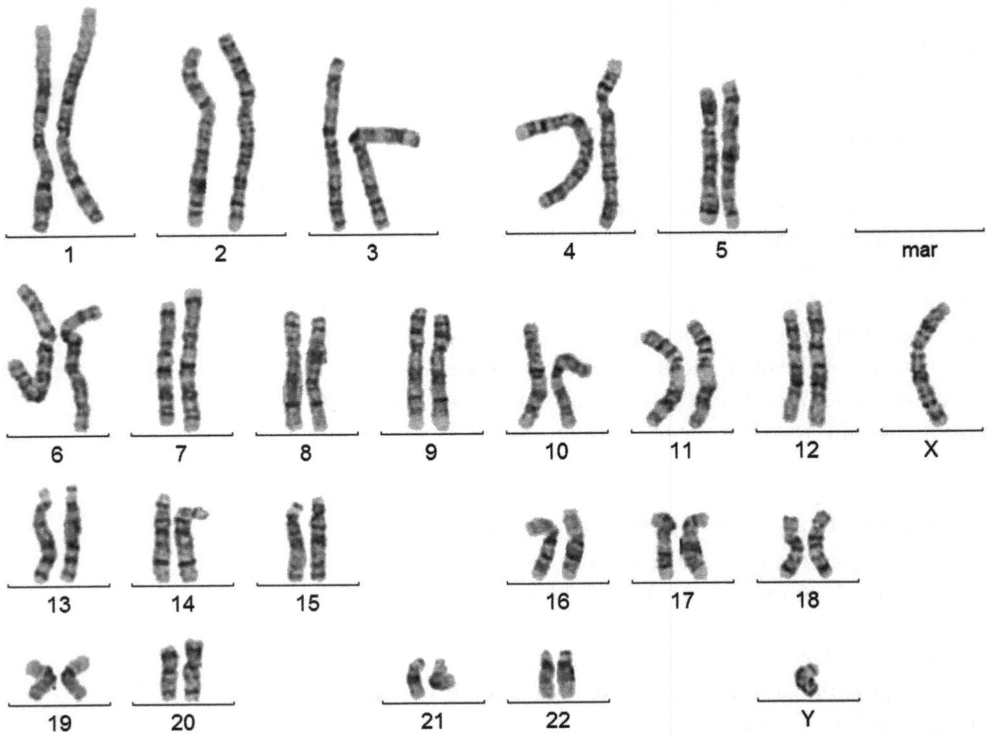

b

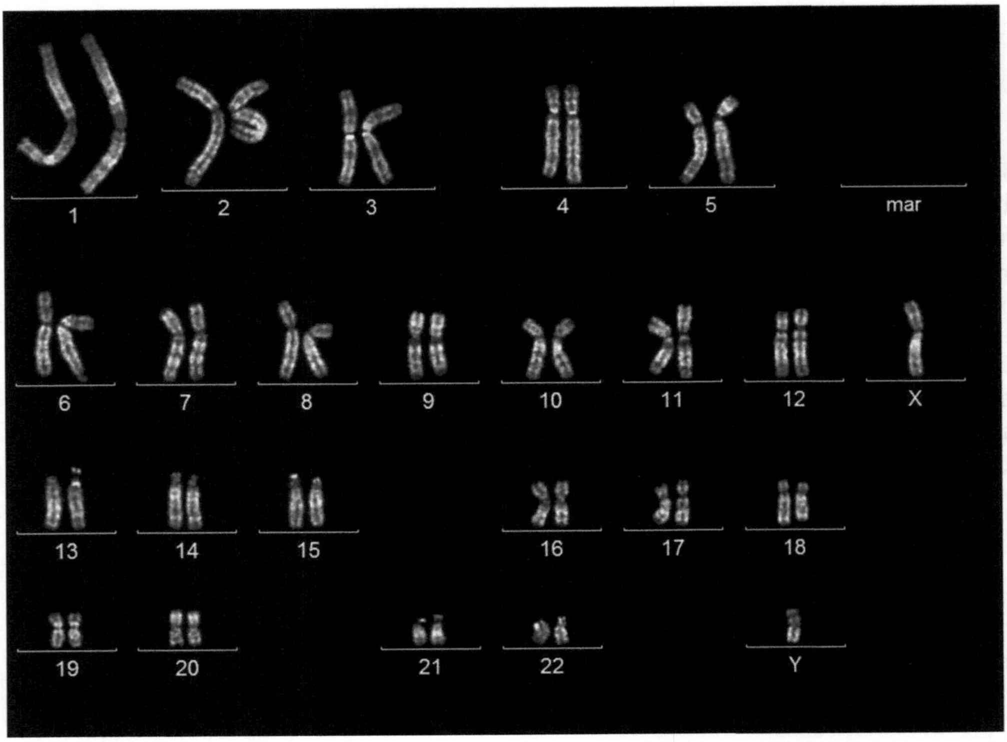

Fig. 6.1 Karyotype 46,XY,inv(1)(p12q12),inv(1)(p13.1q21). **a** GTG-banding, **b** QFQ-banding, 550- and 400-bands per genome, **c** CBG-banding, **d** QFQ-banding, **e** GTG-banding, chromosomes from different mitoses; *left* inv(1)(p12q12), *right* inv(1)(p13.1q21), **f** ideograms of Case 1. Both homologues show a pericentric inversion, *left* breakpoints in heterochromatic regions, *right* breakpoints in the euchromatin

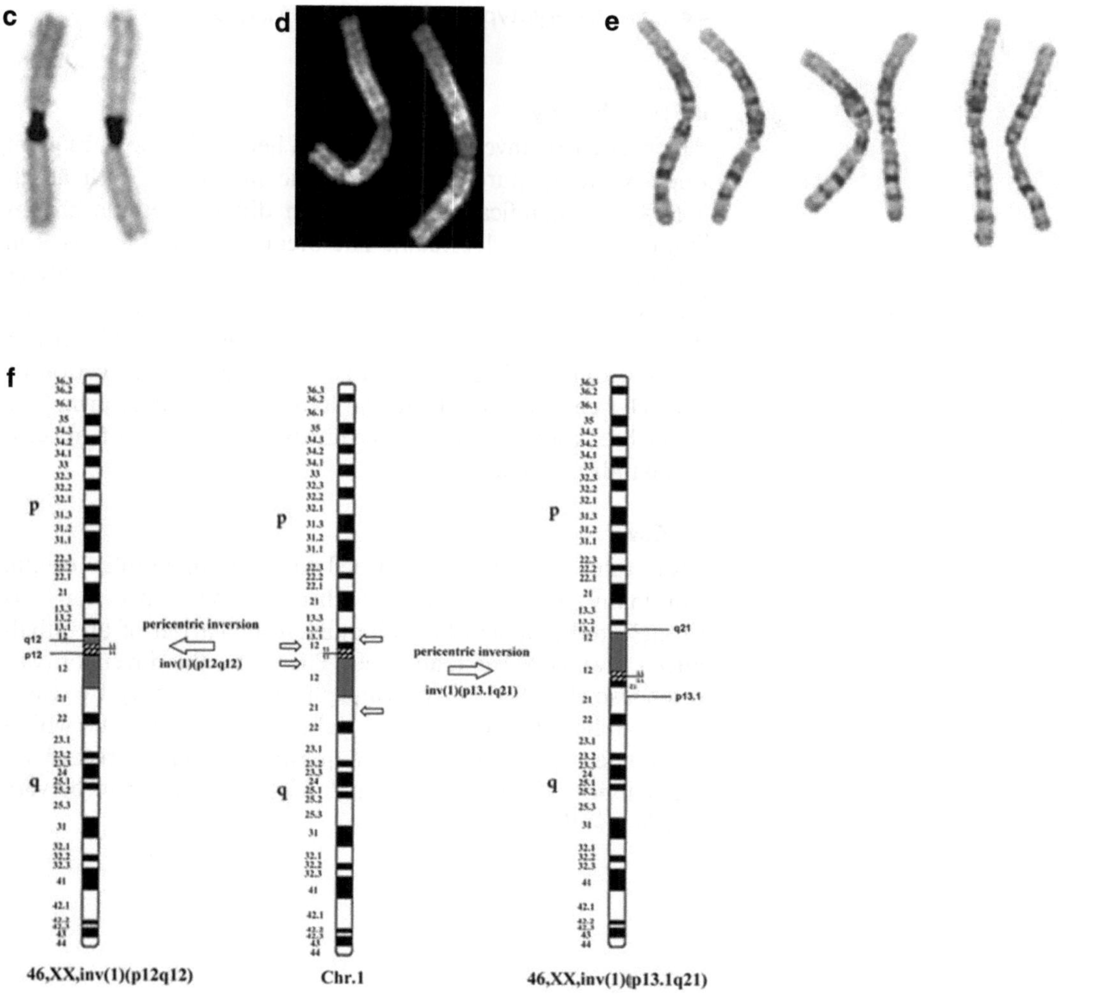

Fig. 6.1 (continued)

▪▪ Case 2: Karyotype: 46,XY,inv(2)(p11.2q13)

▪ Morphology

The pericentric inversion 2 presented here is the most frequent one, excluding pure heterochromatic inversions, with about 1:1300 but significant differences in different ethnic groups (lowest frequency 1:10.000). The breakpoints are localised in regions of common fragile sites. Molecular genetic analyses have shown that the breakpoints are not always quite identical. The inverted region comprises about 5% of the total length of chromosome 2. This inversion is recognised by dislocation of the AT-rich band q13 from the long into the short arm. The exact breakpoint localisation affords a resolution of at least 550 bands per genome.

▪ Cave

More than 95% of the published cases are of familial origin. For many years, this type of chromosome 2 inversion was thought to be stable when inherited, as no carrier of an unbalanced inversion was diagnosed. Therefore, this inversion was often also classified as a polymorphism. But recently, two cases were published, one with a partial duplication and one with a partial deletion of the region in question. A genetic risk therefore cannot be excluded although it can be defined to be very low (<1%) (◻ Fig. 6.2).

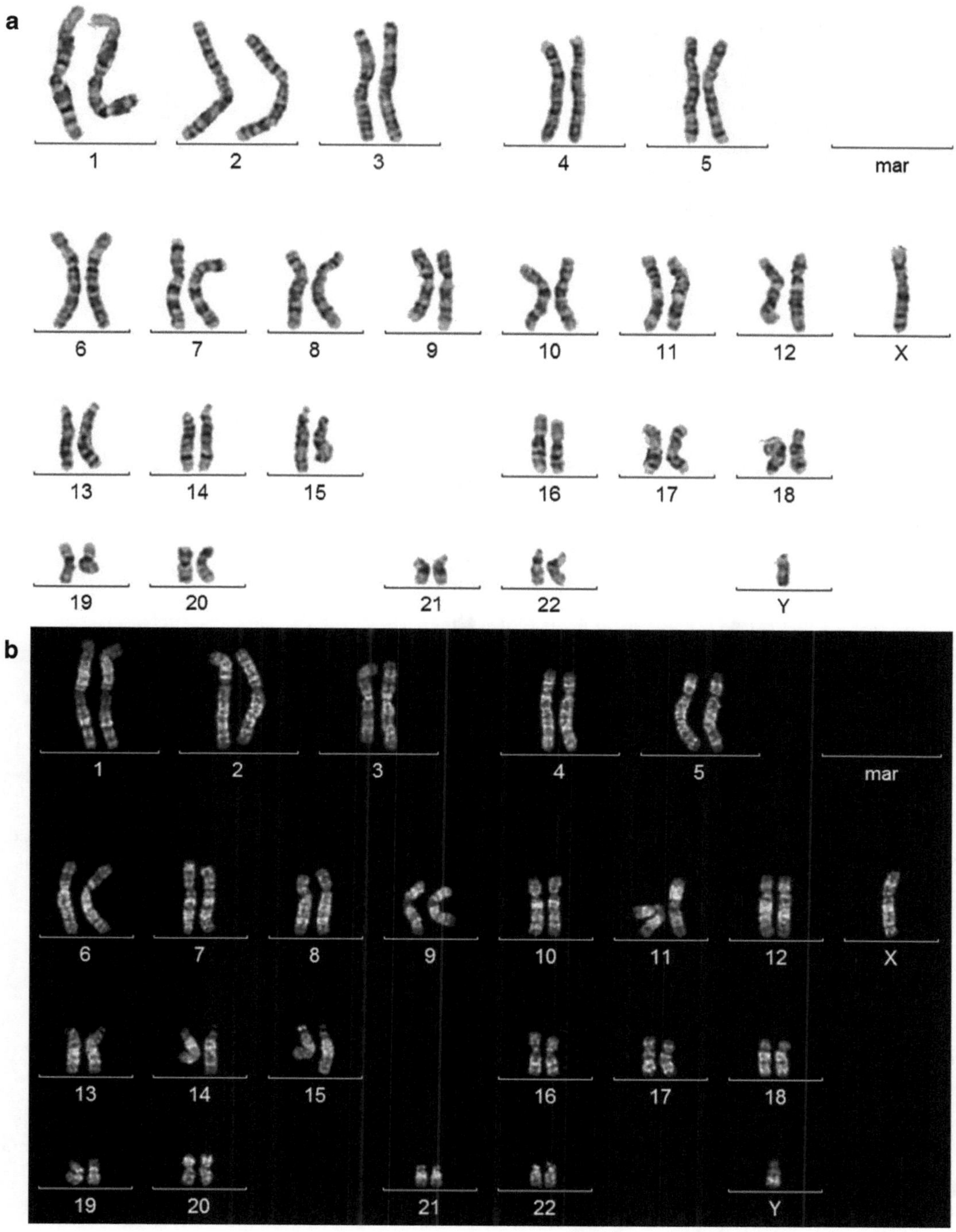

Fig. 6.2 Karyotype 46,XY,inv(2)(p11.2q13). **a** GTG-, **b** QFQ-banding, 550 and 400 bands per genome, **c** QFQ-, **d** GTG-banding, chromosomes from different mitoses, *left* normal, *right* inverted chromosome 2, **e** Ideogram of Case 2, small proximal inversion

6

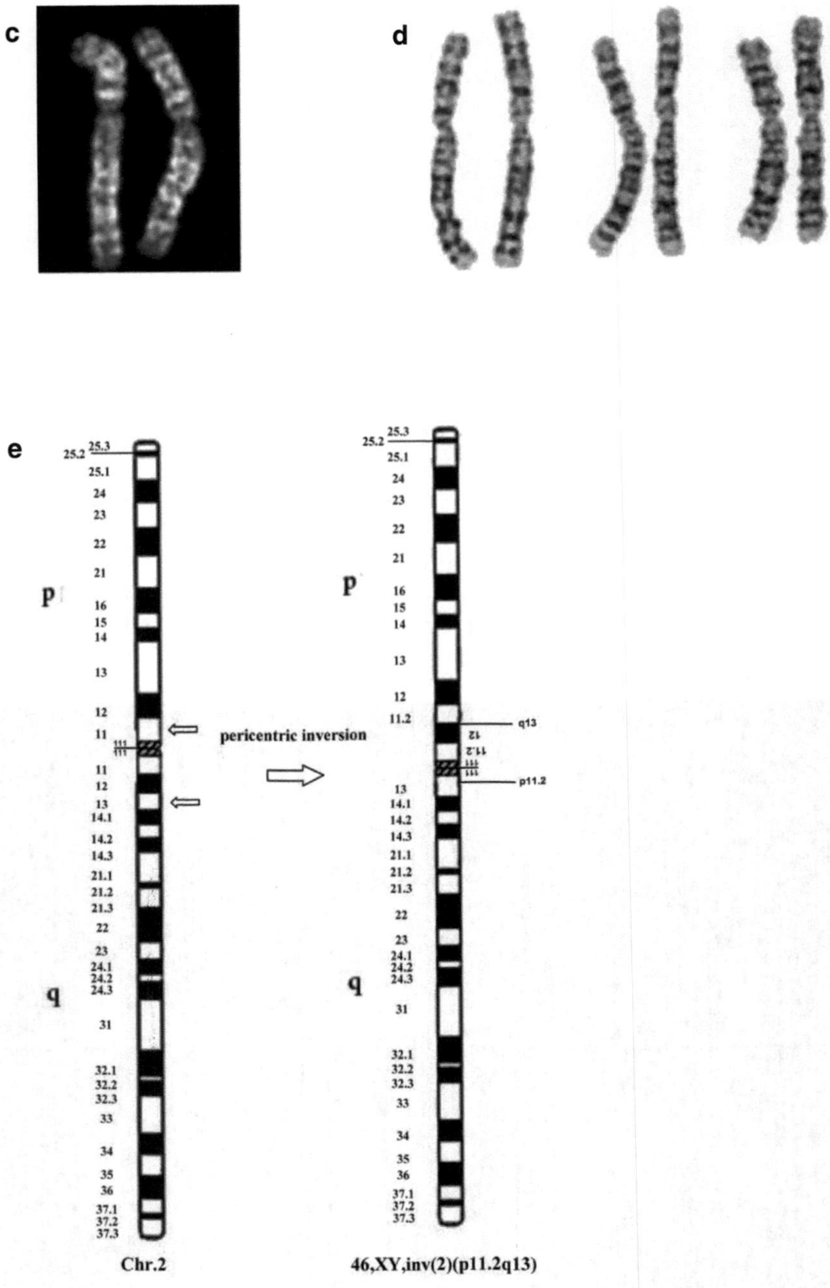

Fig. 6.2 (continued)

Case 3: Karyotype: 46,XY,inv(3)(p12.2q26.32)

Morphology

This pericentric inversion comprises about 50% of the total length of chromosome 3. The most prominent change in the chromosome morphology is an altered centromere index from a proportion of short and long arm of 48:52 to 17:83%. The broad AT-band p12, proximal located in the shortened long arm is a further relevant criterion for identification. A resolution of 550 bands per genome enables a correct diagnosis. In the presented case, an additional CBG-banding improves the investigation. Chromosome 3 is one of the three chromosomes which, according to the literature, present with the highest frequency of pericentric inversions.

Cave

The inverted region comprises about 50% of the total length of chromosome 3, according to literature this causes a genetic risk of average probability of about 10%. Therefore, potential inversion carriers in the family should be investigated and the indication for prenatal diagnostics is given (◻ Fig. 6.3).

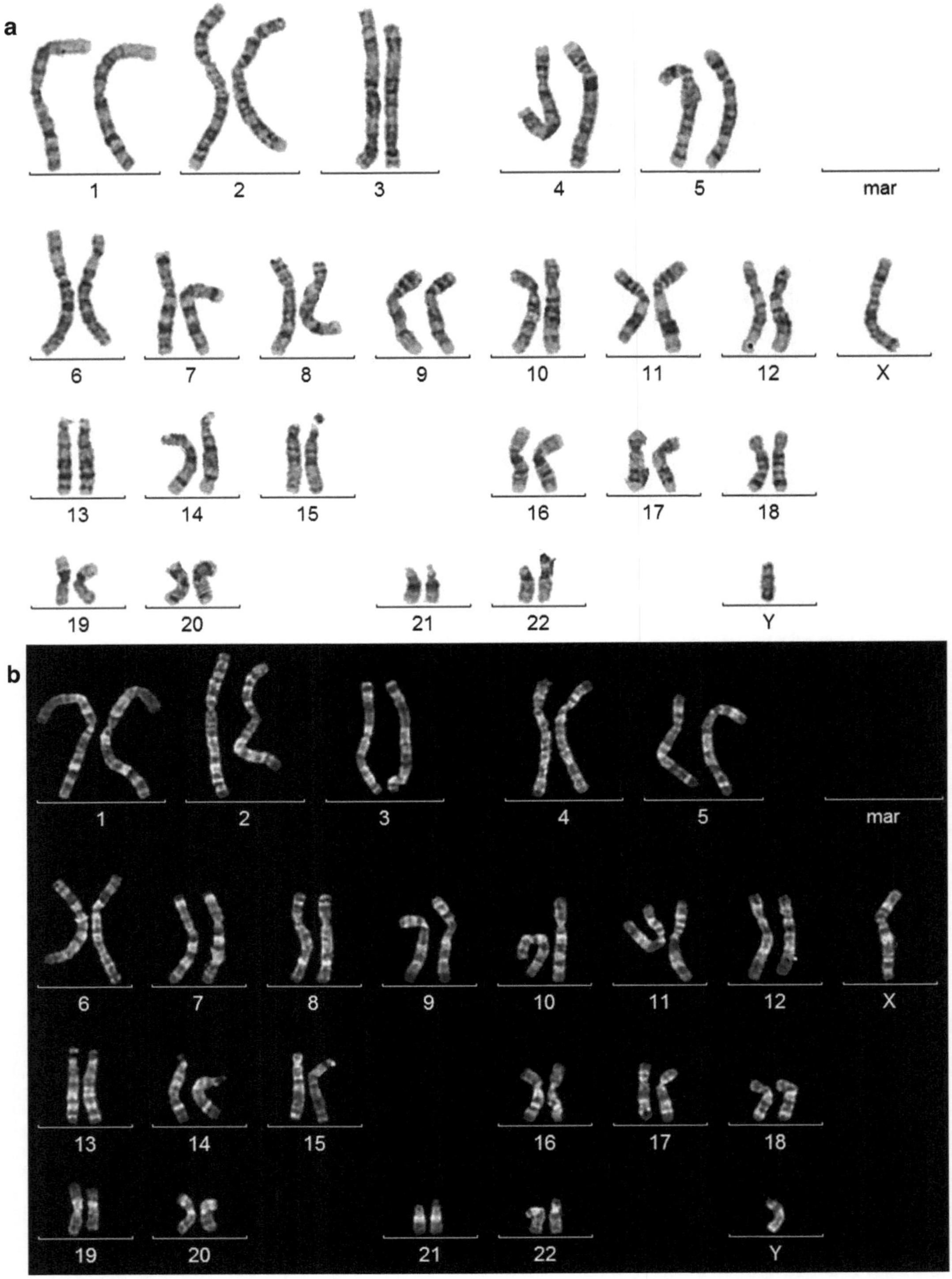

Fig. 6.3 Karyotype 46,XY,inv(3)(p12.2q26.32). **a** GTG-, **b** QFQ-banding, 550 and 400 bands per genome, **c** CBG-, **d** QFQ-, **e** GTG-banding, chromosomes from different mitoses, *left* normal, *right* inverted chromosome 3, **f** ideograms of Case 3, large inverted segment

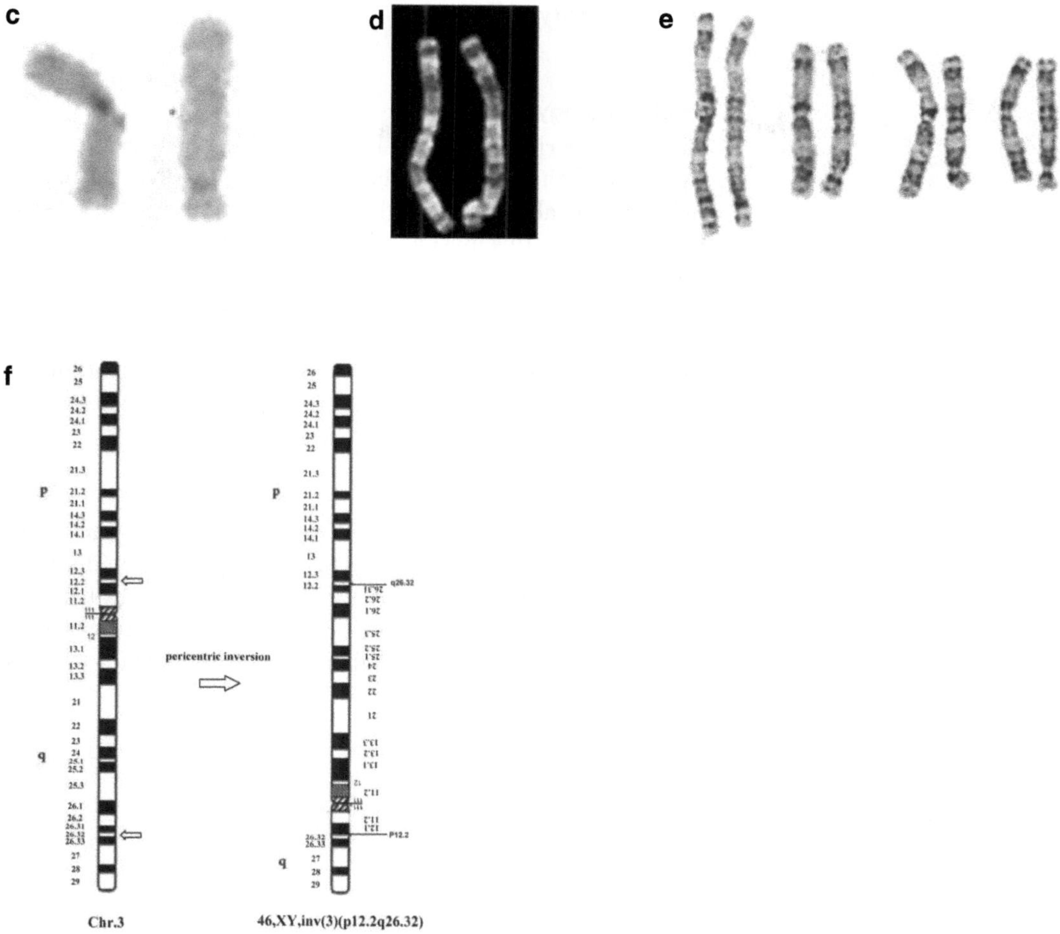

Fig. 6.3 (continued)

■■ Case 4: Karyotype: 46,XX,inv(4)(p12q21.1)

·

■ Morphology

The small pericentric inversion is localised in the proximal segments of the short and the long arm. The inverted region includes only about 16% of the total length of chromosome 4. The analysis of this inversion is easy to perform, as the centromeric index has changed and as the characteristic AT-rich double band q13.1 to q13.3 has changed its position from the long into the short arm. Of additional diagnostic relevance in this pericentric inversion is its appearance after QFQ-banding. It shows the shifting of the brilliant fluorescent heterochromatic band 4q11.2 from the long into the short arm.

■ Cave

The breakpoints in the short as well as in the long arm are localised in GC-rich regions. Therefore, genetic counselling of the carrier is indicated as well as prenatal investigations including microarray to exclude an unbalanced karyotype in the conceptus caused by an unequal crossing over in meiosis (◘ Fig. 6.4).

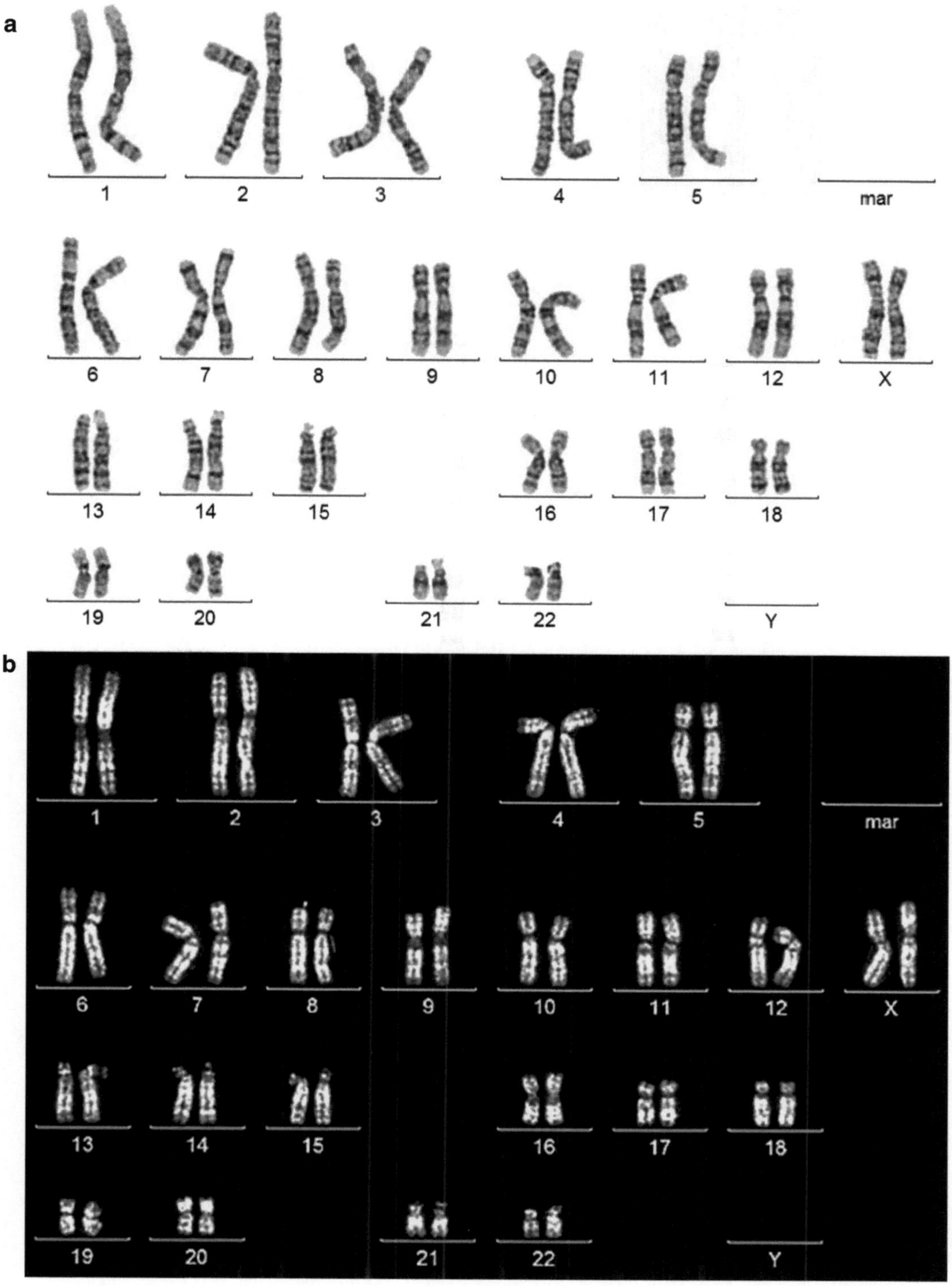

Fig. 6.4 Karyotype 46,XX,inv(4)(p12q21.1). **a** GTG-, **b** QFQ-banding, 550 and 400 bands per genome, **c** QFQ-, **d** GTG-banding, chromosomes from different mitoses, *left* normal chromosome, *right* inverted chromosome 4, **e** ideograms of Case 4, small proximal segment inverted

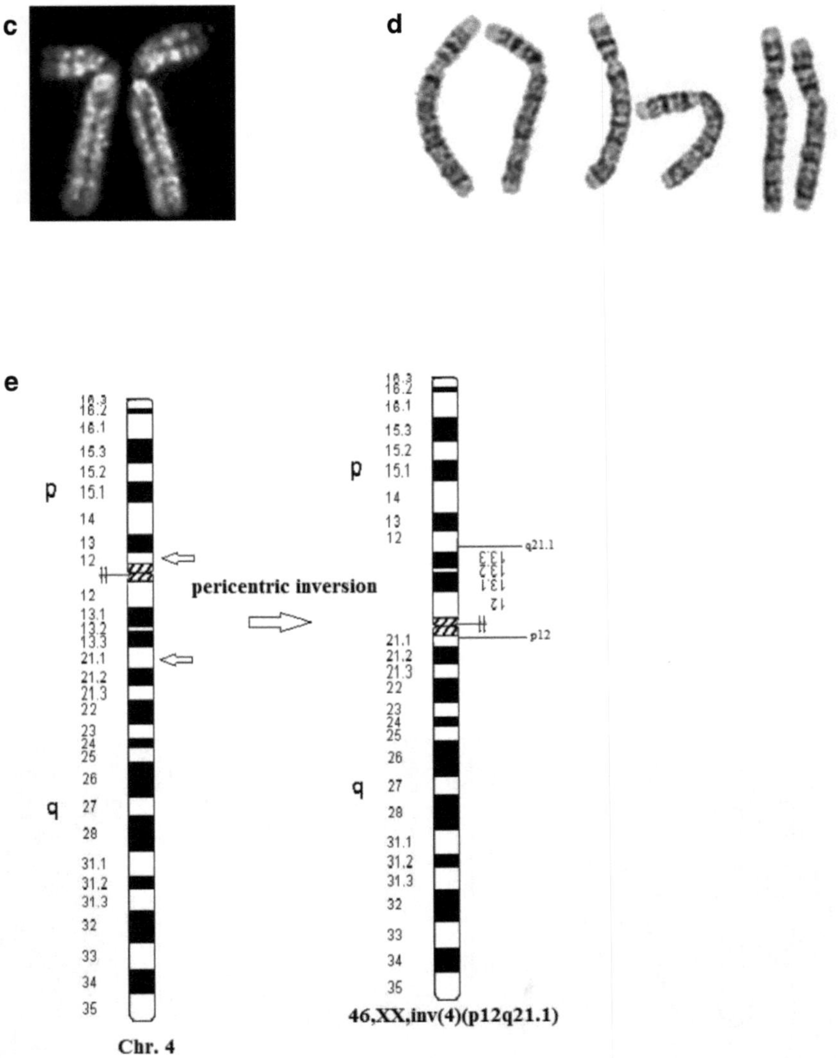

Fig. 6.4 (continued)

▪▪ Case 5: Karyotype: 46,XY,inv(5)(p15.1q33.2)

▪ Morphology

The pericentric inversion 5 presented here encloses a region of about 80% of the total length of the chromosome. By the displacement of the centromere, the aberrant chromosome has a more metacentric morphology with the arm-ratio changing from 1:28 to 1:2.2. The shifting of the AT-rich band q34 from the long into the short arm leads to a very characteristic change of the banding pattern and thus improves the analysis of the inverted chromosome. The resolution should be at least 550 bands per genome to insure an exact localisation of the breakpoints.

▪ Cave

Caused by the size of the inverted segment the risk for the formation of an unbalanced karyotype has to be regarded as high with approximately 30%. Prenatal diagnosis including microarray analysis should be recommended. Relatives of the carrier should be offered chromosome investigations. The inversion includes the critical region for the Cri-du-chat syndrome (▪ Fig. 6.5).

6

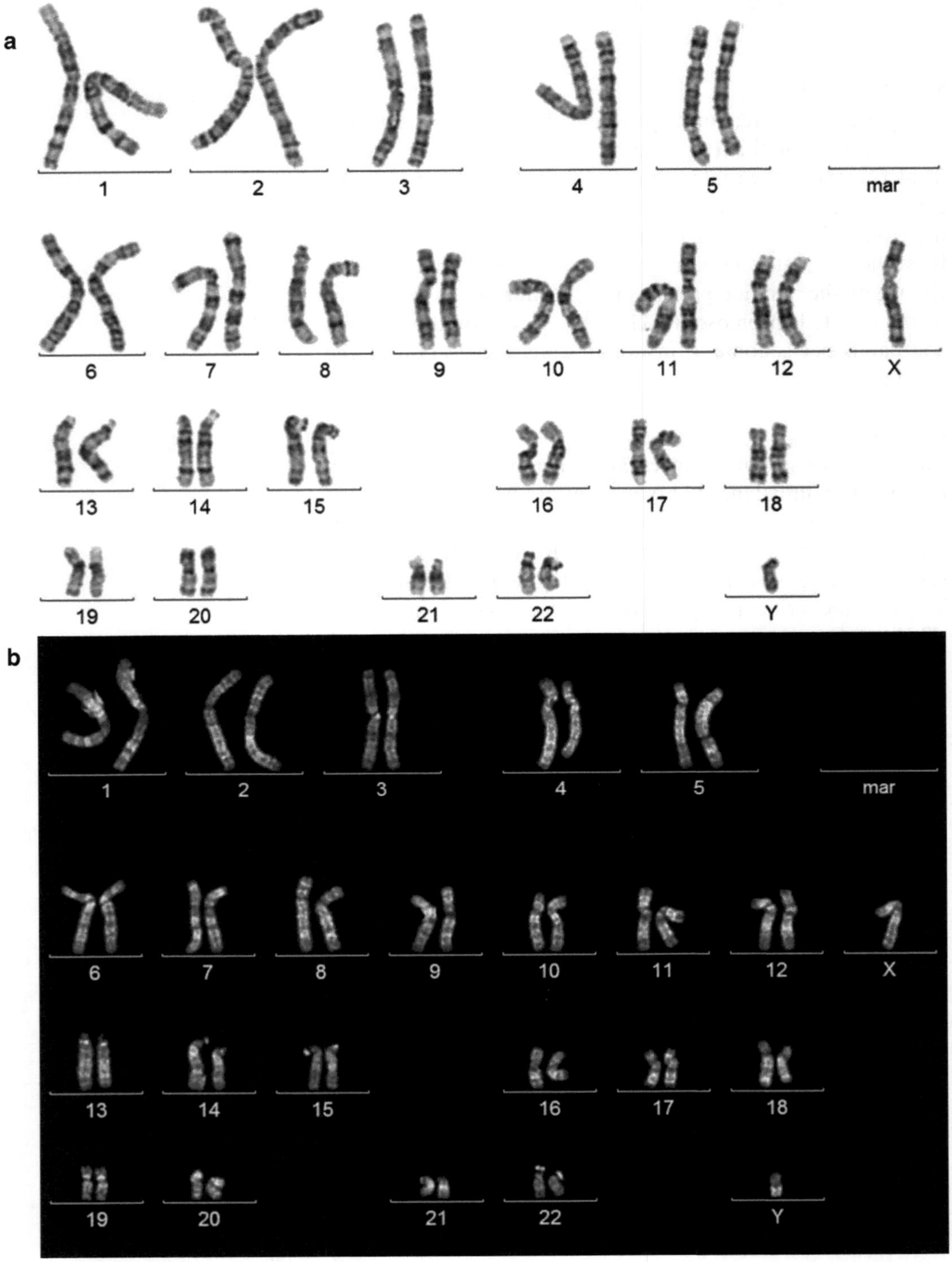

Fig. 6.5 Karyotype 46,XY,inv(5)(p15.1q33.2). **a** GTG-, **b** QFQ-banding, 550 and 400 bands per genome, **c** QFQ-, **d** GTG-banding, chromosomes from different mitoses, *left* normal, *right* inverted chromosome 5, **e** ideograms of Case 5, extreme large inverted segment

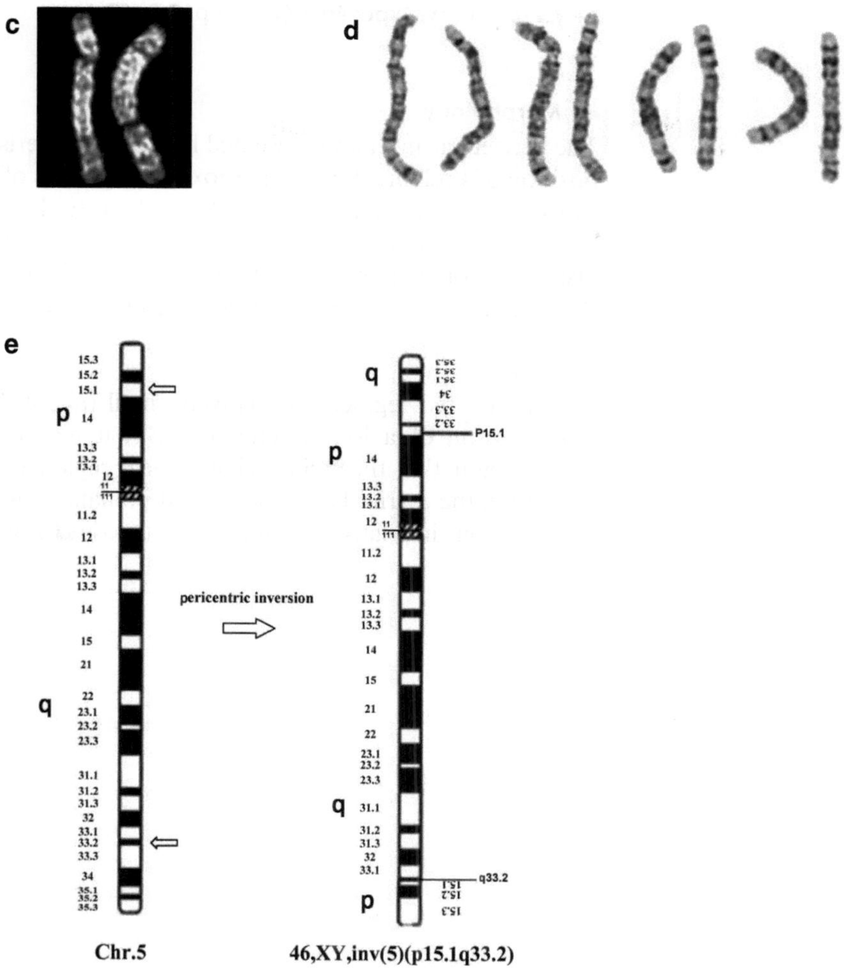

Fig. 6.5 (continued)

▪▪ Case 6: Karyotype: 46,XY,inv(5)(p13.2q13.2)

.

▪ Morphology

The pericentric inversion presented here is characterised by the position of both breakpoints in proximal segments of long and short arm and includes about 18% of the total length of the chromosome. As the breakpoint in the long arm has a more distal position than in the short one, the arm ratio has changed, the inverted chromosome appears more metacentric.

▪ Cave

As the inverted segment is relatively small (about 20%), this is the reason of a low genetic risk. But it has to be taken into account that the Cri-du-chat critical region is included. Therefore, the carrier has to be offered prenatal diagnosis and investigations in relatives should be discussed (◘ Fig. 6.6).

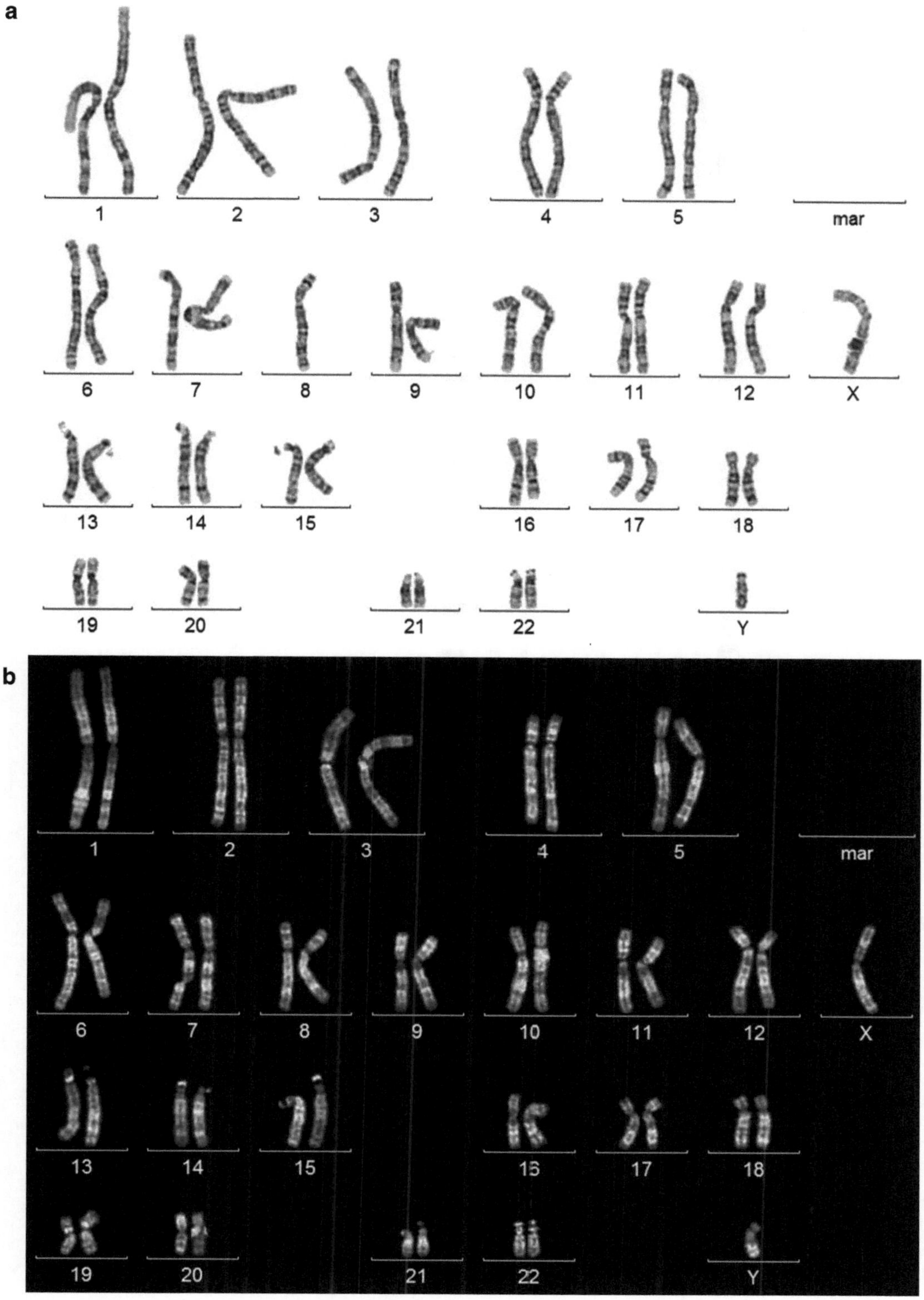

Fig. 6.6 Karyotype 46,XY,inv(5)(p13.2q13.2). **a** GTG-, **b** QFQ-banding, 550 and 400 bands per genome, **c** QFQ-, **d** GTG-banding, chromosomes from different mitoses, *left* normal, *right* inverted chromosome 5, **e** ideograms of Case 6, inversion in proximal segments

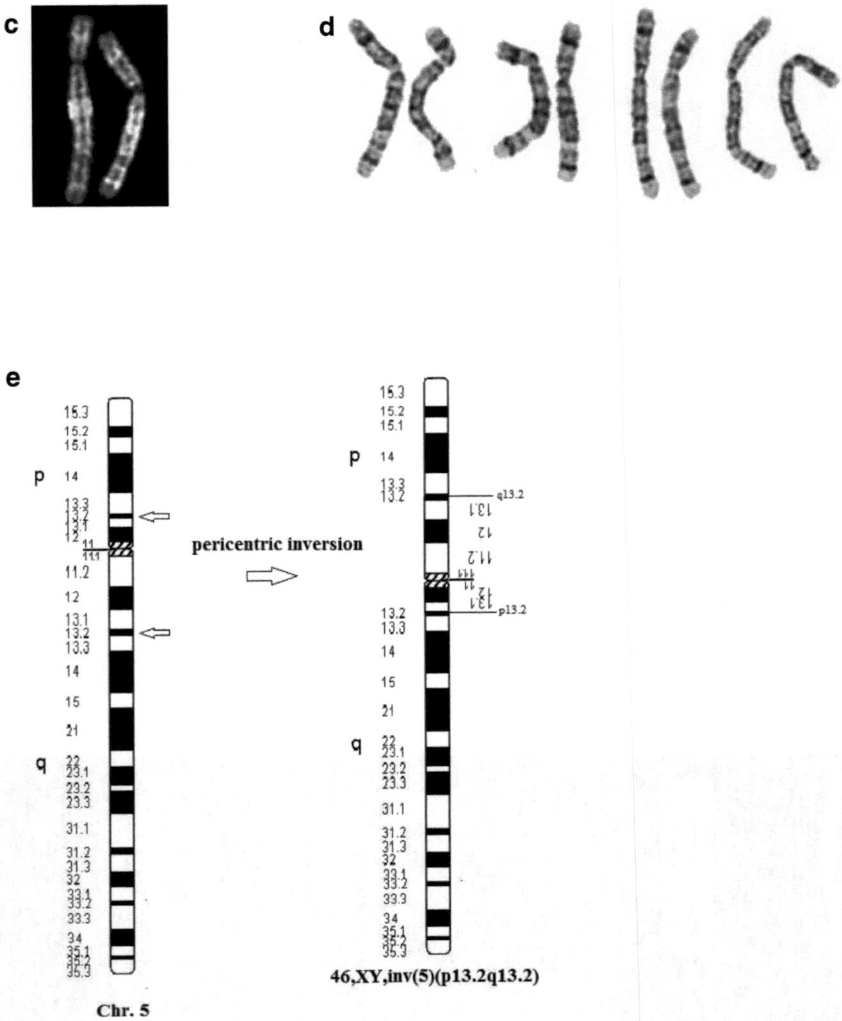

Fig. 6.6 (continued)

▪▪ Case 7: Karyotype: 46,XY,inv(7)(p10q11.23)

·

▪ Morphology

The small pericentric inversion has both breakpoints localised in proximal regions. The breakpoint in the short arm is localised in the pericentromeric heterochromatin. The size of the rearranged segment is about 10% of the total length of chromosome 7. The centromere index has changed from an arm ratio of 34:66 to 48:52%, the inverted chromosome appearing almost metacentric. The inversion is easy to diagnose by the altered centromere index and banding pattern.

▪ Cave

As one of the breakpoints is localised in the heterochromatic centromere region, the risk of an unequal crossing over is lower than in cases with both breakpoints in euchromatic regions (◘ Fig. 6.7).

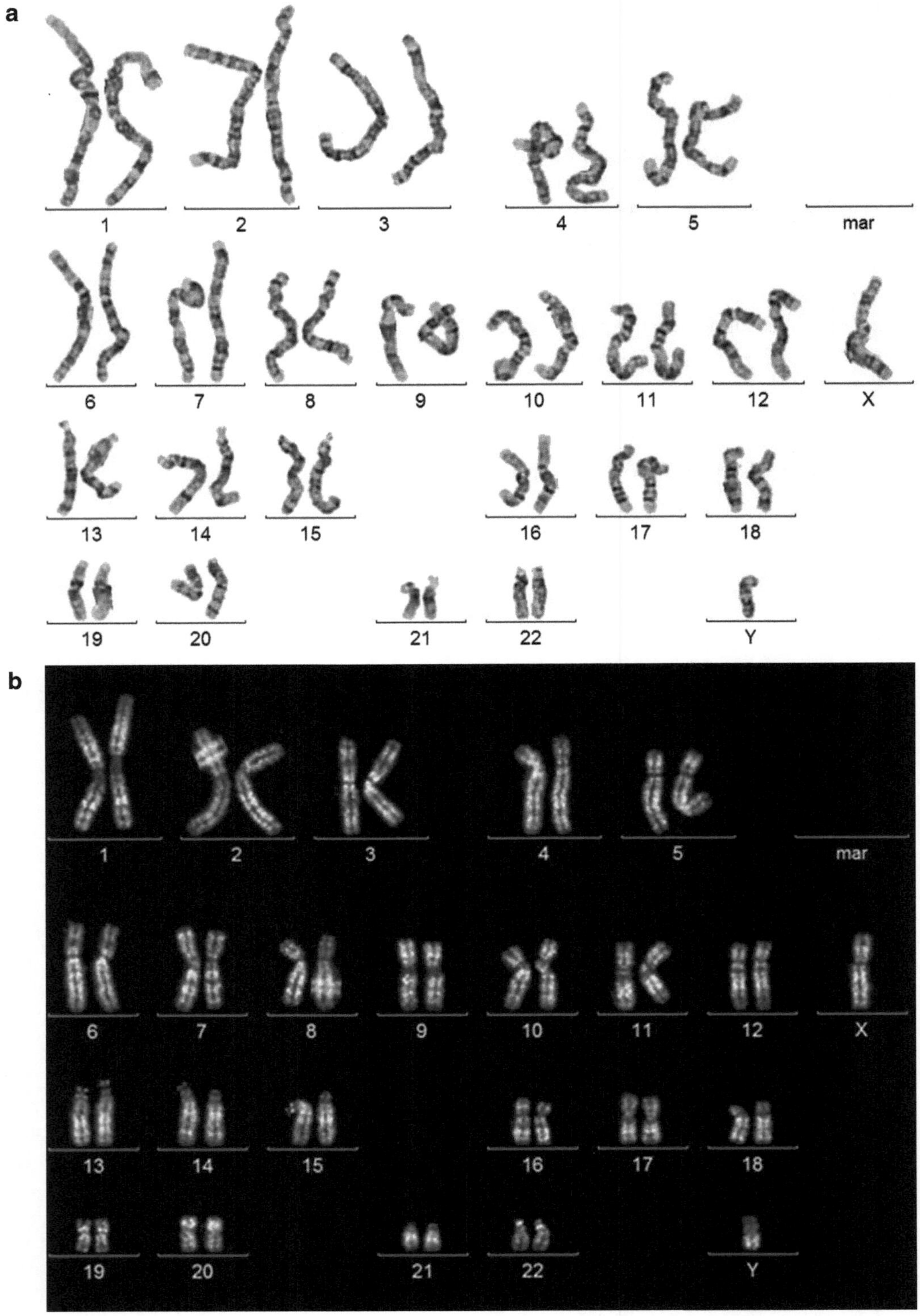

◻ Fig. 6.7 Karyotype 46,XY,inv(7)(p10q11.23). **a** GTG-, **b** QFQ-banding, 700 and 400 bands per genome, **c** QFQ-, **d** GTG-banding, chromosomes from different mitoses, *left* normal, *right* inverted chromosome 7, **e** ideograms of Case 7, small inversion with 1 breakpoint in the heterochromatic centromere region

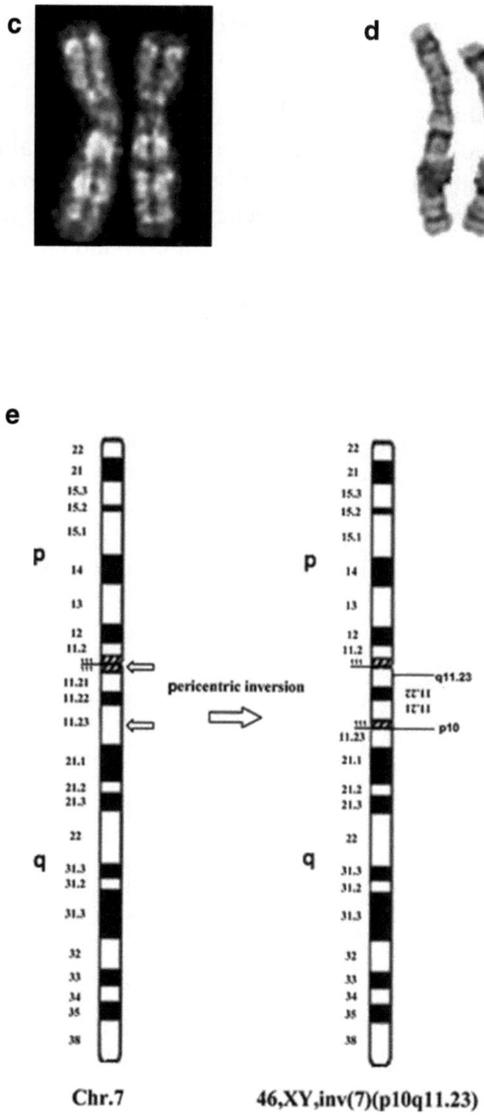

Fig. 6.7 (continued)

▪▪ Case 8: Karyotype: 46,XX,inv(7)(p22q34)

.

▪ Morphology

The inverted region includes about 90% of the total length of chromosome 7. Chromosome 7 belongs to the three chromosomes with the highest frequency of pericentric inversions, the other two are chromosome 3 (Case 3) and chromosome 8 (Case 9). As it is well known that the risk of an unbalanced rearrangement increases with increasing length of the inverted region, it must be estimated to be about 30%.

▪ Cave

Because of the high genetic risk chromosome investigations should be offered to relatives of the carrier. To all carriers, prenatal diagnoses should be recommended and chromosome analyses are to be combined with microarray investigation. In cases of an aberrant karyotype findings from the literature can be included in genetic counselling (◘ Fig. 6.8).

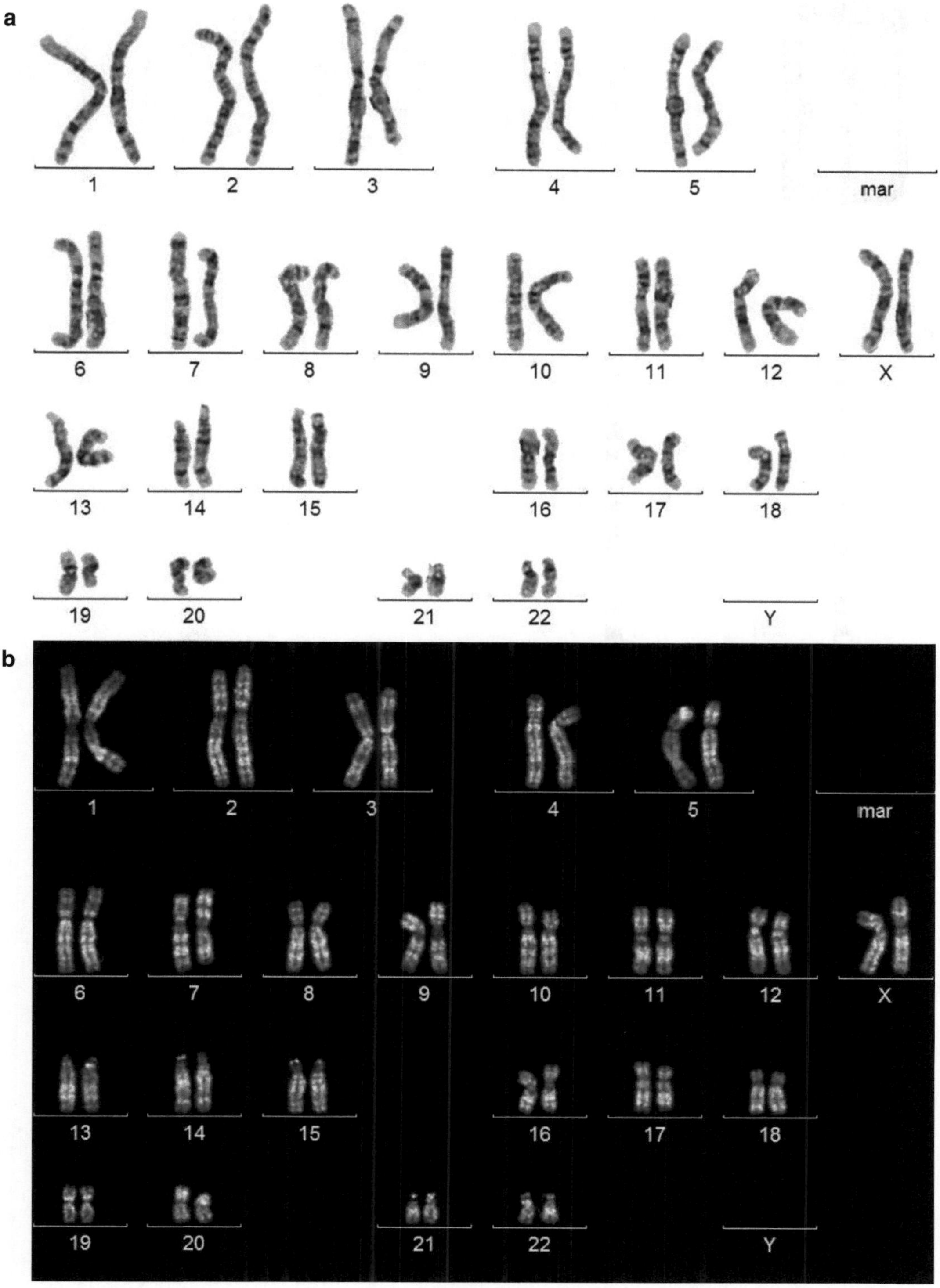

Fig. 6.8 Karyotype 46,XX,inv(7)(p22q34). **a** GTG-, **b** QFQ-banding, 550 and 400 bands per genome, **c** QFQ-, **d** GTG-banding, chromosomes from different mitoses, *left* normal, *right* inverted chromosome 7, **e** ideograms of Case 8, extreme large inverted segment

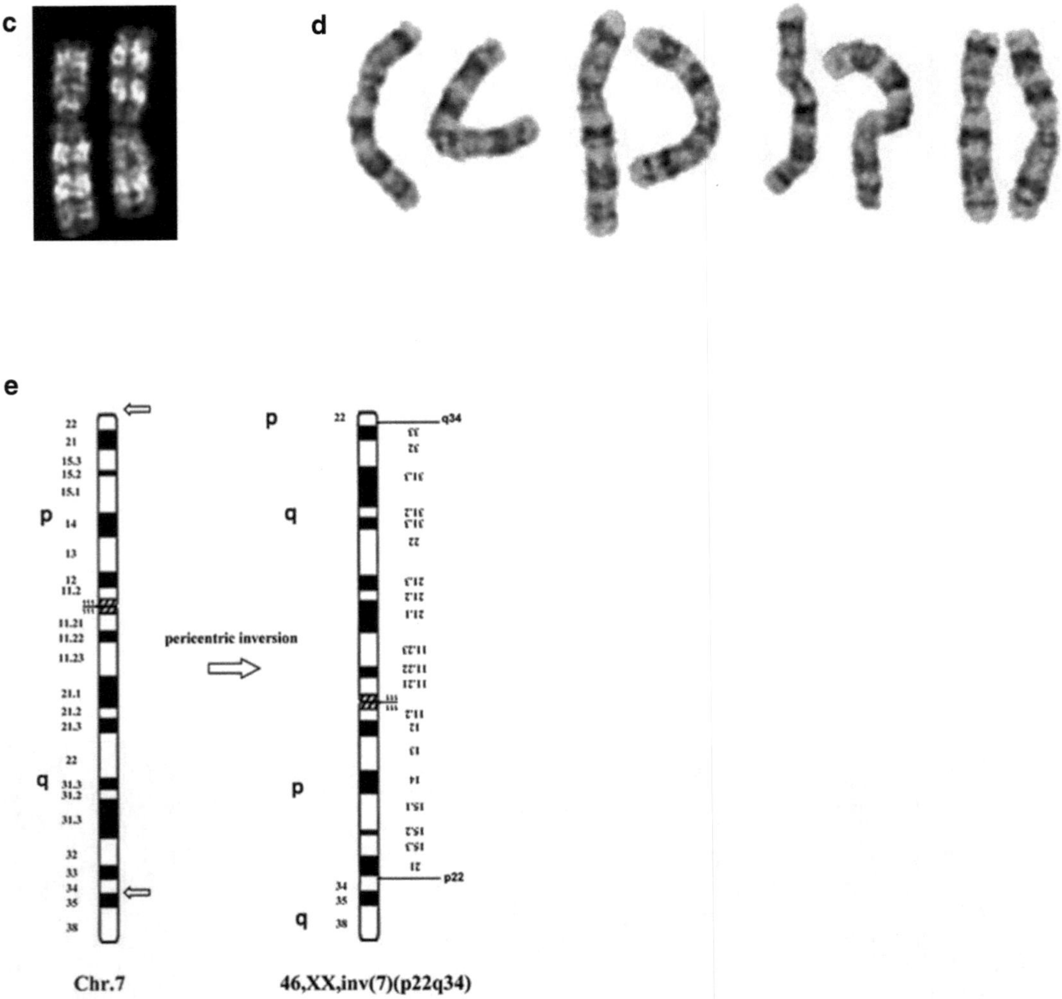

■ **Fig. 6.8** (continued)

■■ Case 9: Karyotype: 46,XX,inv(8)(p11.2q21.12)

.

■ Morphology

The inversion 8 presented here comprises about 25% of the total length of the chromosome. The breakpoint in the short arm is localised in p11.2, a proximal GC-rich band which is neighbouring the pericentromeric heterochromatin. The second breakpoint is intermediate in q21.2. Therefore, the arm ration p:q has changed from 30:70 to 50:50%. The inverted chromosome appears metacentric. The exact breakpoints can only be determined at a resolution of about 700 bands per genome. This case is a good example that a high banding resolution is required in difficult chromosome rearrangements.

■ Cave

Investigations on pericentric inversions of comparable size lead to an intermediate genetic risk for the carrier. Genetic counselling and prenatal diagnoses are indicated (◘ Fig. 6.9).

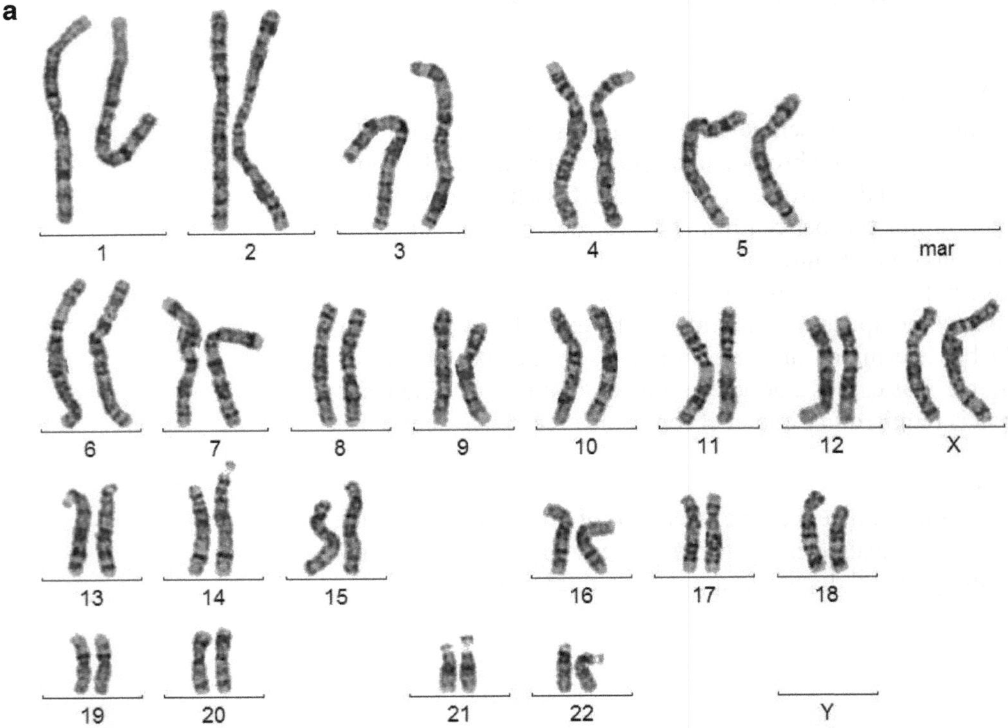

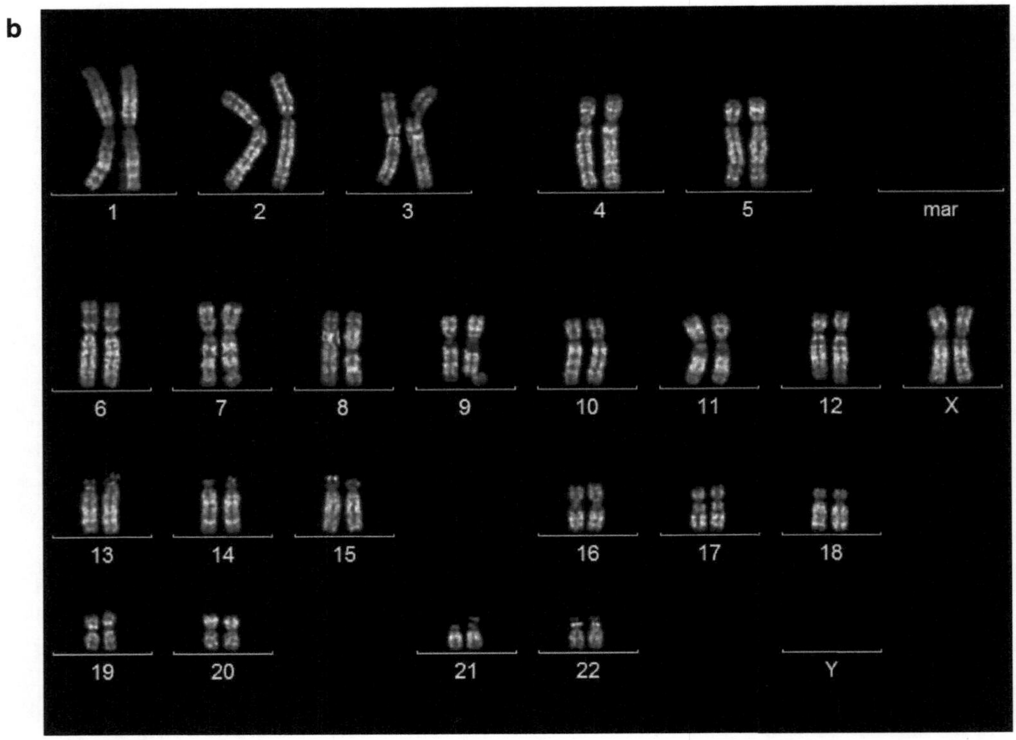

Fig. 6.9 Karyotype 46,XX,inv(8)(p11.2q21.12). **a** GTG-, **b** QFQ-banding, 700 and 400 bands per genome, **c** QFQ-, **d** GTG-banding, chromosomes from different mitoses, *left* normal, *right* inverted chromosome 8, **e** ideograms of Case 9, inverted segment is medium in size

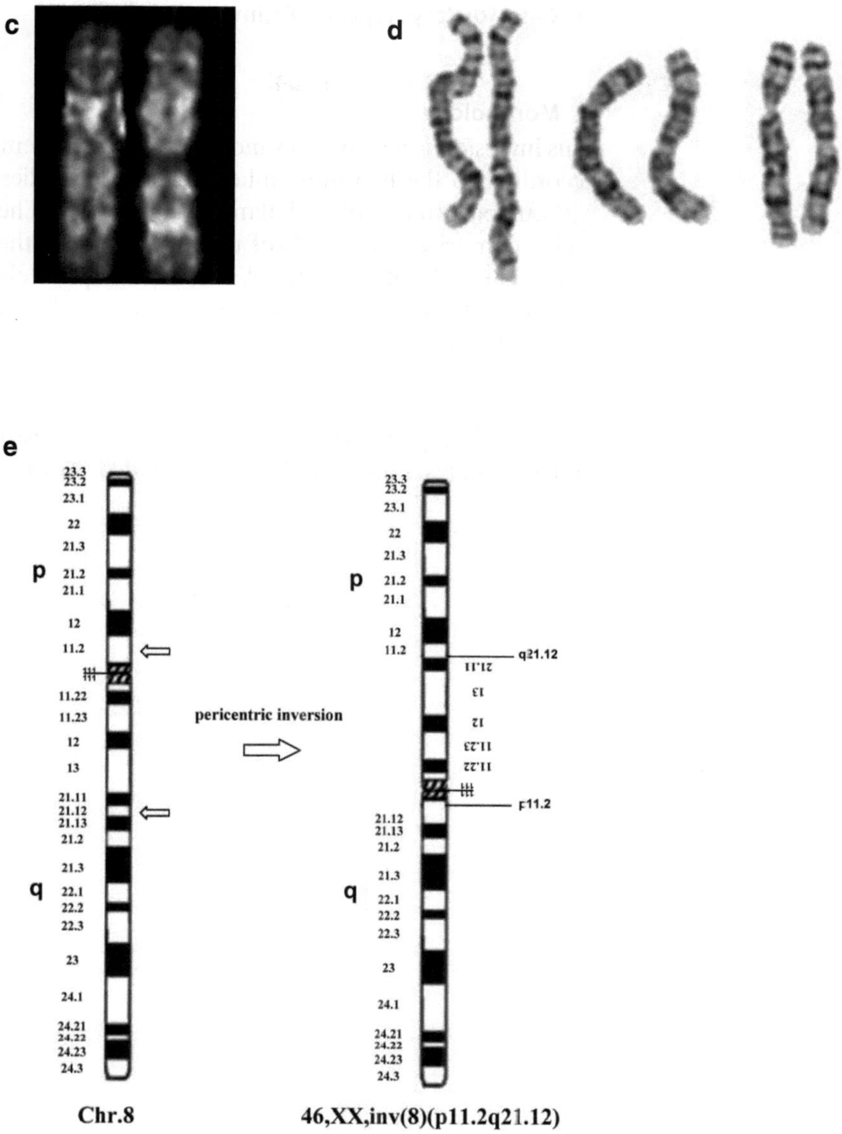

Fig. 6.9 (continued)

▪▪ Case 10: Karyotype: 46,XX,inv(10)(p11.2q21.2)

.

▪ Morphology

This inversion is a stable chromosome rearrangement which is, according to the literature, inherited in the families involved without occurrence of unbalanced karyotypes. The inverted region comprises about 20% of the total length of the chromosome. The dislocation of the AT-rich band q21.1 changes the banding pattern in a characteristic way, so that this inversion is easy to diagnose.

▪ Cave

When analysing this type of inversion, family investigations or prenatal analyses must not be recommended (◘ Fig. 6.10).

a

b

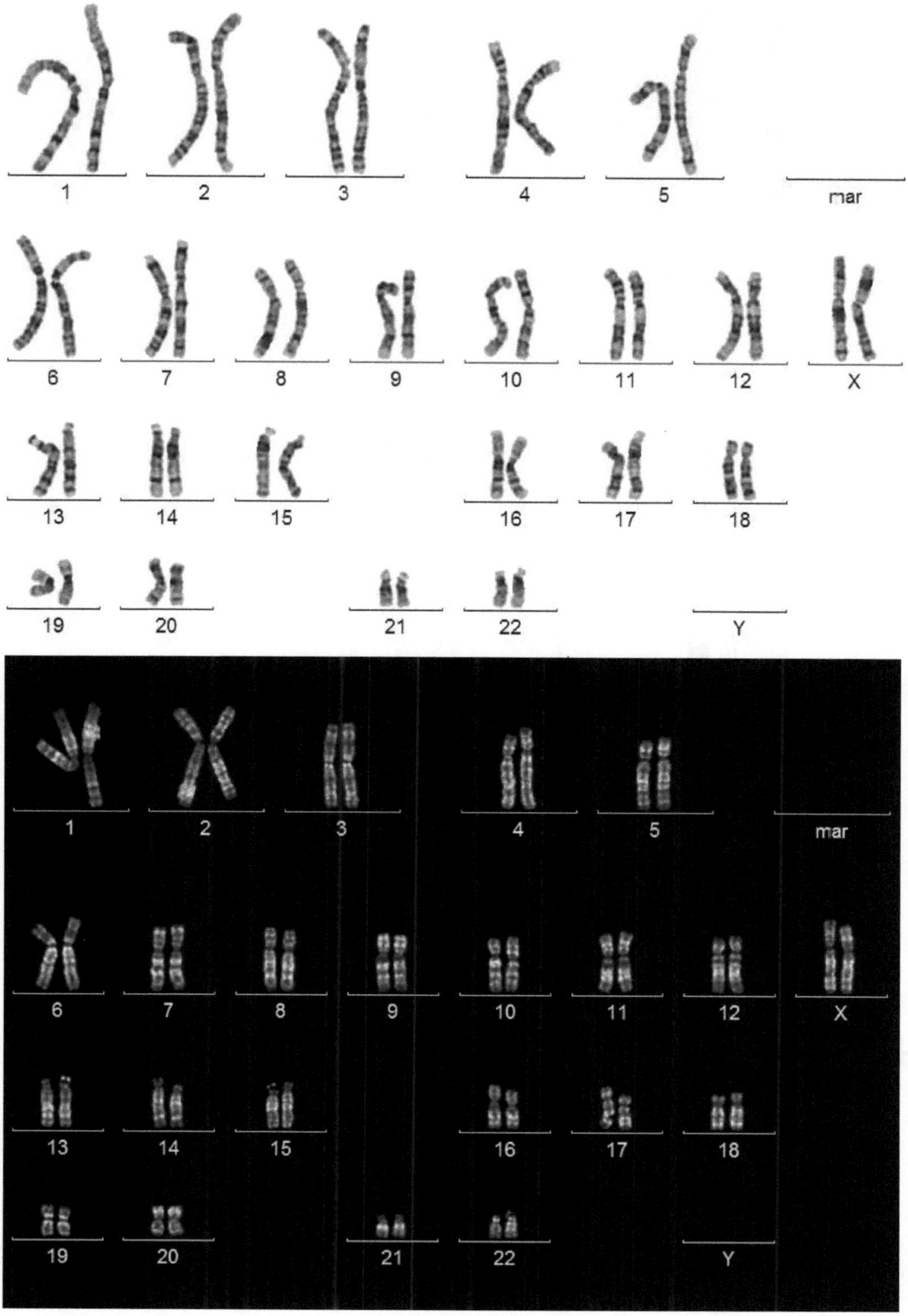

◘ Fig. 6.10 Karyotype 46,XX,inv(10)(p11.2q21.2). **a** GTG-, **b** QFQ-banding, 600 and 400 bands per genome, **c** QFQ-, **d** GTG-banding, chromosomes from different mitoses, *left* normal, *right* inverted chromosome 10, **e** ideograms of chromosome 10, proximal breakpoints of the inverted segments

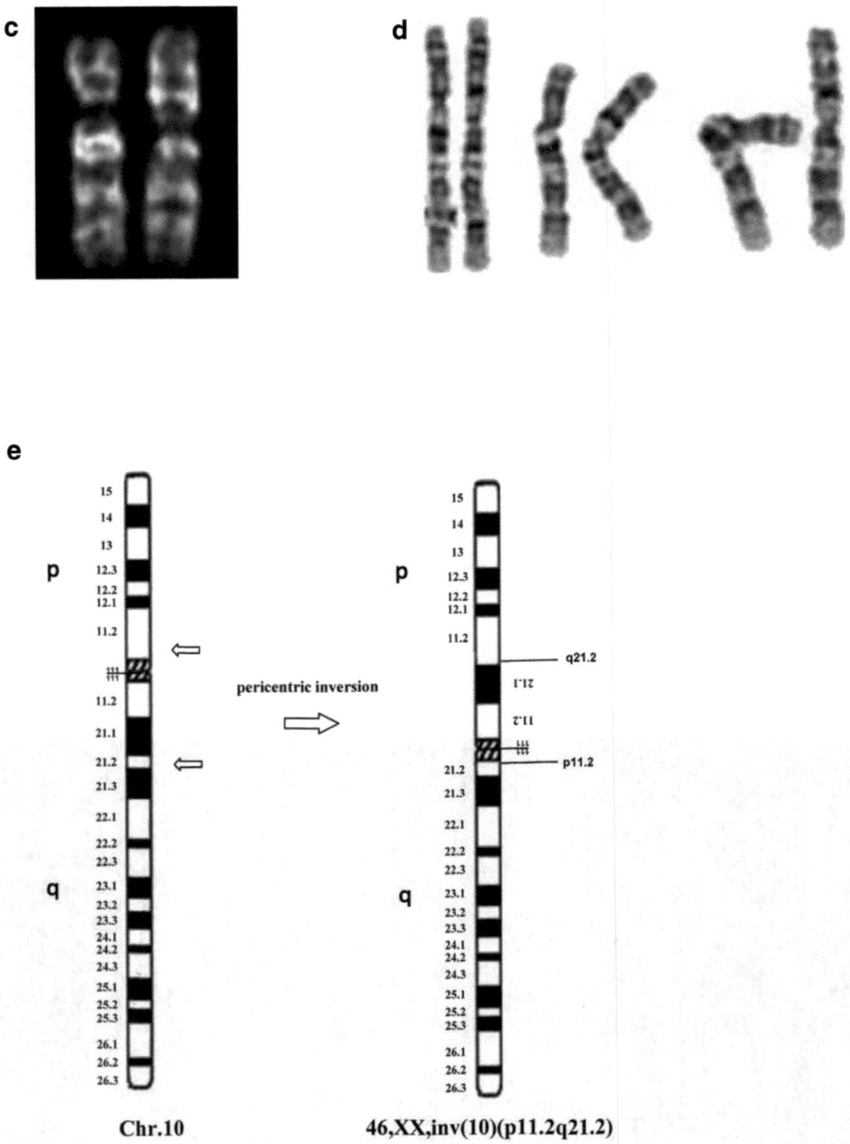

■ **Fig. 6.10** (continued)

▪▪ Case 11: Karyotype: 46,XY,inv(11)(p13q21),inv(9)(p12q12)
The proband presents with two pericentric inversions.

▪ Morphology

1. inv(9)(p12q12): Type II,
   The breakpoints of this pericentric inversion are both localised in heterochromatic segments. The heterochromatic block q12 has been shifted into the proximal short arm region. This type of inversion is defined as polymorphism without genetic relevance. Its frequency in the European population is about 1% (see: ► Sect. 15.6).
2. inv(11)(p13q21).
   This pericentric inversion comprises about 50% of the total length of chromosome 11, both breakpoints are located in intermediate regions. The analysis of the rearrangement is easy to perform because of the dislocation of the AT-rich band q14 from the long into the short arm.

▪ Cave

Genetic counselling after investigation of the relatives includes recommendation of prenatal investigations. But according to the literature, this inversion seems to be stable when inherited and the genetic risk might be small (◘ Fig. 6.11).

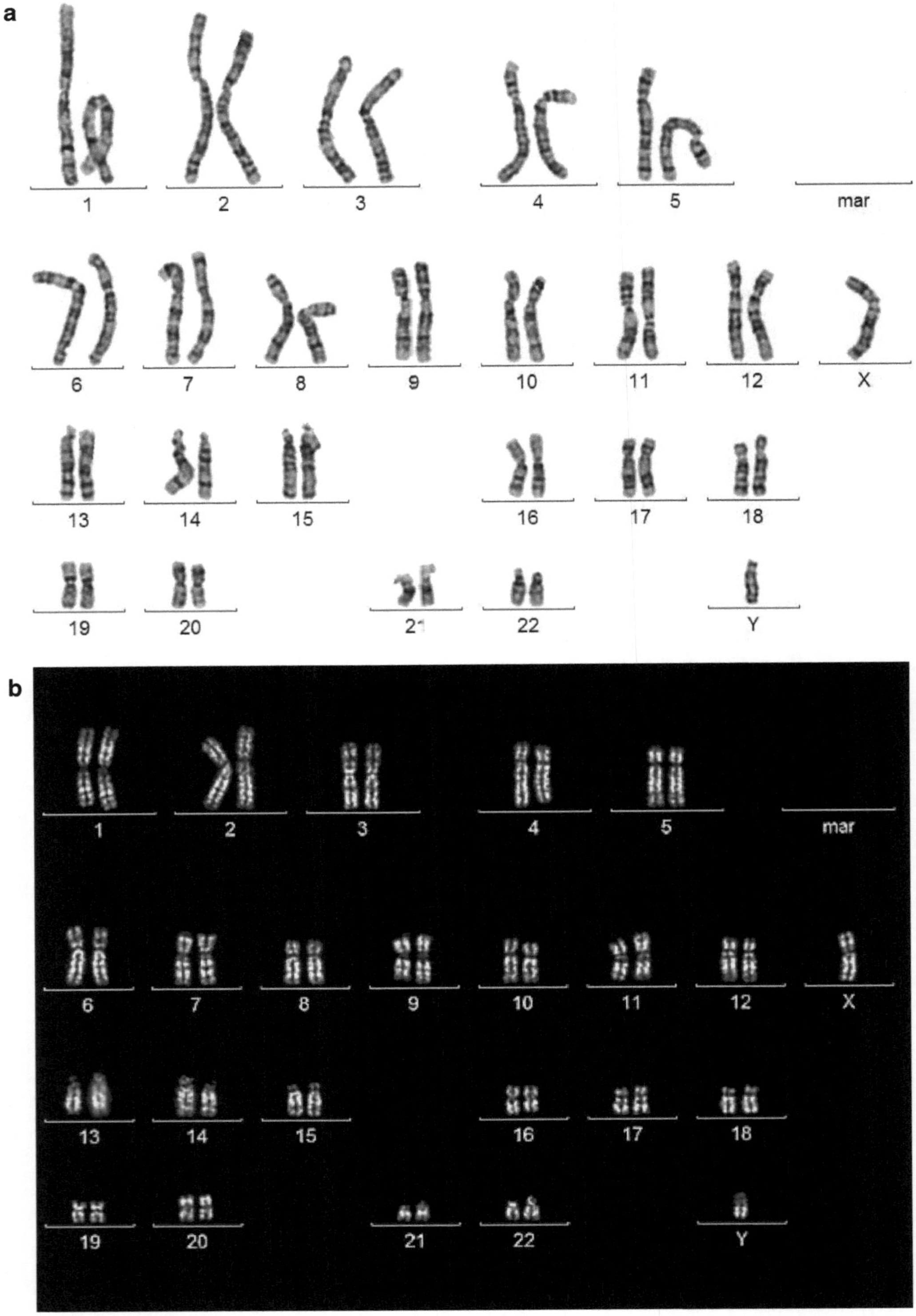

Fig. 6.11 Karyotype 46,XY,inv(11)(p13q21),inv(9)(p12q12). **a** GTG-, **b** QFQ-banding, 600 and 400 bands per genome, **c** CBG-, **d** QFQ-, **e** GTG-banding, chromosomes from different mitoses, chromosome 9: *left* normal, *right* partial inverted chromosome, Chromosome 11: *left* normal, *right* inverted, chromosome 11, **f** ideograms of chromosomes 9 and 11, inverted chromosomes right of the normal ones

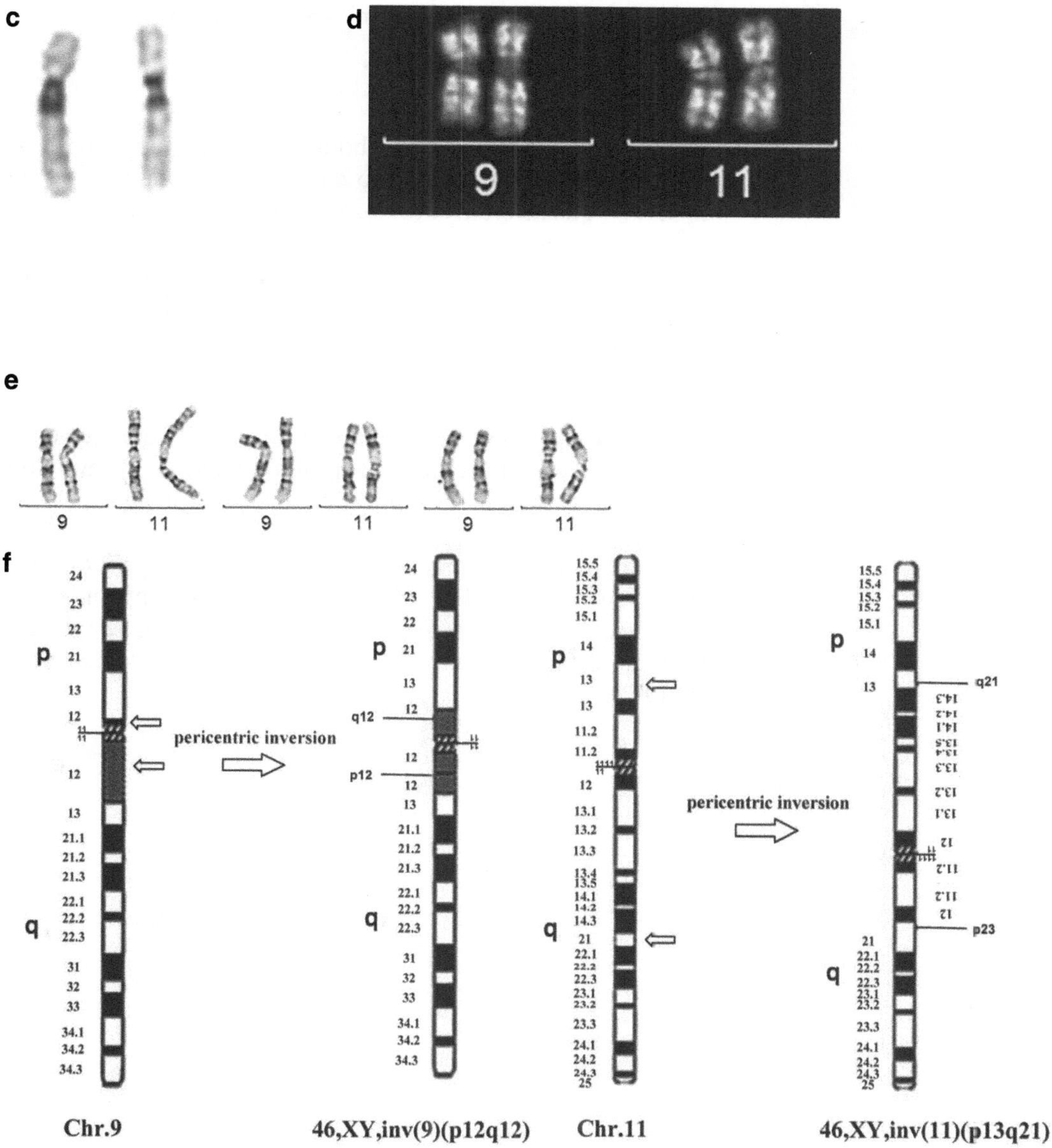

Fig. 6.11 (continued)

Case 12: Karyotype: 46,XX,inv(16)(p13.2q11.1)

.

Morphology

The inverted region encloses about 40% of the total length of chromosome 16. The arm ratio has changed from p:q = 40:60 to 3:97%. The inverted chromosome 16 appears telocentric like the chromosomes 13, 14, 15. In this case, CBG banding is of great diagnostic relevance, as only by this technique the position of the centromere and the heterochromatic block q11.2 can be clearly defined.

Cave

According to our knowledge, comparable cases from the literature do not exist. It would be helpful to find as many inversion carriers as possible in the family and to look at pregnancies and offspring of all carriers.

To all carriers prenatal investigations, including microarray, should be recommended. In a pathologic course of pregnancy—abortion, stillbirth—the possible unbalanced karyotype should be analysed and the phenotype of the conceptus documented (Fig. 6.12).

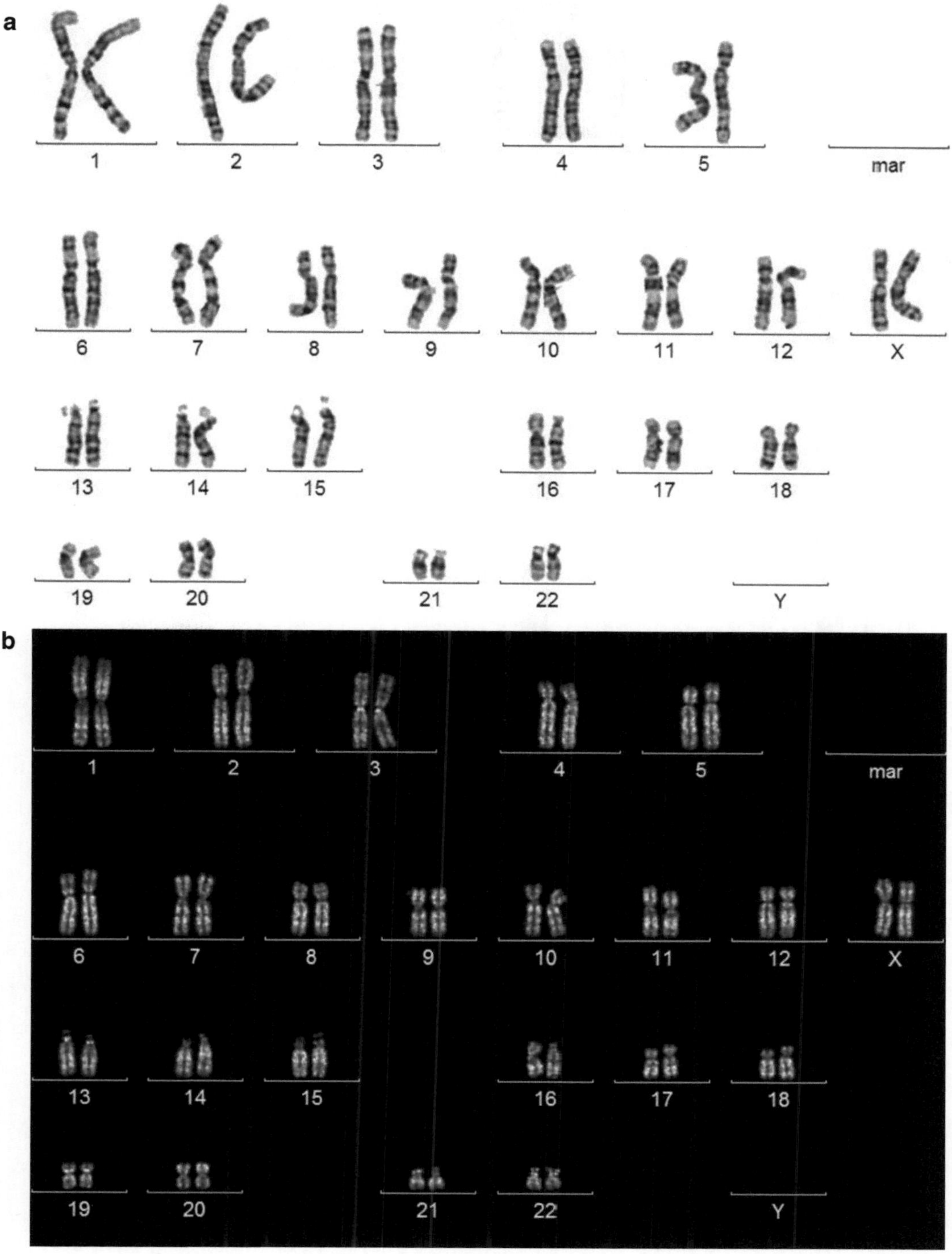

Fig. 6.12 Karyotype 46,XX,inv(16)(p13.2q11.1). **a** GTG-banding, 600 bands per genome, **b** QFQ-banding 400 bands per genome, **c** CBG-, **d** QFQ-, **e** GTG-banding, chromosomes from different mitoses: *left* normal, *right* inverted chromosome 16, **f** ideograms of Case 12, partial dislocation of the centromeric region

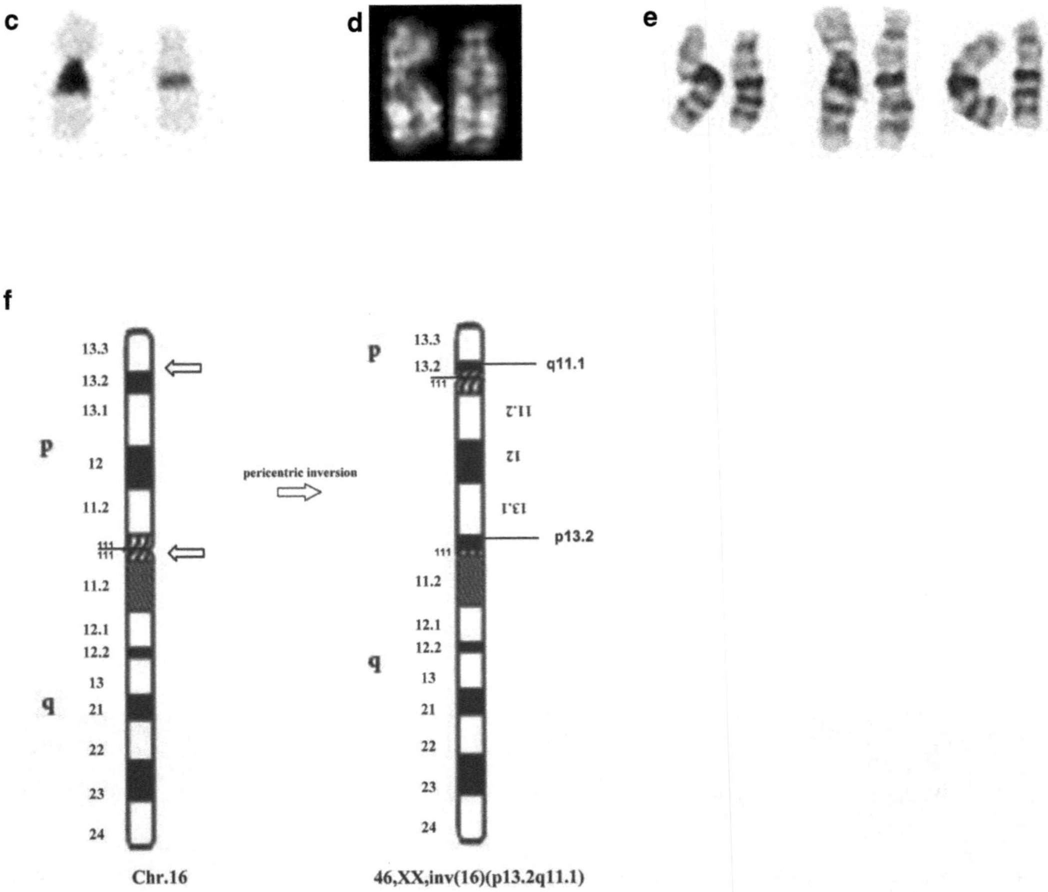

Fig. 6.12 (continued)

■■ Case 13: Karyotype: 46,XY,inv(16)(p12q12.2)

■ Morphology

As in the previous case (Case 12), the complete heterochromatic block q11.2 has changed its position from the long into the short arm. The inverted region encloses about 40% of the total length of the chromosome, the arm ratio has not changed significantly. Both breakpoints are located in small, AT-rich bands. CBG-banding illustrates this change.

■ Cave

The risk of meiotic unequal crossing over causing an unbalanced karyotype is given. Therefore, genetic counselling and prenatal diagnosis should be offered (■ Fig. 6.13).

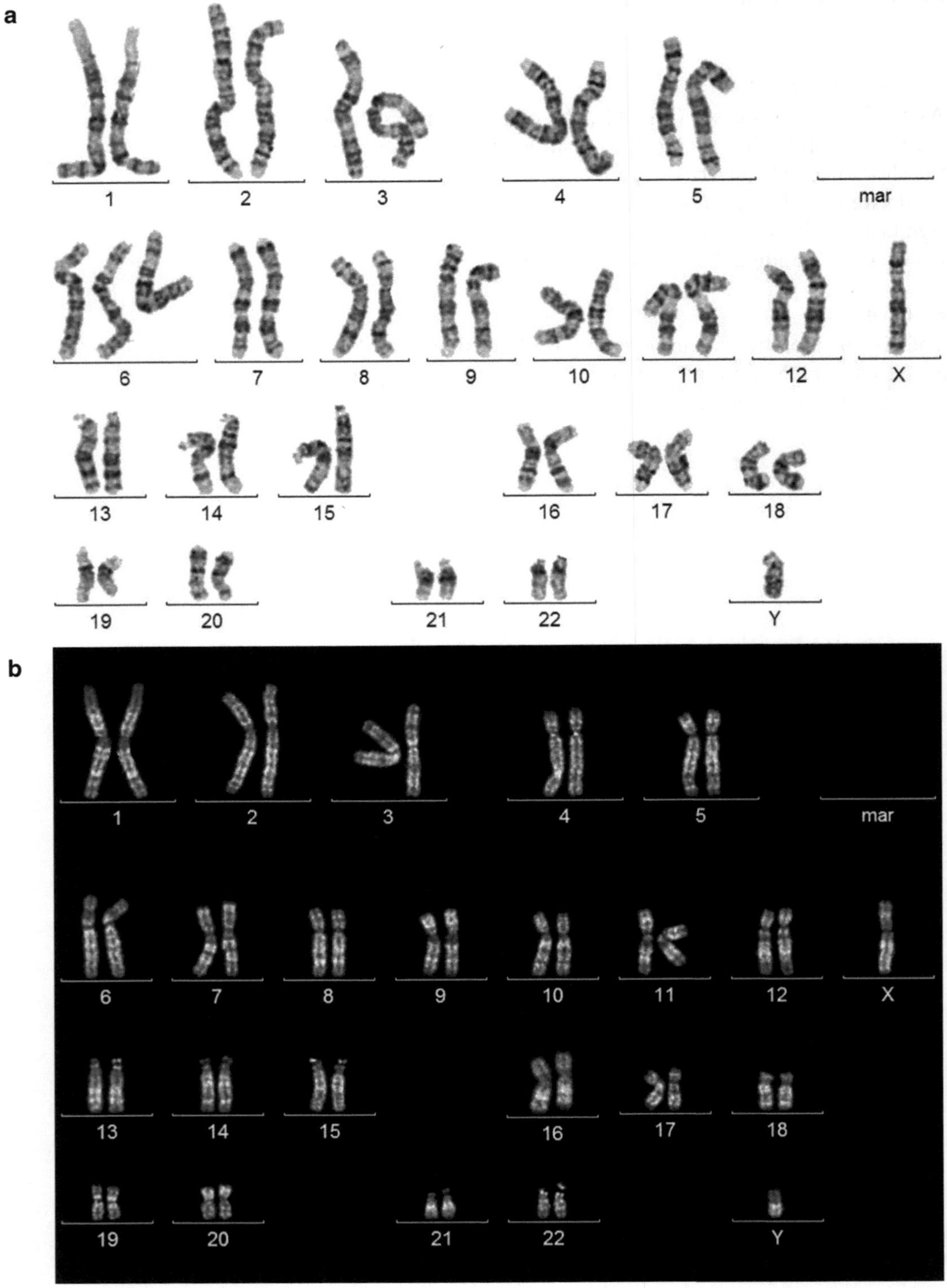

Fig. 6.13 Karyotype 46,XY,inv(16)(p12q12.2). **a** GTG-banding, 700 bands per genome, single pathologic cell with additional chromosome 6, **b** QFQ-banding, 400 bands per genome, **c** QFQ-, **d** GTG-banding, chromosomes from different mitoses, *left* normal, *right* inverted chromosome 16, **e** ideograms of chromosome 16. Transposition of the heterochromatin block into the short arm

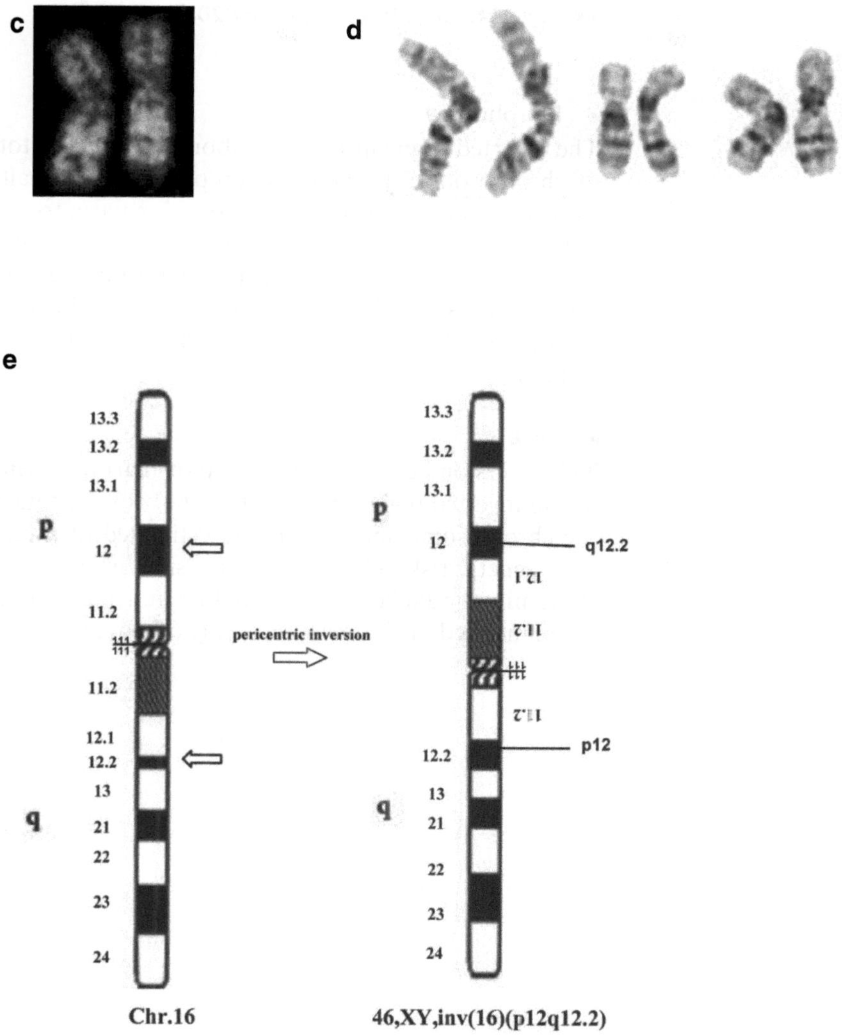

Fig. 6.13 (continued)

Case 14: Karyotype: 46,XX,inv(20)(p13q13.2)

.

Morphology

The inverted segment encloses about 70% of the total length of chromosome 20. The inversion product has an altered centromere position and thus appears submetacentric. The main band of the short arm (p12) has been translocated into the proximal region of the long arm. The morphology of the aberrant chromosome shows an altered banding which can be easily recognised even at a resolution of less than 550 bands per genome.

Cave

This inversion belongs to the group of rearrangements enclosing an inverted region of more than 60% of the total length of the chromosome and is therefore estimated to present with a high genetic risk. Therefore, all possible inversion carriers in the family should be investigated. Prenatal diagnoses have to be recommended, including microarray (◘ Fig. 6.14).

a

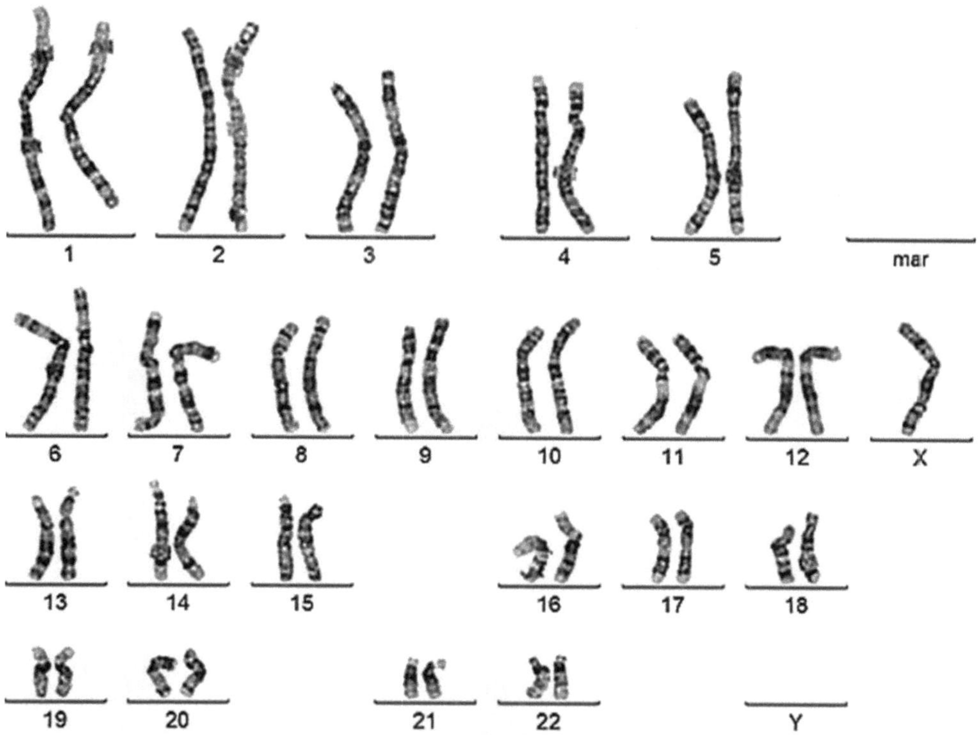

b

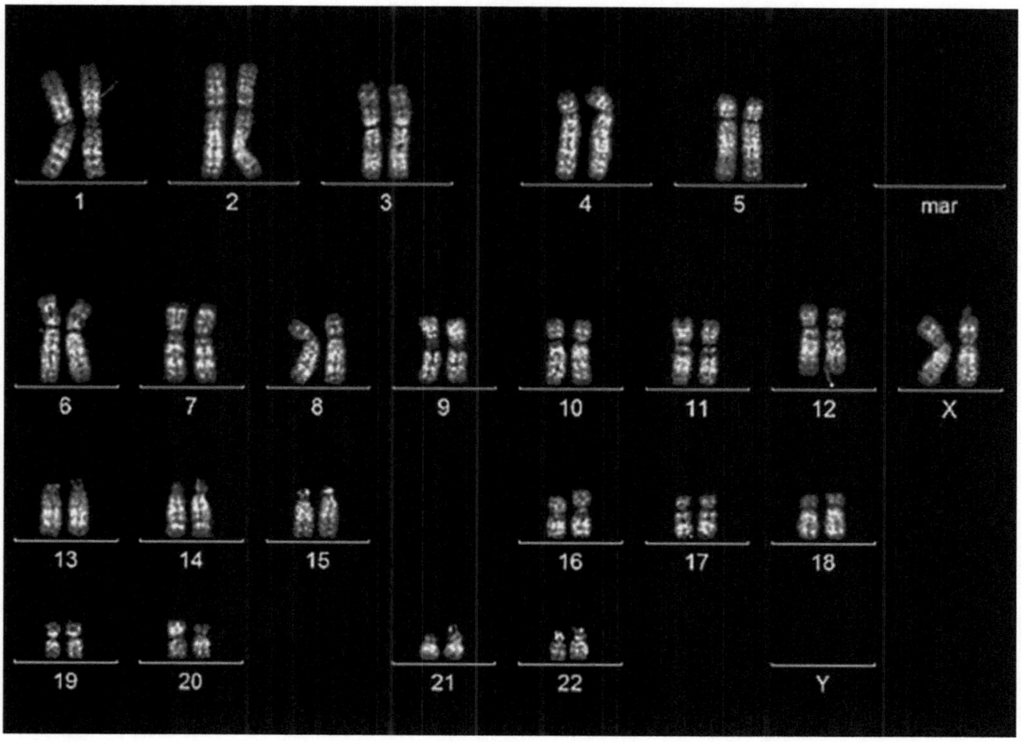

◘ Fig. 6.14 Karyotype 46,XX,inv(20)(p13q13.2). **a** GTG-banding, 1 chromosome X lost by preparation. **b** QFQ-banding, 700 and 400 bands per genome, **c** GTG-banding, chromosomes from different mitoses, *left* normal, *right* inverted chromosome 20, **d** ideograms of chromosome 20, large inverted segment

c

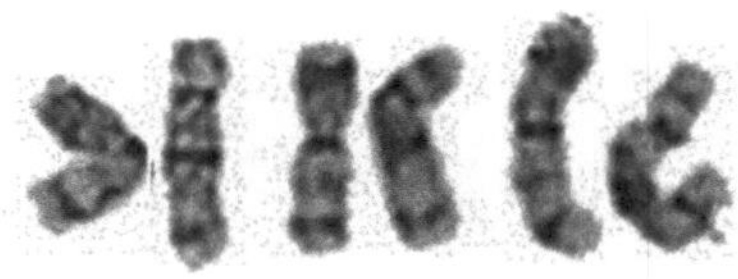

d

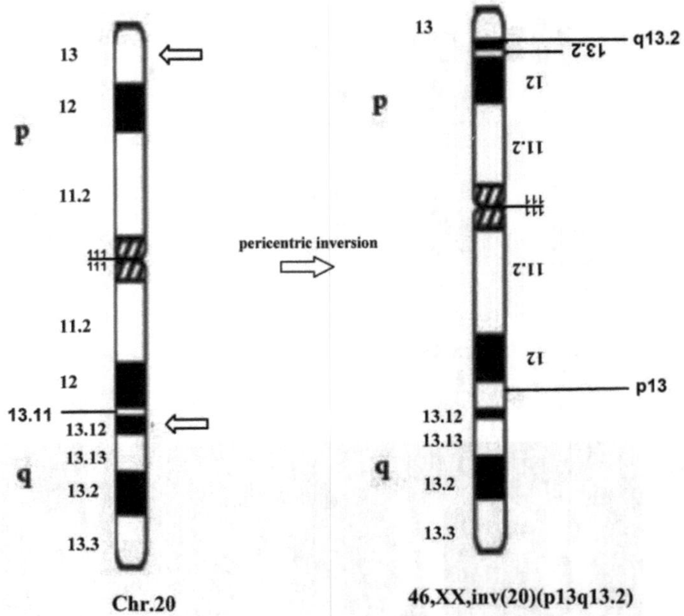

■ Fig. 6.14 (continued)

Case 15: Karyotype: 46,X,inv(X)(p22.1q26)

Morphology

The inverted region contains about 80% of the total length of the X-chromosome. The arm ratio changed from about 35:65 to 80:20%. This and the translocation of the AT-rich band q21 into the short arm make the diagnosis quite easy.

Cave

The risk of an unequal crossing over in meiosis is high because of the size of the inverted segment. The phenotypic alterations will be more severe in males (1 X-chromosome) inheriting an aberrant X-chromosome than in females (2 X-chromosomes). In the latter group, the clinical symptoms caused by an unbalanced karyotype can be expected to show great variability caused by a possible screwed inactivation of the second X-chromosome.

An additional risk factor is given when the breakpoint in Xq is localised between q13 and q26 as this can induce primary or secondary amenorrhoea in the female carrier (◘ Fig. 6.15).

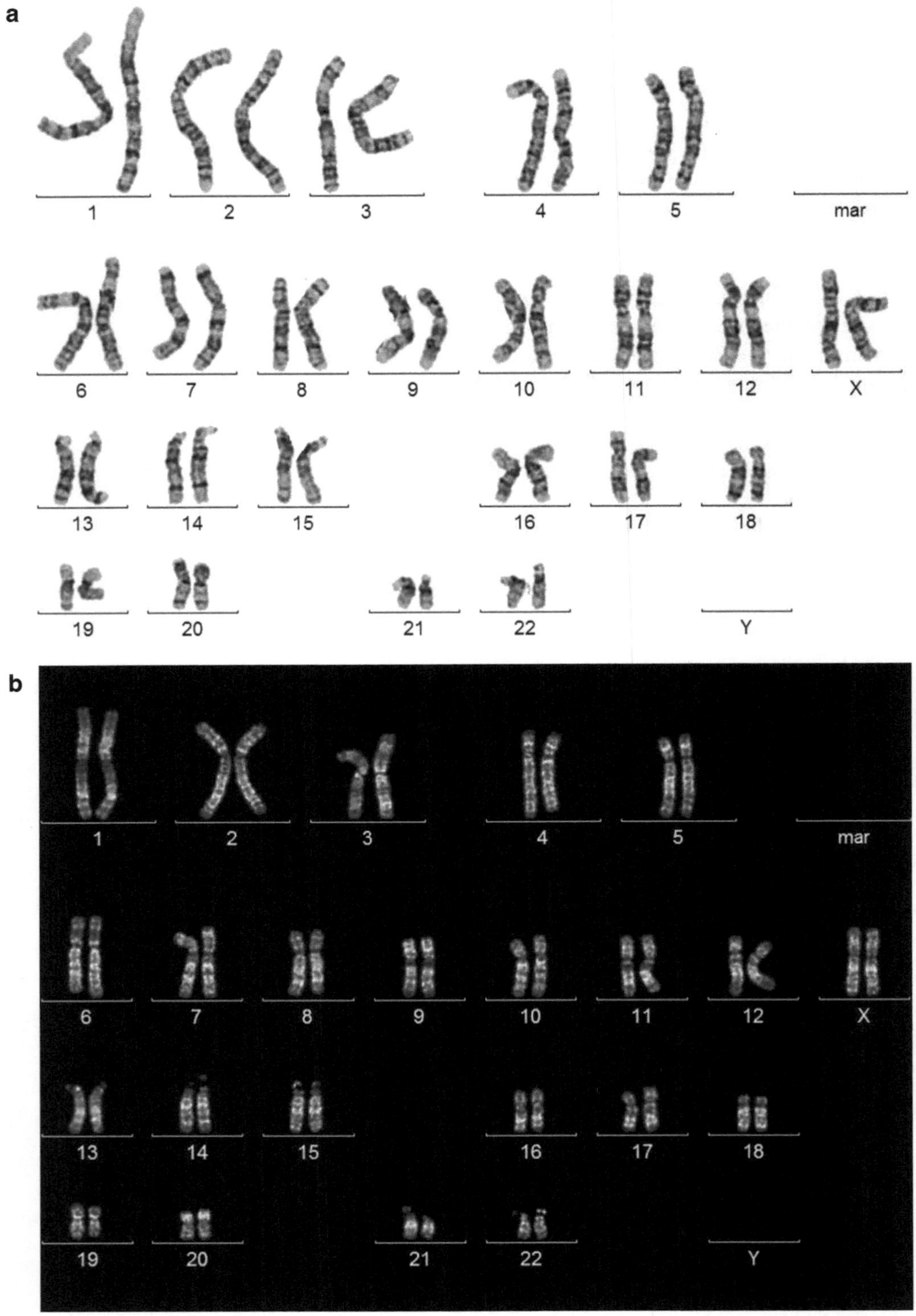

Fig. 6.15 Karyotype 46,X,inv(X)(p22.1q26). **a** GTG-, **b** QFQ-banding, 700 and 400 bands per genome, **c** QFQ-, **d** GTG-banding, chromosomes from different mitoses, *left* normal, *right* inverted X-chromosome, **e** ideograms of the X-chromosomes, large inverted segment

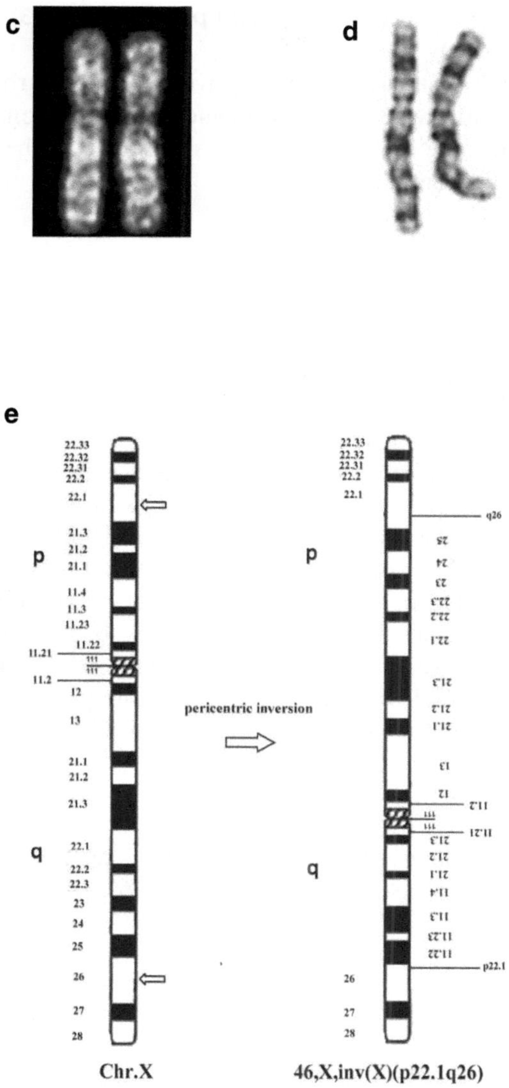

Fig. 6.15 (continued)

▪▪ Case 16: Karyotype: 46,X,inv(X)(p22.13q24),inv(3)(p11.2q12)

The proband has two pericentric inversions: one in the X chromosome with two breakpoints in euchromatic regions and one in chromosome 3 which inversion is restricted to the constitutive heterochromatin.

▪ Morphology

1. inv(3)(p11.2q12),

   The pericentric inversion 3 includes no euchromatic segments and is therefore defined as polymorphism. It can only be diagnosed after QFQ-banding. Its frequency in the European population is about 2% (see: ▶ Sect. 15.4).
2. inv(X)(p22.13q24).

   The breakpoints are both located in the distal regions of the short and the long arm. The large inverted segment encloses about 70% of the total length of chromosome X. The arm ratio has changed as in the previous case (Case 15) from 35:65 to 55:45% and the aberrant X appears almost metacentric. Besides the altered banding pattern, it is characterised by the translocation of the AT-rich bands p22.1 to 22.3 from the short into the long arm and q21.1, q21.3 from the long into the short arm. Therefore, the diagnosis is easy to perform.

▪ Cave

The same genetic risk factors have to be taken into account as in the previous case (Case 15) (◘ Fig. 6.16).

a

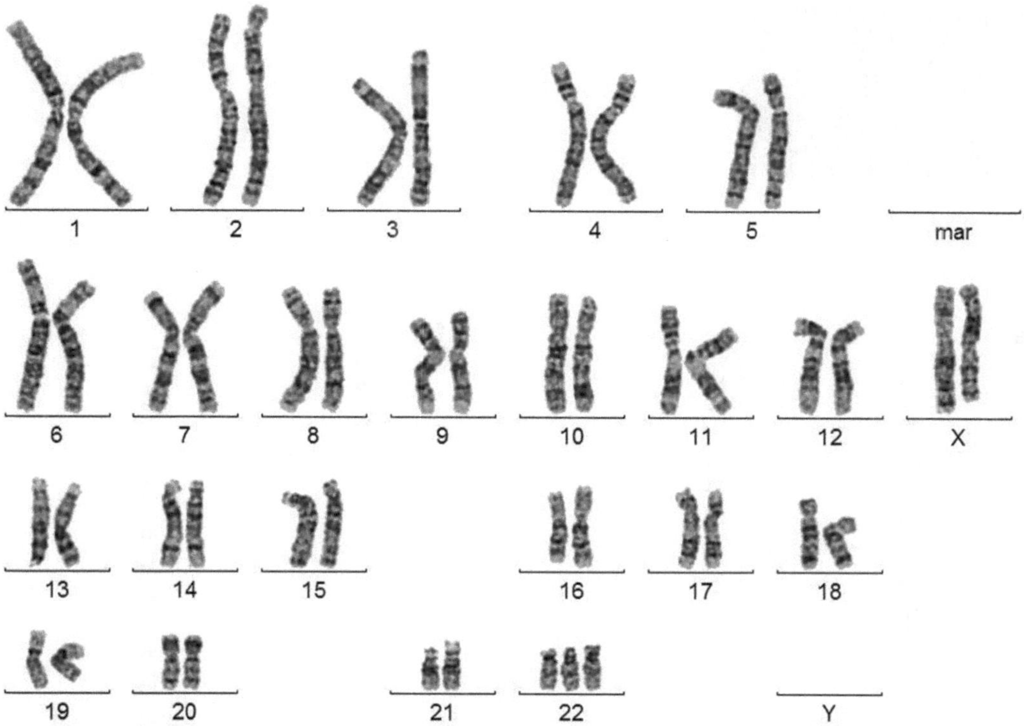

b

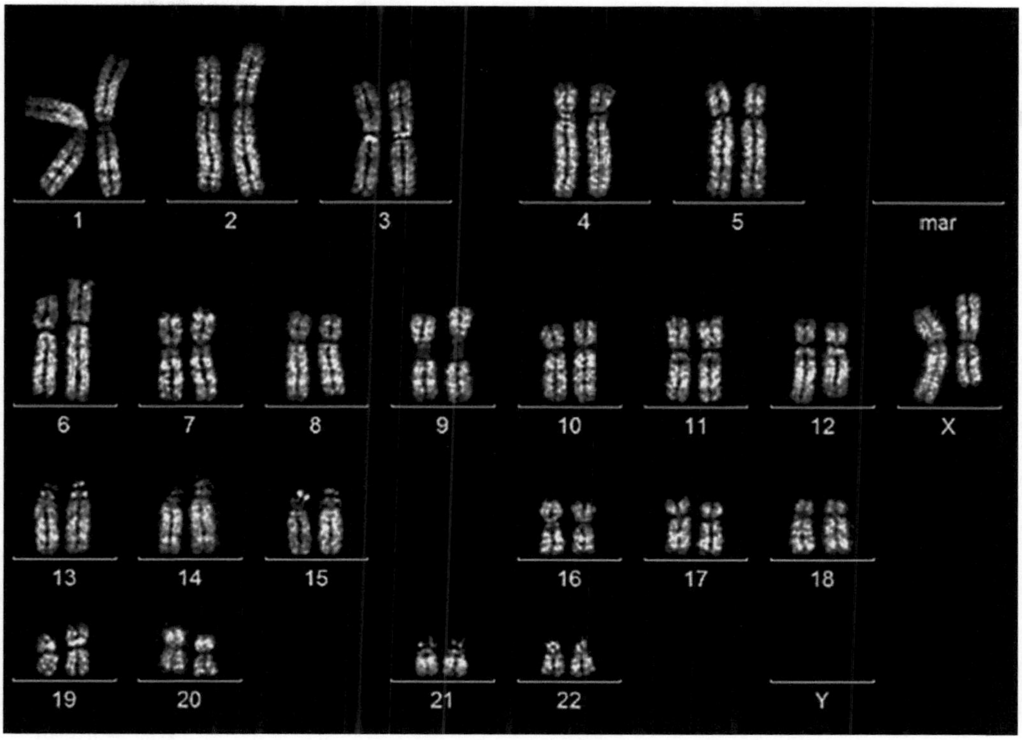

Fig. 6.16 Karyotype 46,X,inv(X)(p22.13q24),inv(3)(p11.2q12). **a** GTG-banding, single pathologic cell with additional chromosome 22, **b** QFQ-banding, 700 and 400 bands per genome, **c** QFQ-, **d** GTG-banding, chromosome 3 and X from different mitoses, *left* normal, *right* inverted chromosomes, **e**, **f** normal and inverted ideograms of chromosome 3 and X

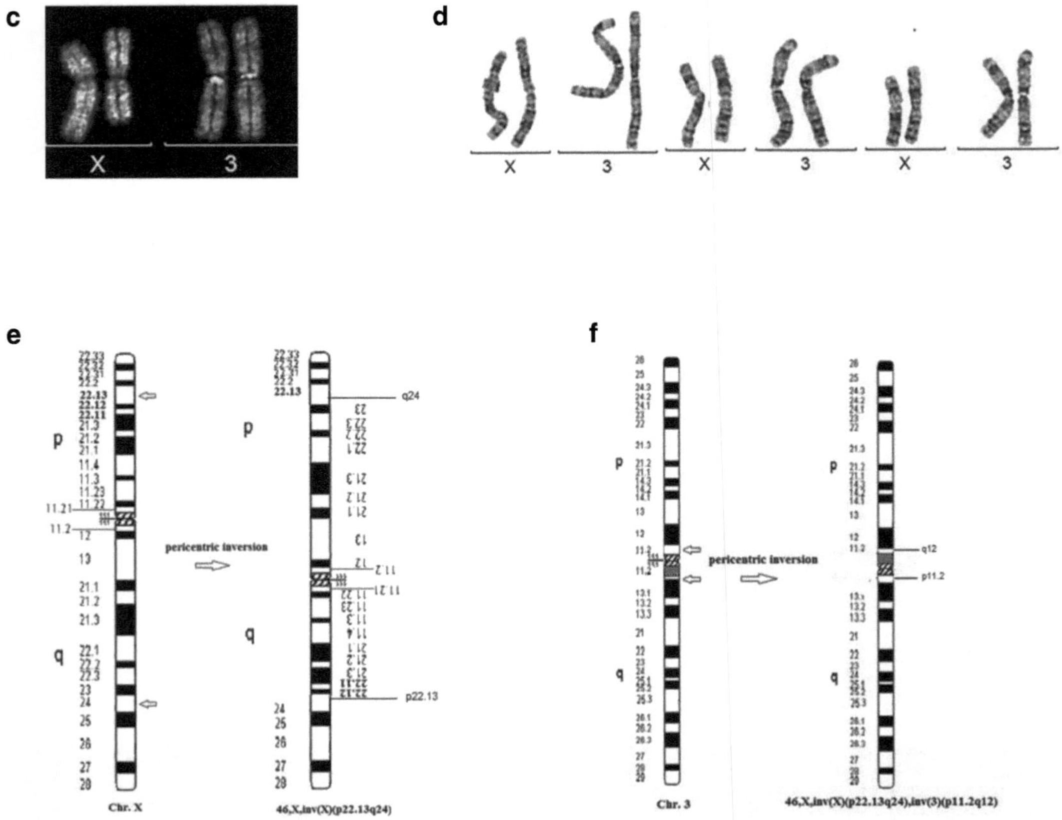

Fig. 6.16 (continued)

■■ Case 17: Karyotype: 46,X,inv(Y)(p11.2q11.23)

·

■ Morphology

The pericentric inversion presented here demonstrates a complete transfer of the distal heterochromatic band q12 from the long into the short arm. The aberrant chromosome is metacentric. In this case, the inverted region encloses about 50% of the total length of the Y-chromosome. Caused by the variation in length of q12 in different probands, this index can only be determined for the individual carrier. The specific presentation of the inverted chromosome should be done not only by QFQ but also by DA/DAPI-banding as the proximal heterochromatic region of Yq12 can only be analysed by DA/DAPI. This is of diagnostic relevance because a breakpoint in the proximal heterochromatin is of no genetic importance in contrast to a breakpoint in the following euchromatin.

■ Cave

The Y-inversion presented here has a frequency of 1–2‰ with differences in single ethnic groups. There is a large variability in phenotypic peculiarities caused by possible pairing abnormalities of X and Y that can lead to oligo- and azoospermia in the carrier (◘ Fig. 6.17).

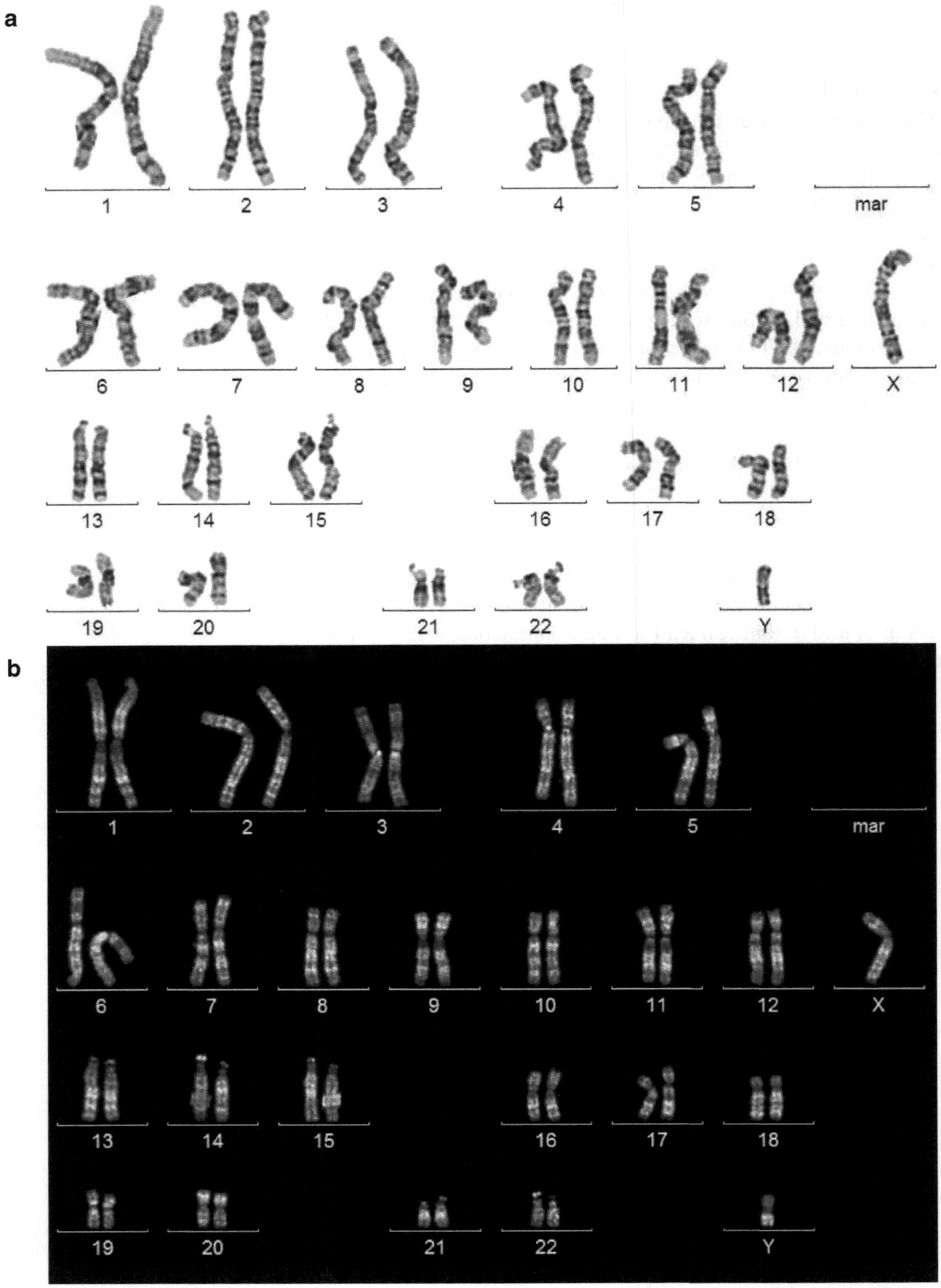

Fig. 6.17 Karyotype 46,X,inv(Y)(p11.2q11.23). **a** GTG-, **b** QFQ-banding, 700 and 400 bands per genome, **c** CBG-, **d** QFQ-, **e** GTG-banding, Y-chromosome from different mitoses, **d/1**: Y-chromosome from a normal control person, **f** ideograms of Y-chromosome, the inverted segment leads to a metacentric morphology

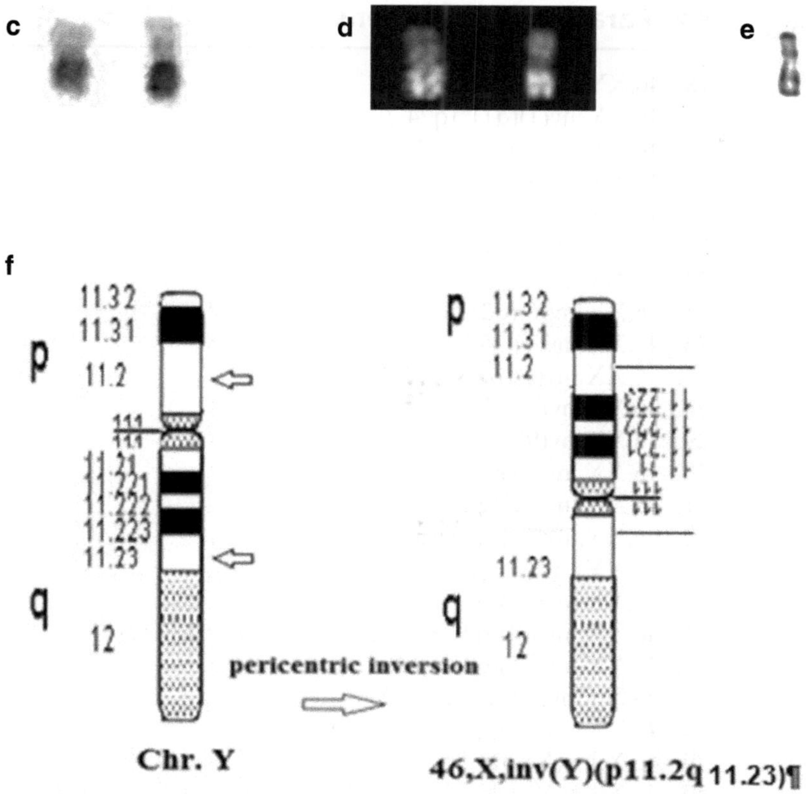

Fig. 6.17 (continued)

6.2 Paracentric Inversions

18. 46,XY,inv(1)(p32.1p36.1)
19. 46,XY,inv(1)(q11.1q24.2)
20. 46,XX,inv(2)(q14.2q21.1)
21. 46,XY,inv(3)(p22.2p25)
22. 46,XY,inv(4)(q13.2q27–q28.2)
23. 46,XY,inv(5)(p11p15.1)
24. 46,XX,inv(6)(q16.2q24.2)
25. 46,XY,inv(6)(q23.3q25.3)
26. 46,XX,inv(6)(q25.1q25.3)
27. 46,XX,inv(7)(p14.2p21.2)
28. 46,XY,inv(8)(q13q24.22)
29. 46,XY,inv(9)(q32q34.3)
30. 46,XX,inv(11)(q21q23.3)
31. 46,XX,inv(12)(q21.1q24.1)
32. 46,XX,inv(13)(q12.1q22)
33. 46,XY,inv(13)(q13q32)
34. 46,XX,inv(13)(q21.2q31)
35. 46,XX,inv(14)(q23.3q32.2),inv(3)(p11.2q12)(x2)
36. 46,XX,inv(15)(q24q26.3)
37. 46,XY,inv(19)(p12p13.13)
38. 46,XX,inv(19)(q13.31q13.42)
39. 46,X,inv(X)(p11.3p22.32)
40. 46,X,inv(Y)(q11.21q12),inv(9)(p12q12),inv(9)(p12q13).

▪▪ Case 18: Karyotype: 46,XY,inv(1)(p32.1p36.1)

.

▪ Morphology

The case presented here shows a small distal inversion in the short arm (p) which comprises only about 10–15% of the total length of chromosome 1. This inversion is in spite of its small extend easy to identify because of the shift of AT-rich bands into the distal GC-rich segment. A further peculiarity of the inverted chromosome is the division of the heterochromatic block q12 into 3 subunits. This is a polymorphism without genetic relevance (see: ► Sect. 15.6).

▪ Cave

The region 1p36 contains an aberration hot spot. If a microdeletion (del < 5 Mb) occurs in this locus the carrier shows a characteristic severe malformation syndrome. On the basis of this observation, all carriers of the aberration type presented here should be recommended prenatal diagnosis in each pregnancy because of the risk of an unequal crossing over in meiosis, and this investigation has to include microarray or FISH (LSI-probes) analyses. Relatives of the carrier should be recommended to perform chromosome analyses (◘ Fig. 6.18).

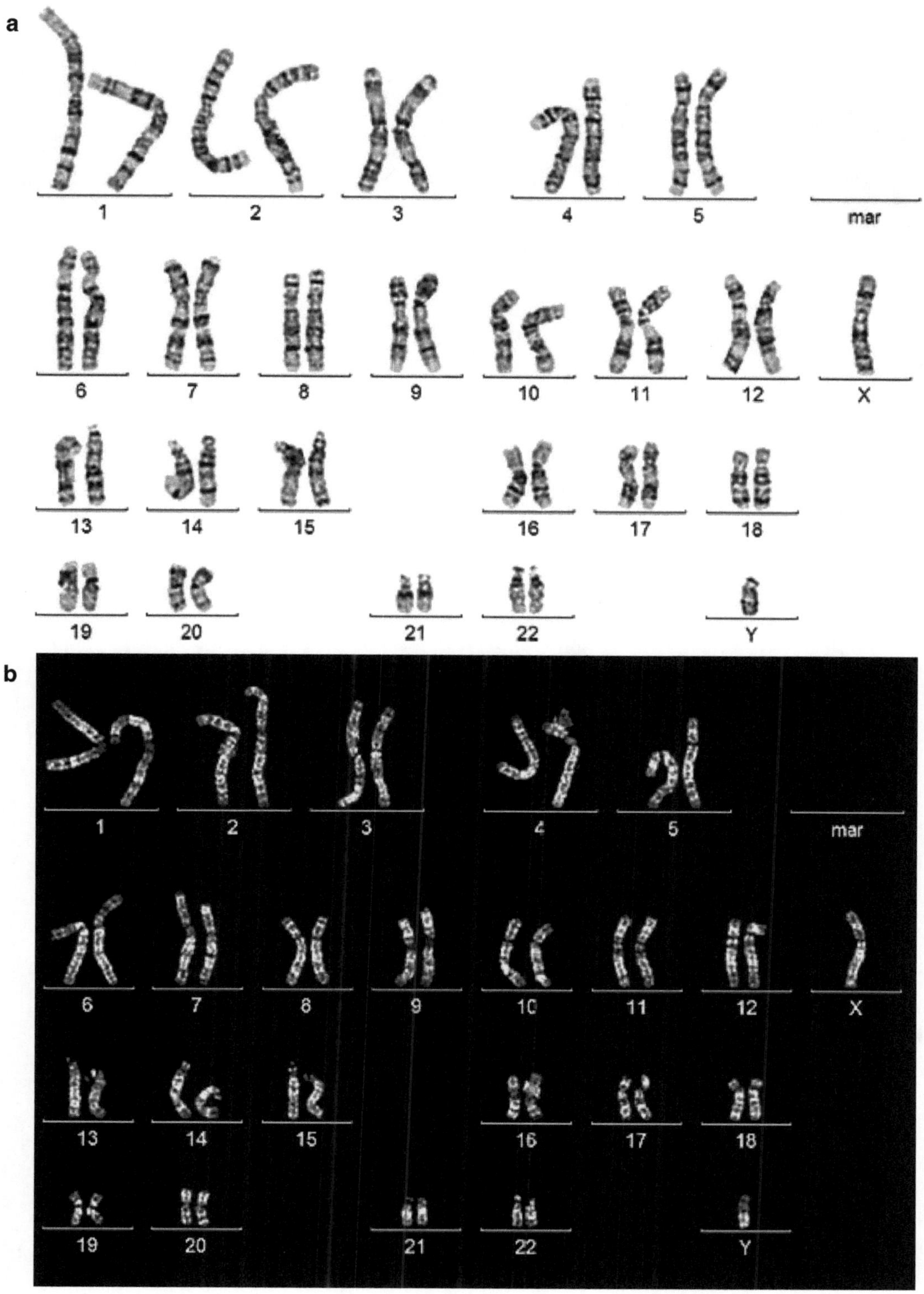

Fig. 6.18 Karyotype 46,XY,inv(1)(p32.1p36.1). **a** GTG-, **b** QFQ-banding, 700 and 500 bands per genome, **c** QFQ-, **d** GTG-banding, chromosomes from different mitoses, *left* normal, *right* inverted chromosome 1, **e** ideograms of chromosome 1, small inverted segment in the short arm

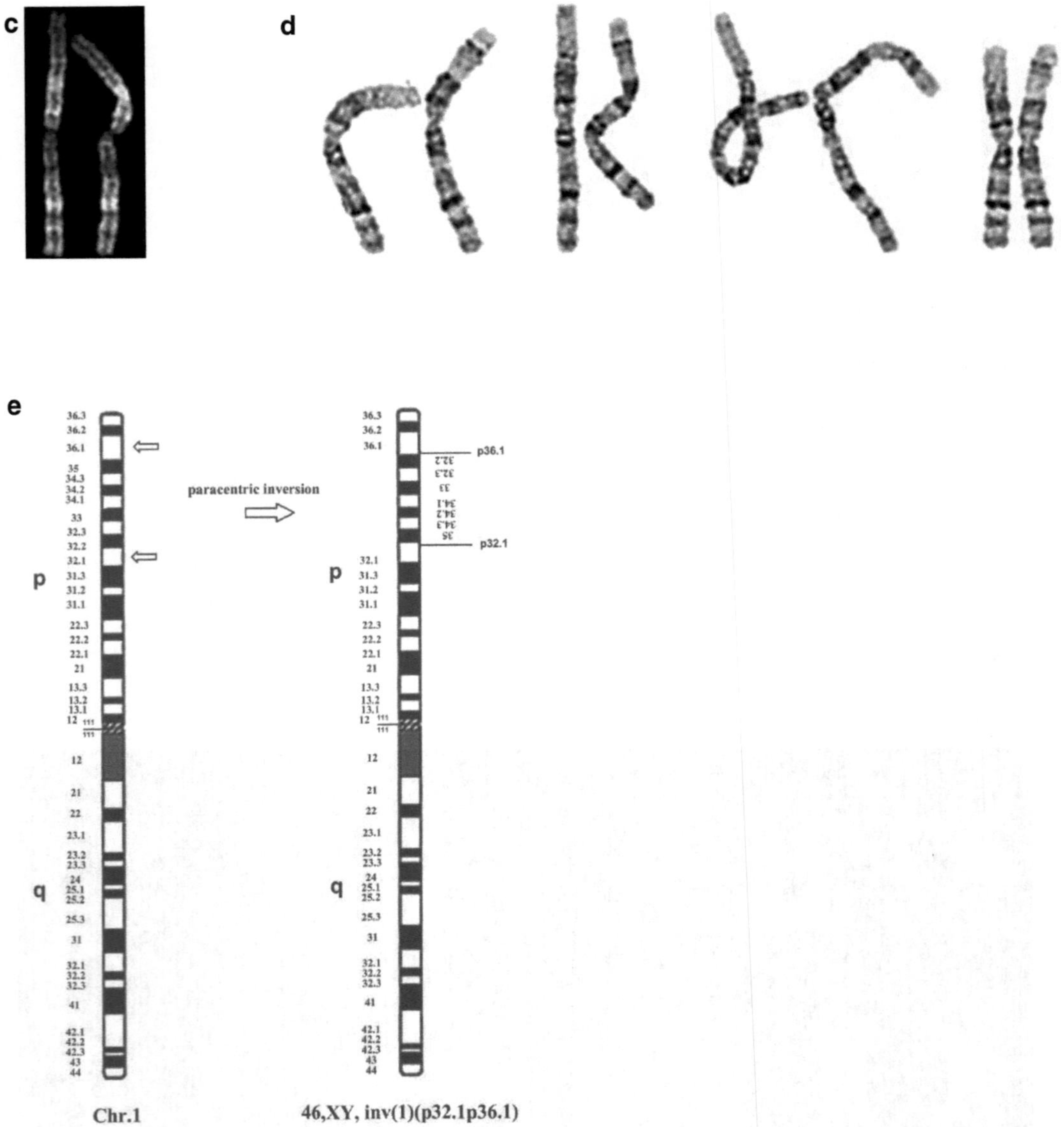

Fig. 6.18 (continued)

Case 19: Karyotype: 46,XY,inv(1)(q11.1q24.2)

Morphology

The inverted segment includes about 50% of the long arm (q) of chromosome 1. Its size corresponds to almost 22% of the length of this chromosome. The proximal breakpoint is located in the peripheral region of the centromere (q11.1) and therefore in a non-coding segment. The phenotype of the inverted chromosome is especially characterised by the dislocation of the heterochromatic block q12 from its pericentromeric position into the intermediate band q24.2. In this case CBG-banding gives the best presentation of the rearranged heterochromatic block.

Cave

The case presented here is a rare type of inversion. Therefore, the genetic risk caused by an unequal meiotic crossing over is not easy to check. Thus, in the course of genetic counselling of the carrier and his family potential inversion carriers should be investigated and prenatal analyses should be recommended to all of them (◘ Fig. 6.19).

6

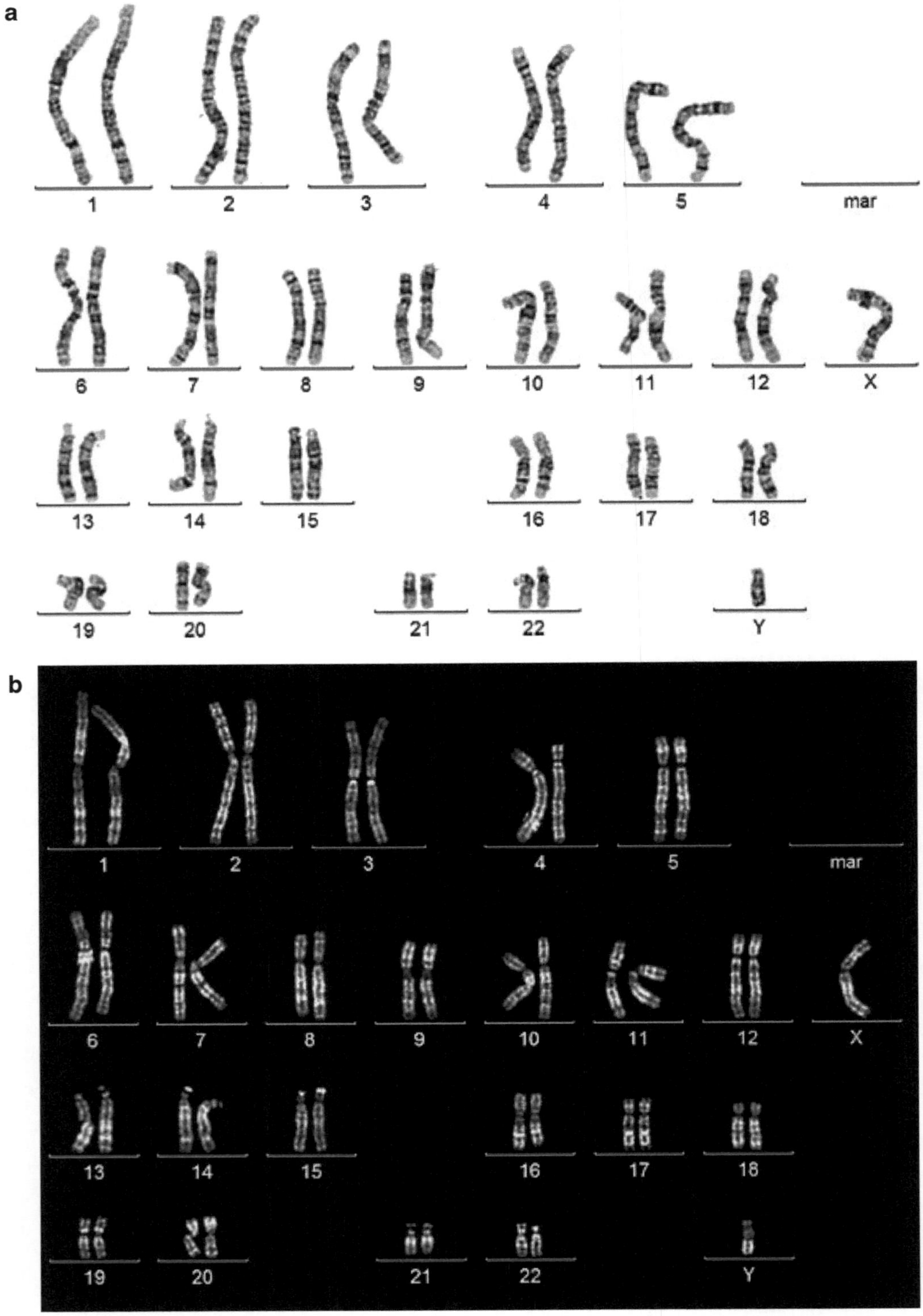

Fig. 6.19 Karyotype 46,XY,inv(1)(q11.1q24.2). **a** GTG-, **b** QFQ-banding, 700 and 550 bands per genome, **c** CBG-, **d** QFQ-, **e** GTG-banding, chromosomes from different mitoses, *left* normal, *right* inverted chromosome 1, **f** ideograms of chromosome 1. Characteristic changed appearance by dislocation of the heterochromatic block

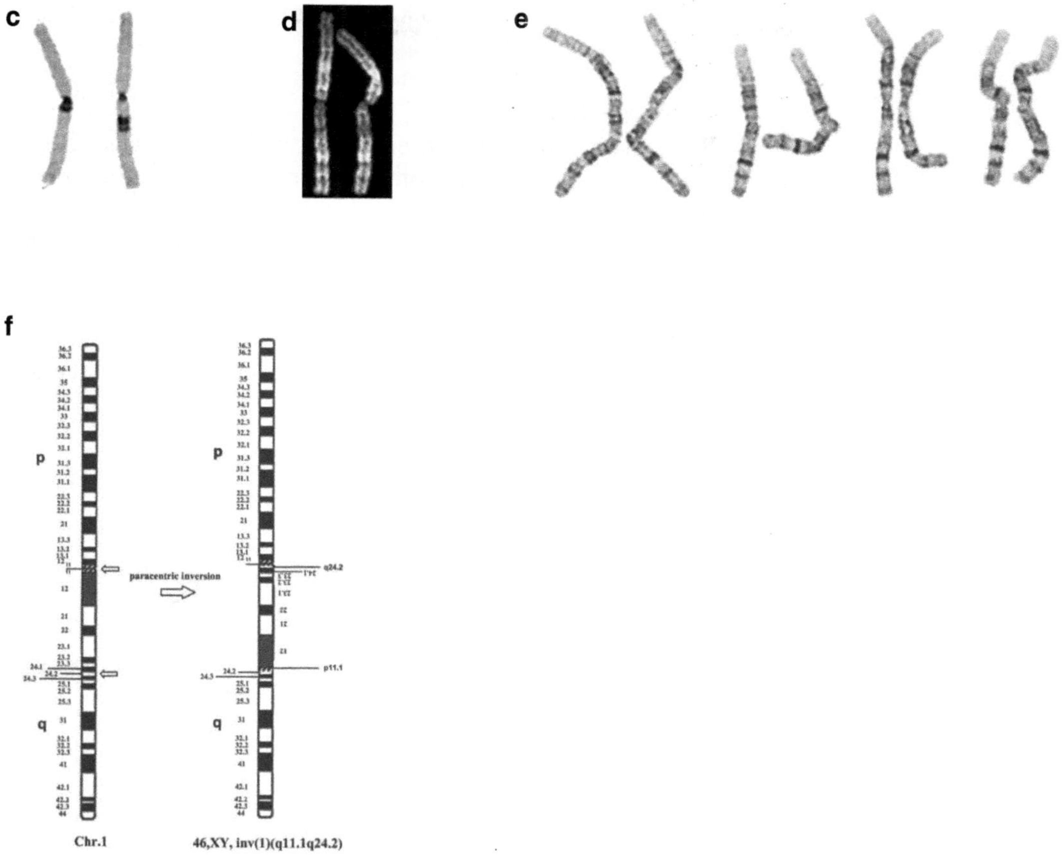

Fig. 6.19 (continued)

■■ Case 20: Karyotype: 46,XX,inv(2)(q14.2q21.1)

.

■ Morphology

The small paracentric inversion located in the intermediate region of the long arm, comprises only about 4% of the total length of chromosome 2. The identification of this inversion requires a high banding resolution of 700 to 850 bands, as can be delineated from the GTG karyotype.

■ Cave

As in the previous cases genetic counselling and prenatal diagnoses are indicated (◘ Fig. 6.20).

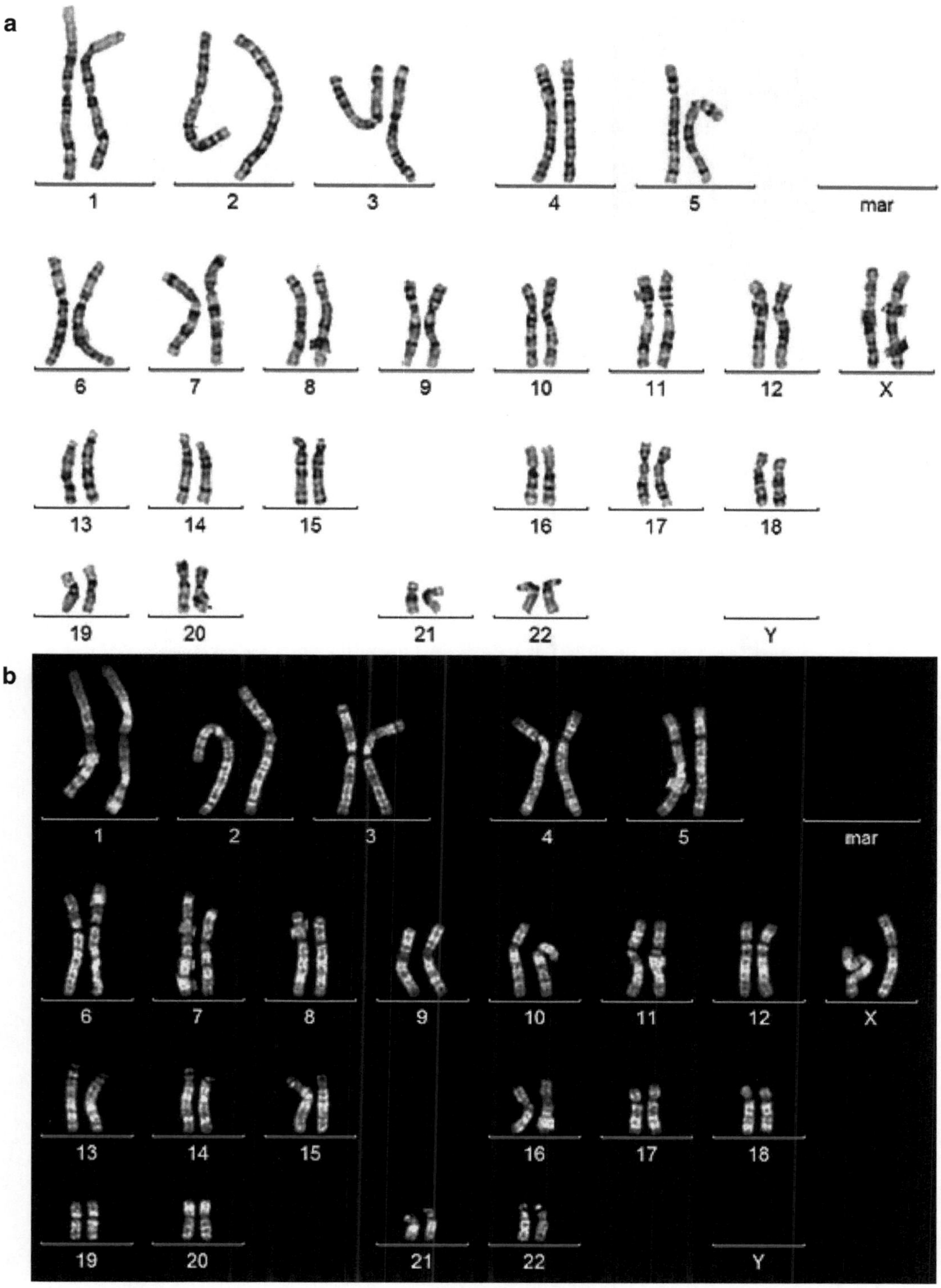

Fig. 6.20 Karyotype 46,XX,inv(2)(q14.2q21.1). **a** GTG-, **b** QFQ-banding, 700 and 550 bands per genome, **c** QFQ-, **d** GTG-banding, chromosomes from different mitoses, *left* normal, *right* inverted chromosome 2, **e** ideograms of chromosome 2, small inverted segment in the proximal long arm

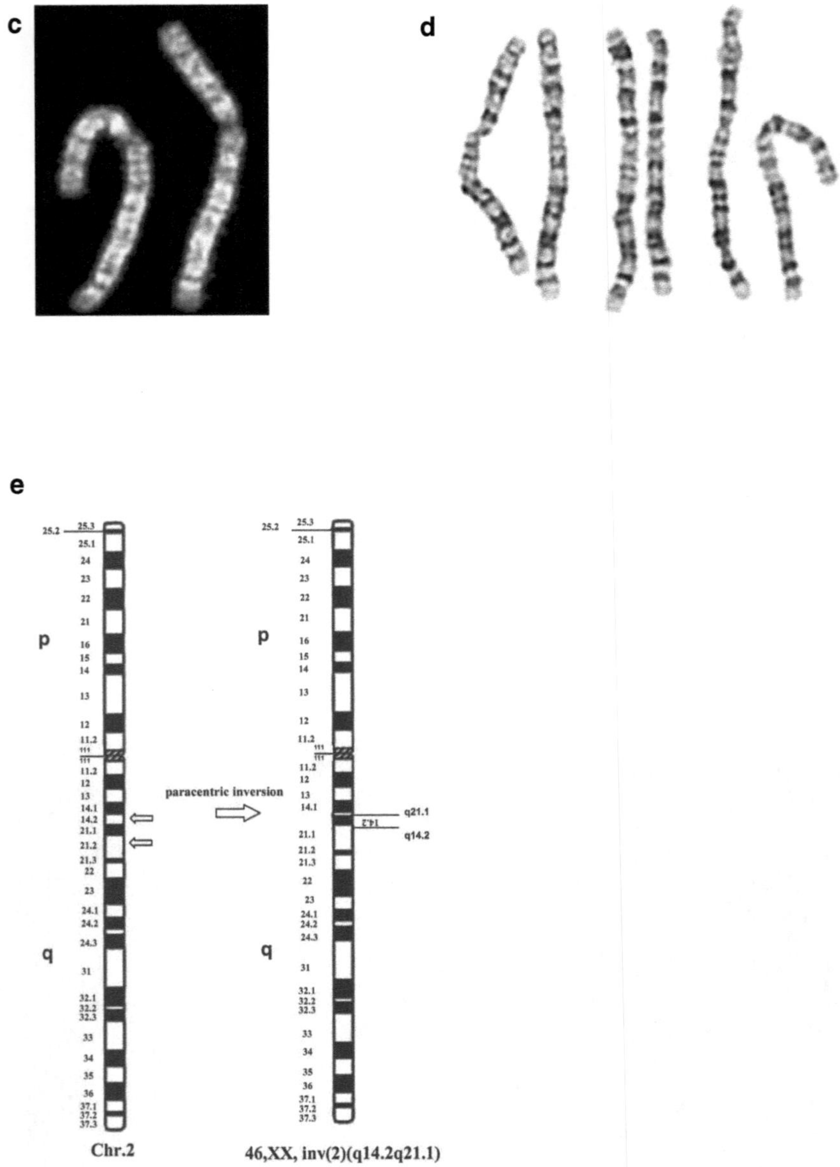

Fig. 6.20 (continued)

Case 21: Karyotype: 46,XY,inv(3)(p22.2p25)

Morphology

The inverted segment is located in the distal part of the short arm of chromosome 3. The length of the inversion is about 12% of the total length of chromosome 3. The exact localisation of the breakpoints is only possible at a resolution of 550–700 bands per genome. Among the paracentric inversions chromosome 3 is frequently involved and the locus p25 is a common breakpoint.

Cave

The genetic risk is estimated to be low, but precise figures can't be given because of the small number of cases published until now (Fig. 6.21).

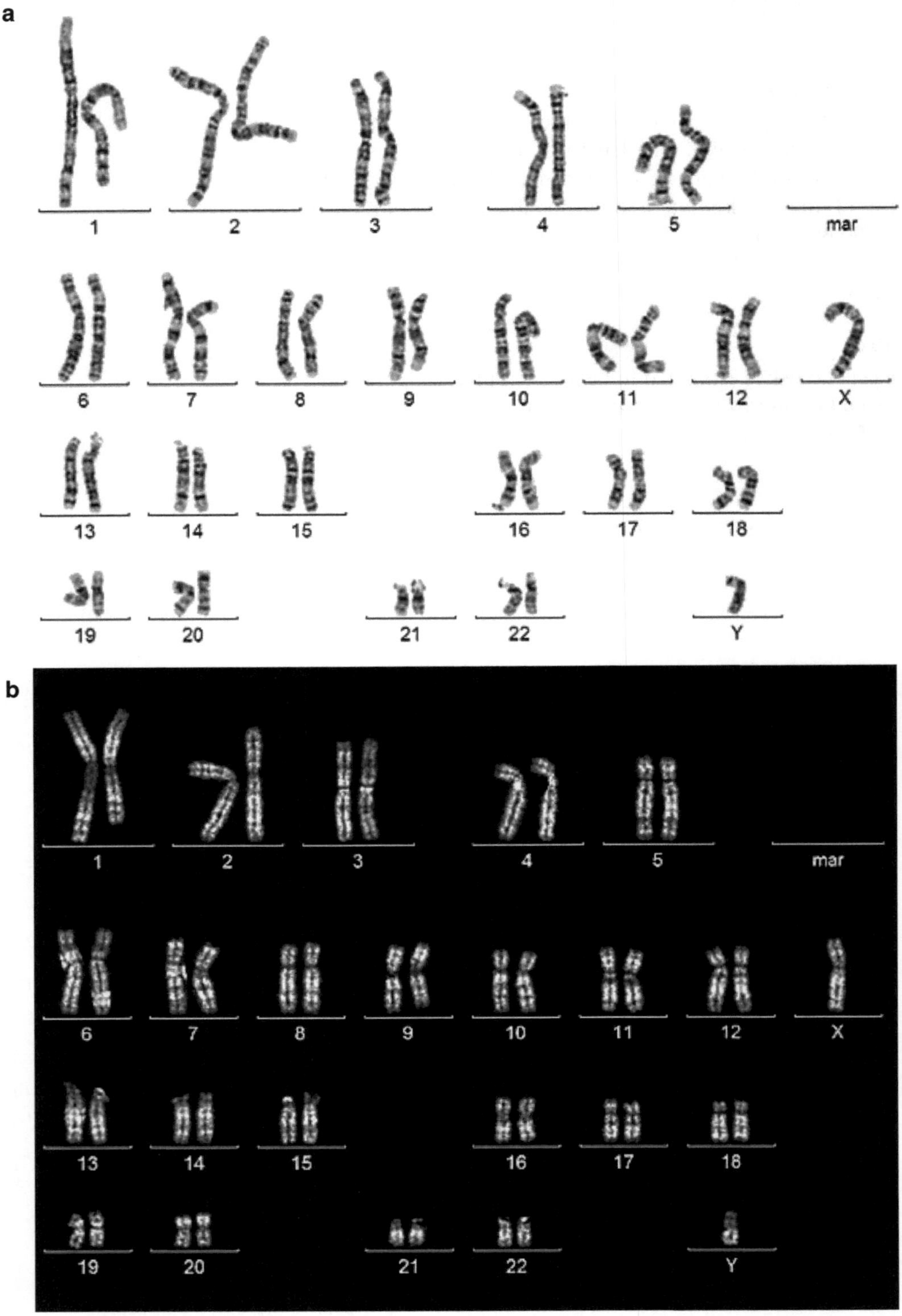

Fig. 6.21 Karyotype 46,XY,inv(3)(p22.2p25). **a** GTG-, **b** QFQ-banding, 600 and 400 bands per genome, **c** QFQ-, **d** GTG-banding, chromosomes from different mitoses, *left* normal, *right* inverted chromosome 3, **e** ideograms of chromosome 3, small, distal segment inverted in the short arm

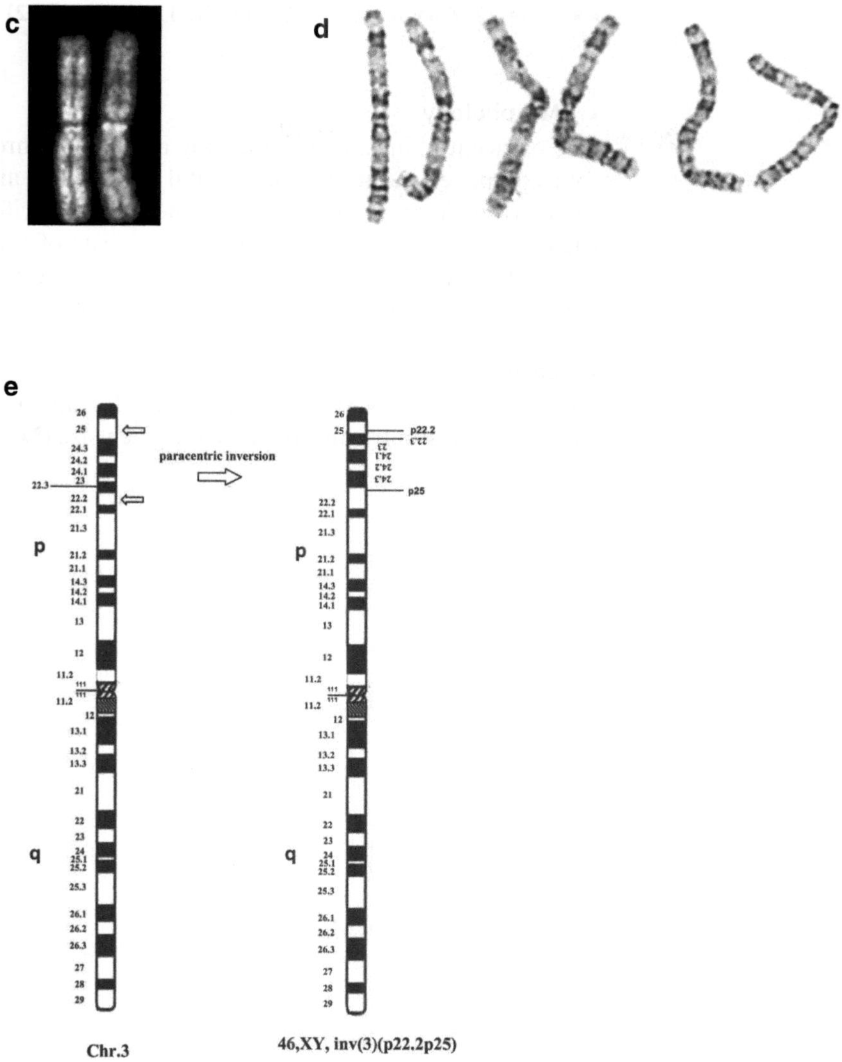

■ **Fig. 6.21** (continued)

■■ Case 22: Karyotype: 46,XY,inv(4)(q13.2q27–q28.2)

.

■ Morphology

The paracentric inversion in the long arm (q) of chromosome four comprises about 30% of its total length. The analysis of the rearrangement is primary performed by the verification of the dislocation of the distal band (q13.2) part of the AT-rich double band q13. For successful diagnosis, a resolution of 550–700 bands per genome has to be recommended.

■ Cave

No specific additional risk factors are documented besides those of a paracentric inversion of the given size (◘ Fig. 6.22).

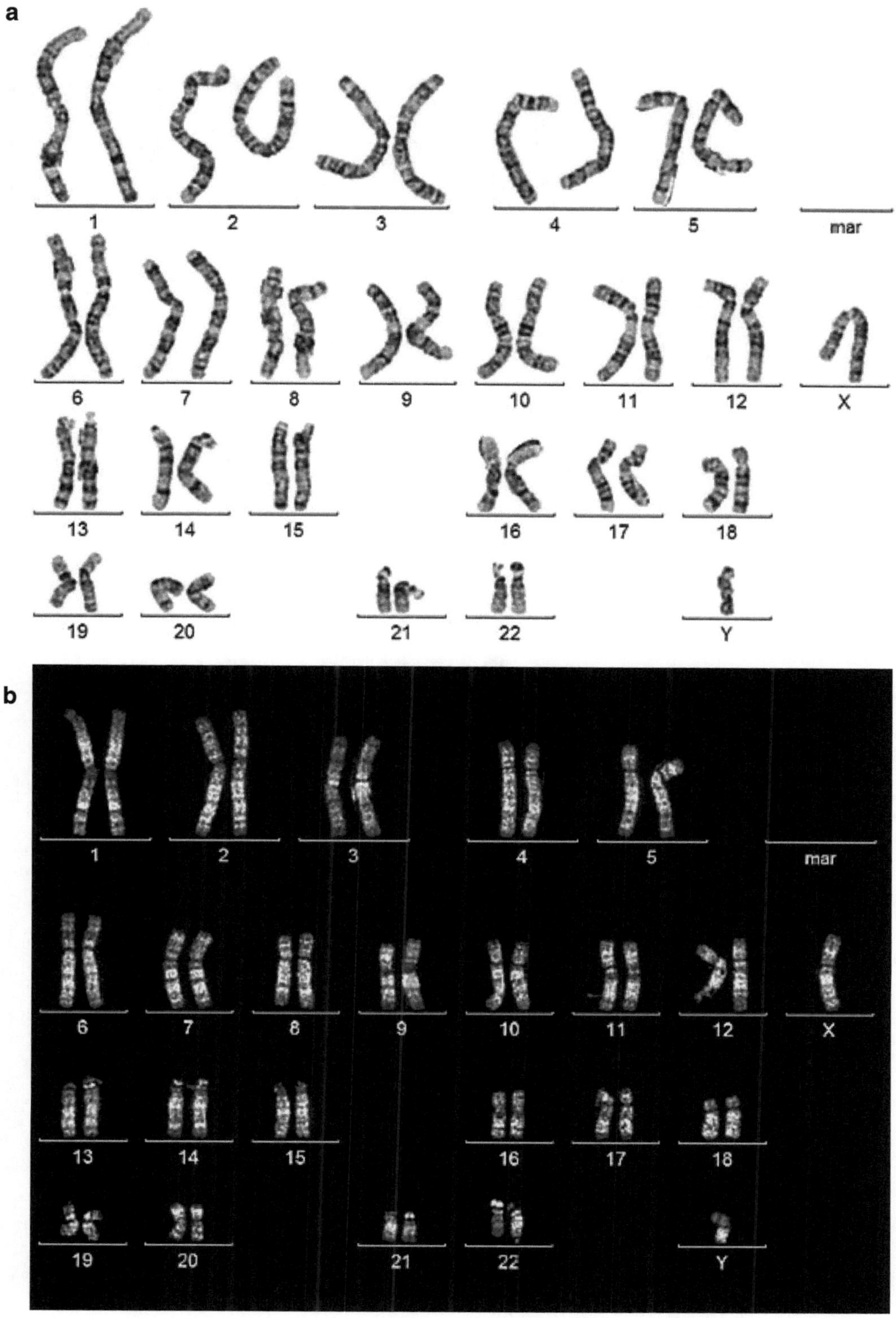

Fig. 6.22 Karyotype 46,XY,inv(4)(q13.2q27–q28.2). **a** GTG-, **b** QFQ-banding, 700 and 400 bands per genome, **c** GTG-banding, chromosomes from different mitoses, *left* normal, *right* inverted chromosome 4, **d** ideograms of chromosome 4, inverted segment in the long arm

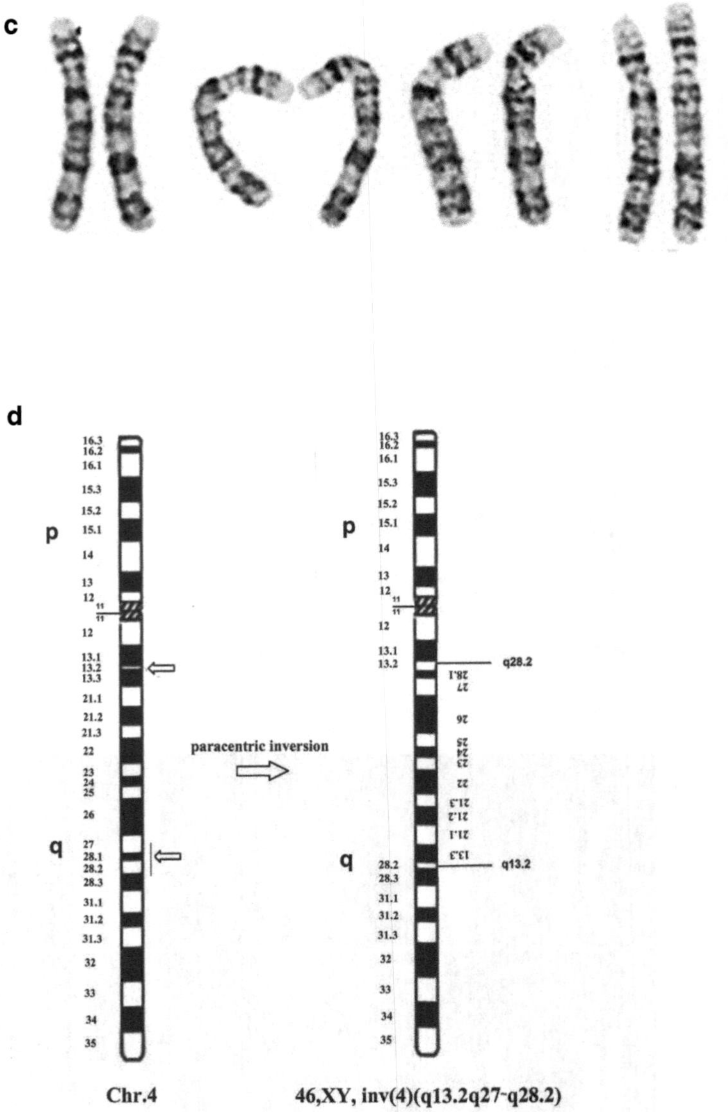

Fig. 6.22 (continued)

Case 23: Karyotype: 46,XY,inv(5)(p11p15.1)

Morphology

The inverted segment comprises about 70% of the short arm and 15% of the total length of chromosome 5. The proximal breakpoint is located in the pericentromeric, heterochromatic region, the distal one in the GC-rich band p15.1. The AT-rich band p14 is now located more proximal and this leads to an abnormal banding pattern and therefore this inversion can even be diagnosed at a resolution of 400–550 bands per genome.

Cave

According to the literature, this small paracentric inversion only presents with a low genetic risk, but it has to be taken into account that the whole Cri-du-chat critical region is included, and this has to be mentioned in the course of genetic counselling (Fig. 6.23).

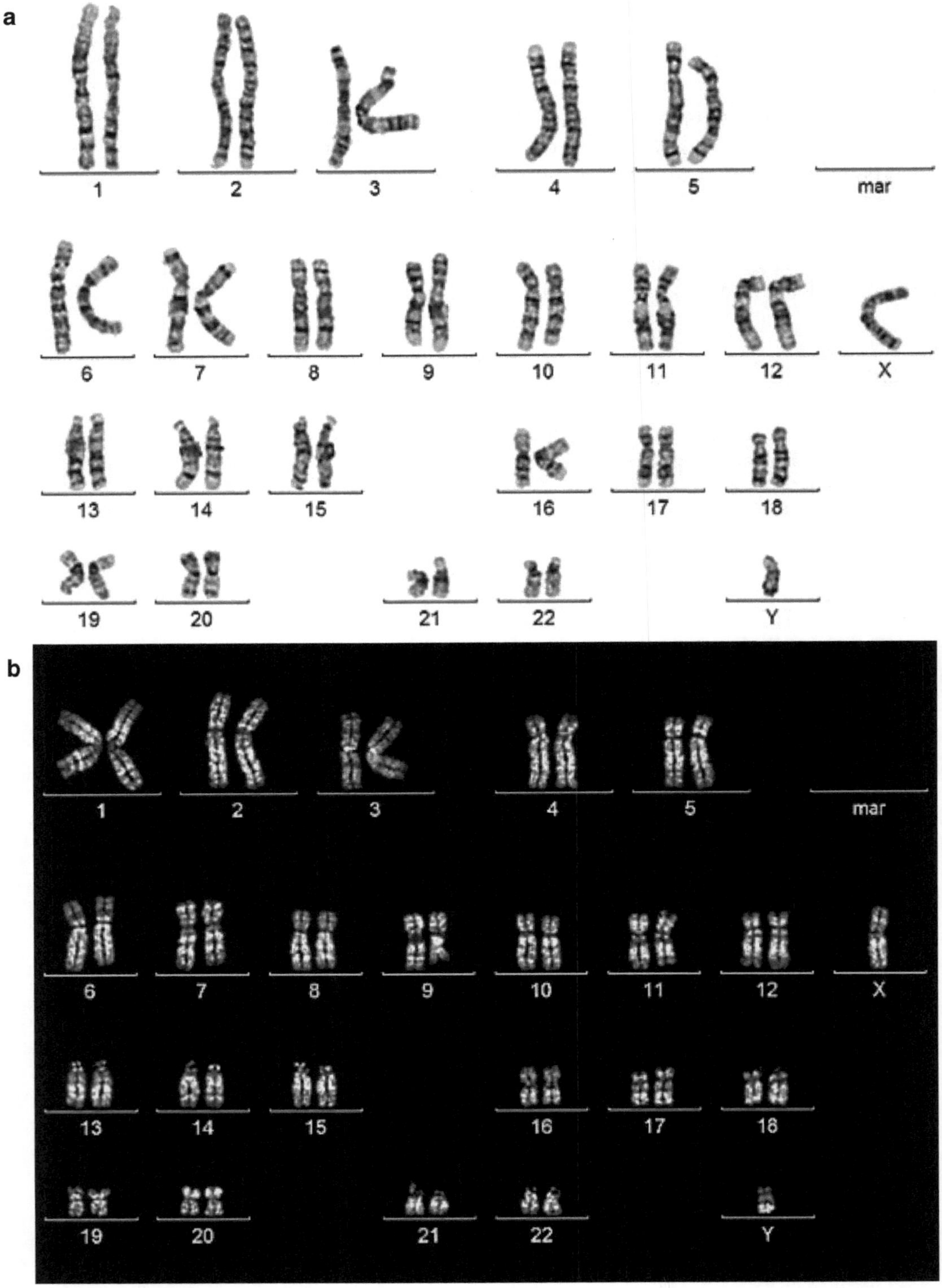

■ **Fig. 6.23** Karyotype 46,XY,inv(5)(p11p15.1). **a** GTG-, **b** QFQ-banding, 550 and 400 bands per genome, **c** QFQ-, **d** GTG-banding, chromosomes from different mitoses, *left* normal, *right* inverted chromosome 5, **e** ideograms of chromosome 5, proximal breakpoint in the pericentromeric heterochromatin

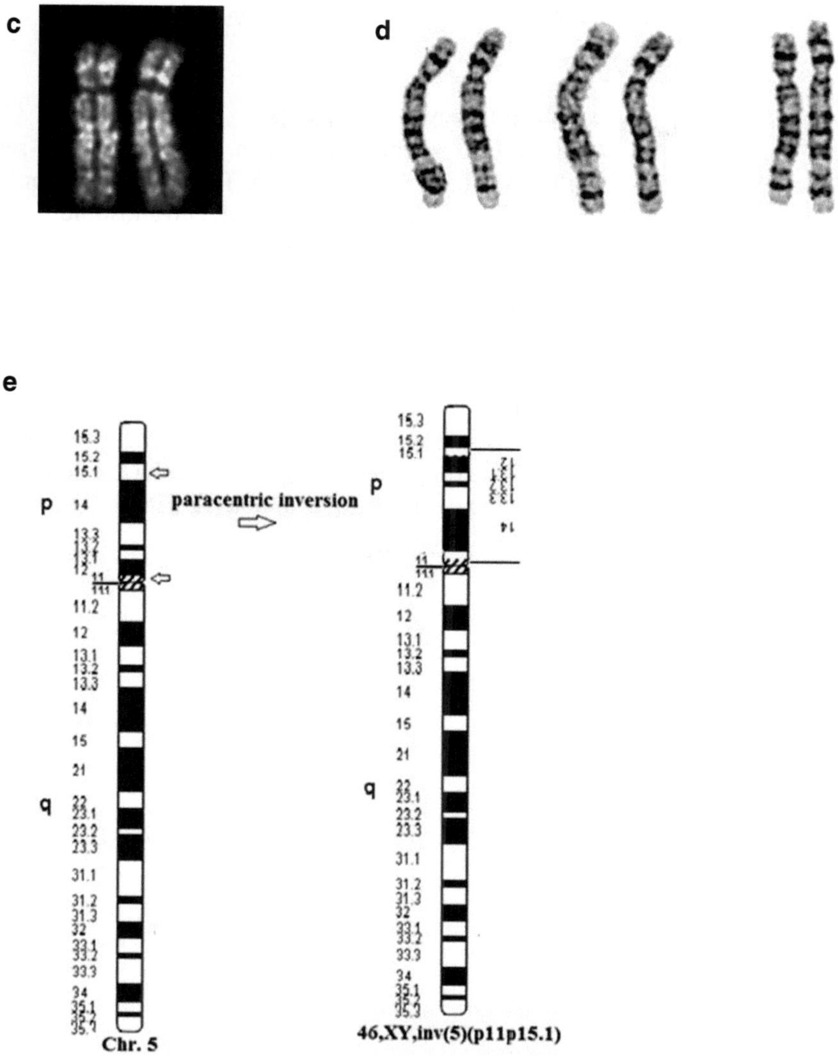

■ **Fig. 6.23** (continued)

▪▪ Case 24: Karyotype: 46,XX,inv(6)(q16.2q24.2)

.

▪ Morphology

The rearrangement occurred in an intermediate segment of the long arm of chromosome 6. The inverted region comprises about 25% of the total length of chromosome 6. This inversion is an excellent example to demonstrate the relevance of a high banding resolution to analyse an inversion in an almost homogenous banded segment. The exact diagnosis affords a resolution of 700–850 bands per genome.

▪ Cave

According to the literature on paracentric inversions, this rare type of rearrangement is estimated to have a low genetic risk. But specific data are not available(◘ Fig. 6.24).

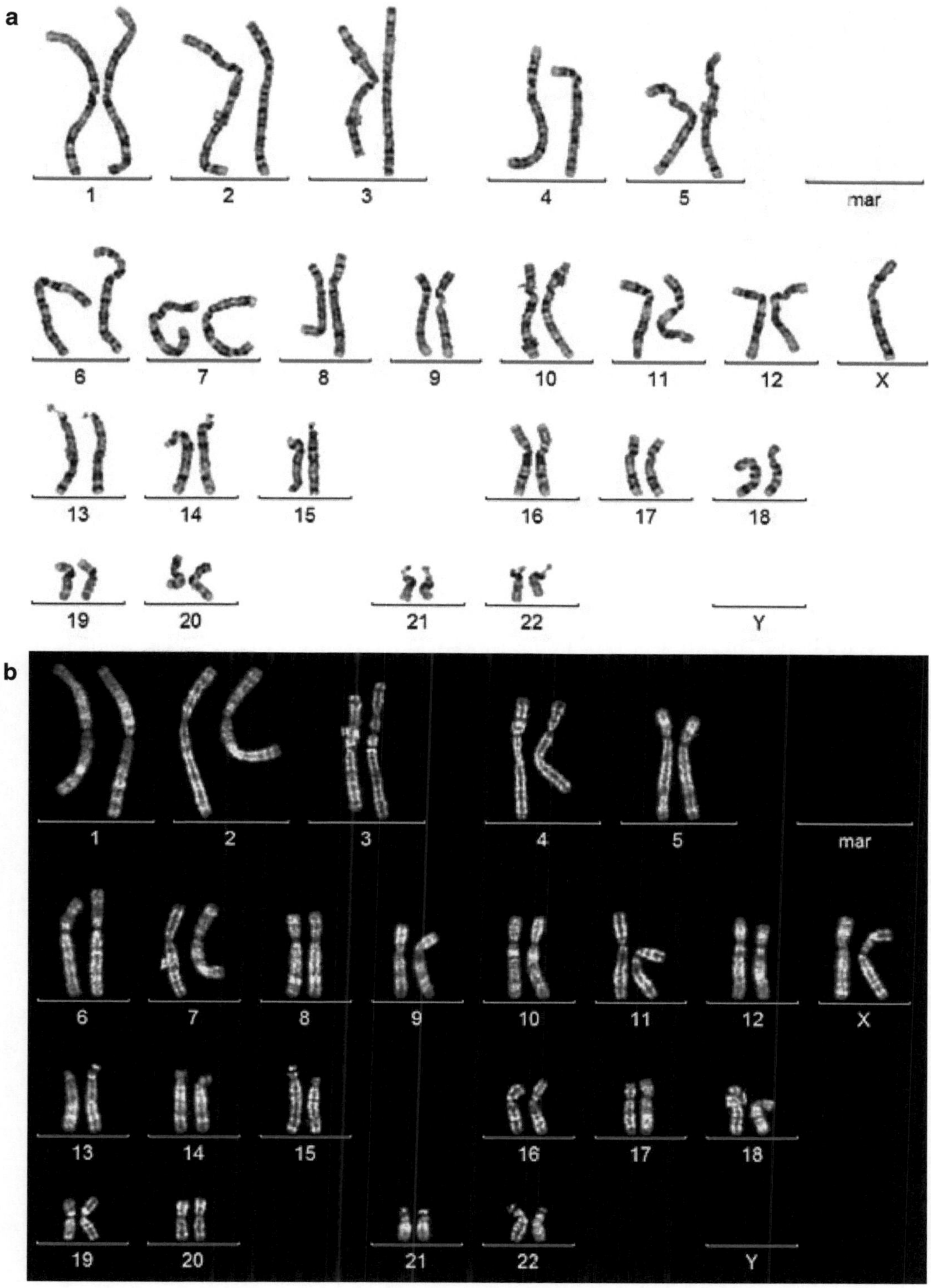

Fig. 6.24 Karyotype 46,XX,inv(6)(q16.2q24.2). **a** GTG-banding, 1 X-chromosome got lost during preparation, **b** QFQ-banding, 700 and 550 bands per genome, **c** QFQ-, **d** GTG-banding, chromosomes from different mitoses, *left* normal, *right* inverted chromosome 6, **e** ideograms of chromosome 6, intermediate position in the long arm of the inverted segment

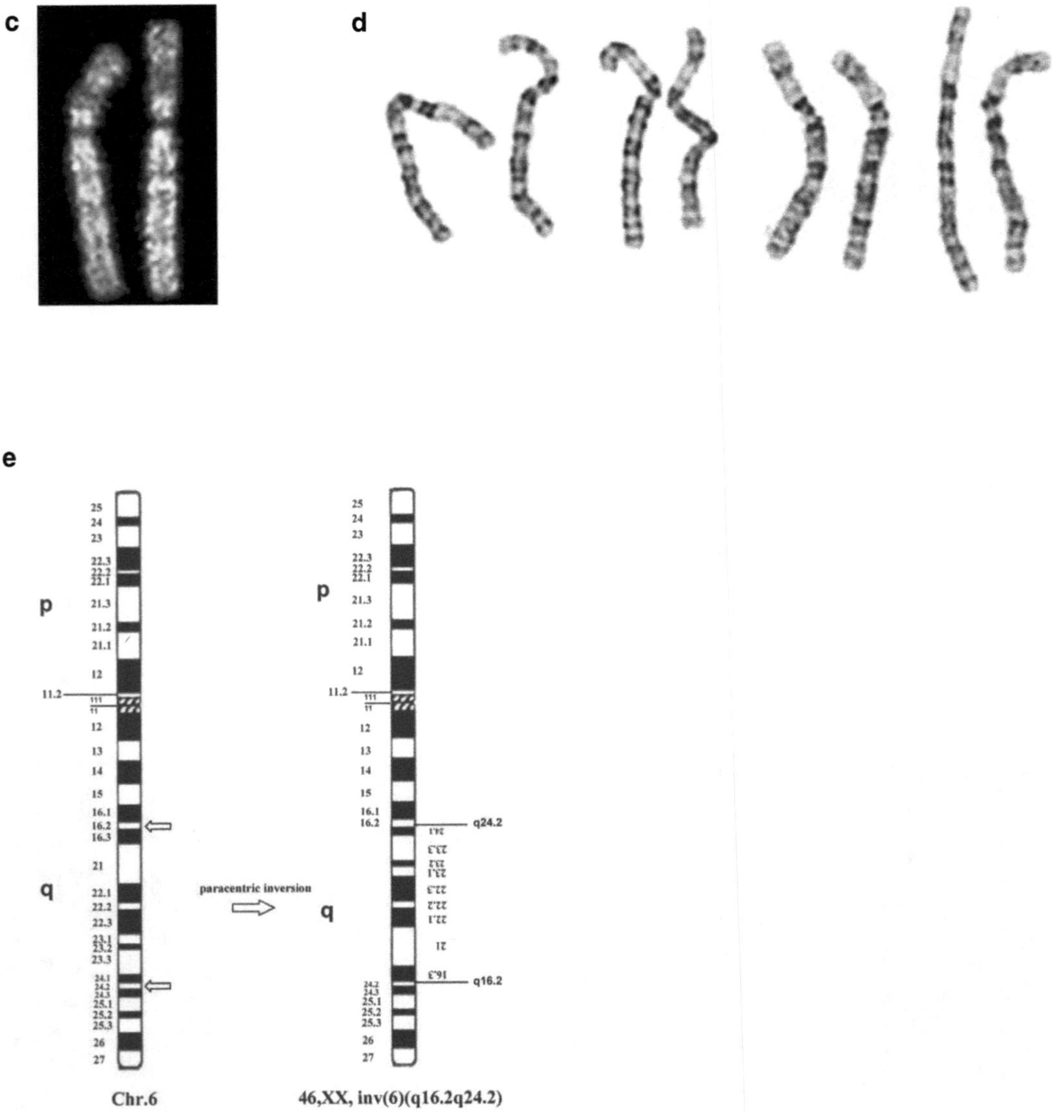

Fig. 6.24 (continued)

Case 25: Karyotype: 46,XY,inv(6)(q23.3q25.3)

Morphology

Presented here is a small inversion in the distal part of the long arm of chromosome 6. The inverted region has a size of about 12% of the total length of the chromosome. For diagnosis, there must be chosen a resolution of at least 550 bands per genome. Both breakpoints in 6q have been repeatedly described in pericentric inversions and in reciprocal translocations.

Cave

Surveys from the literature on paracentric inversions of comparable size show a high genetic stability, and therefore a low genetic risk for the carrier. This should be discussed with the carrier (◘ Fig. 6.25).

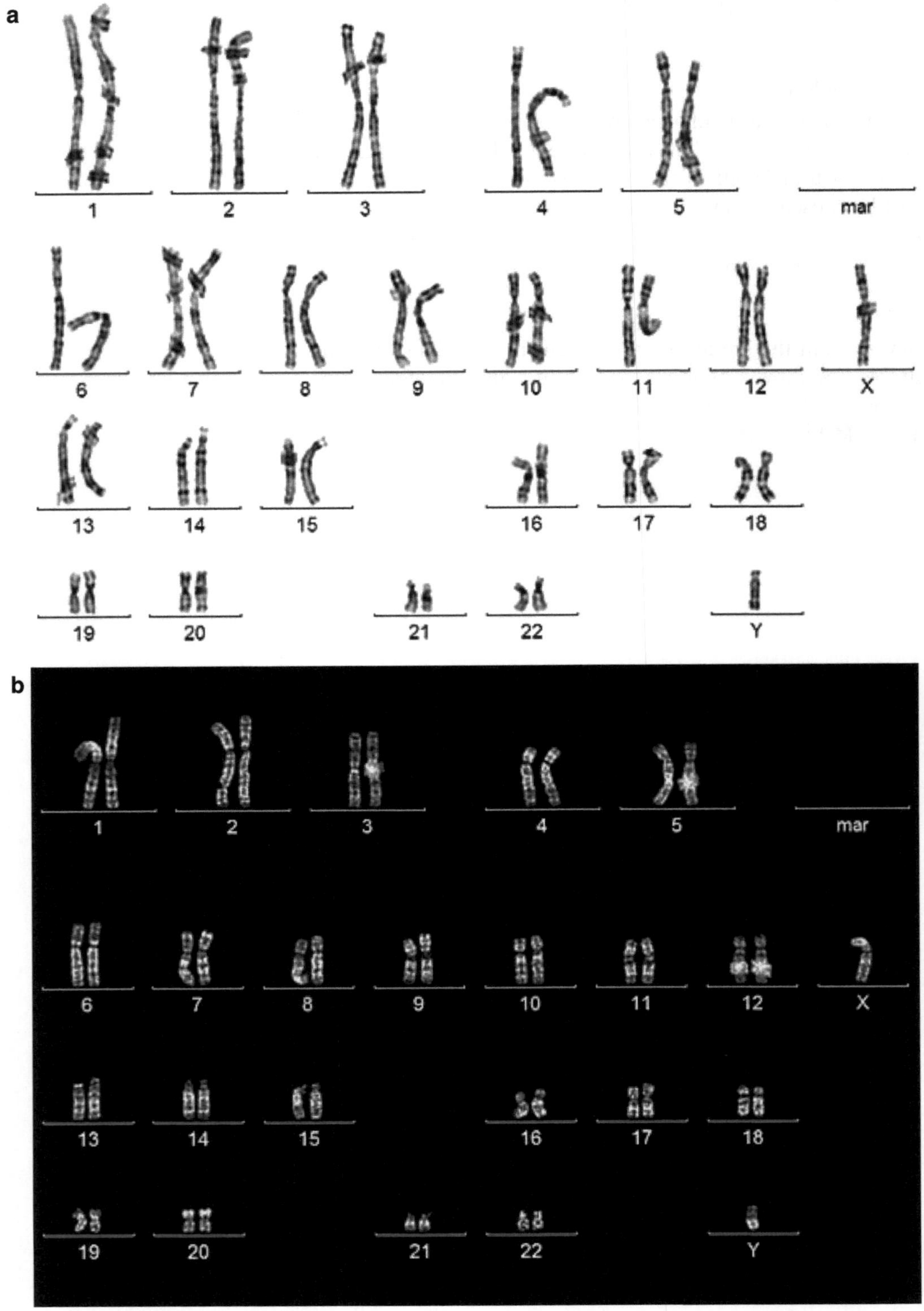

Fig. 6.25 Karyotype 46,XY,inv(6)(q23.3q25.3). **a** GTG-, **b** QFQ-banding, 550 and 400 bands per genome, **c** QFQ-, **d** GTG-banding, chromosomes from different mitoses, *left* normal, *right* inverted chromosome 6, **e** ideograms of chromosome 6, small, distal inversion in the long arm

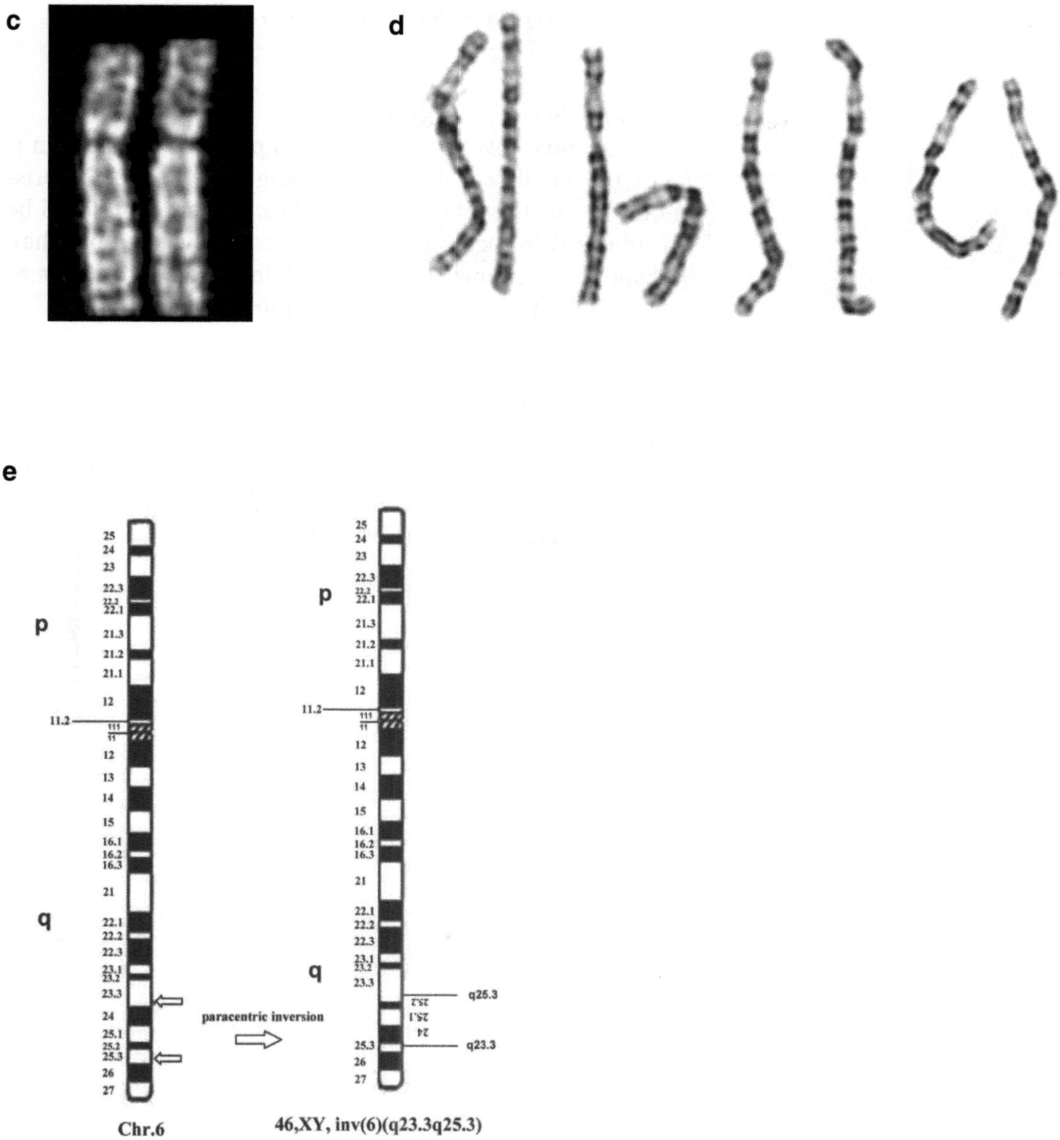

Fig. 6.25 (continued)

▪▪ Case 26: Karyotype: 46,XX,inv(6)(q25.1q25.3)

.

▪ Morphology

This paracentric inversion in the distal part of the long arm is one of the smallest in the own investigation group, comprising only 4% of the total length of chromosome 6. It could be only analysed in metaphases with a resolution of more than 700 bands per genome. The survey of inversion chromosomes 6 from different mitoses makes this quite obvious.

▪ Cave

There is no sufficient information on paracentric inversions of chromosome 6 in this small size. Relying on general risk of chromosome rearrangements with euchromatic breakpoints, this is estimated to be in the order of 3–7% caused by an unequal crossing over in meiosis (◘ Fig. 6.26).

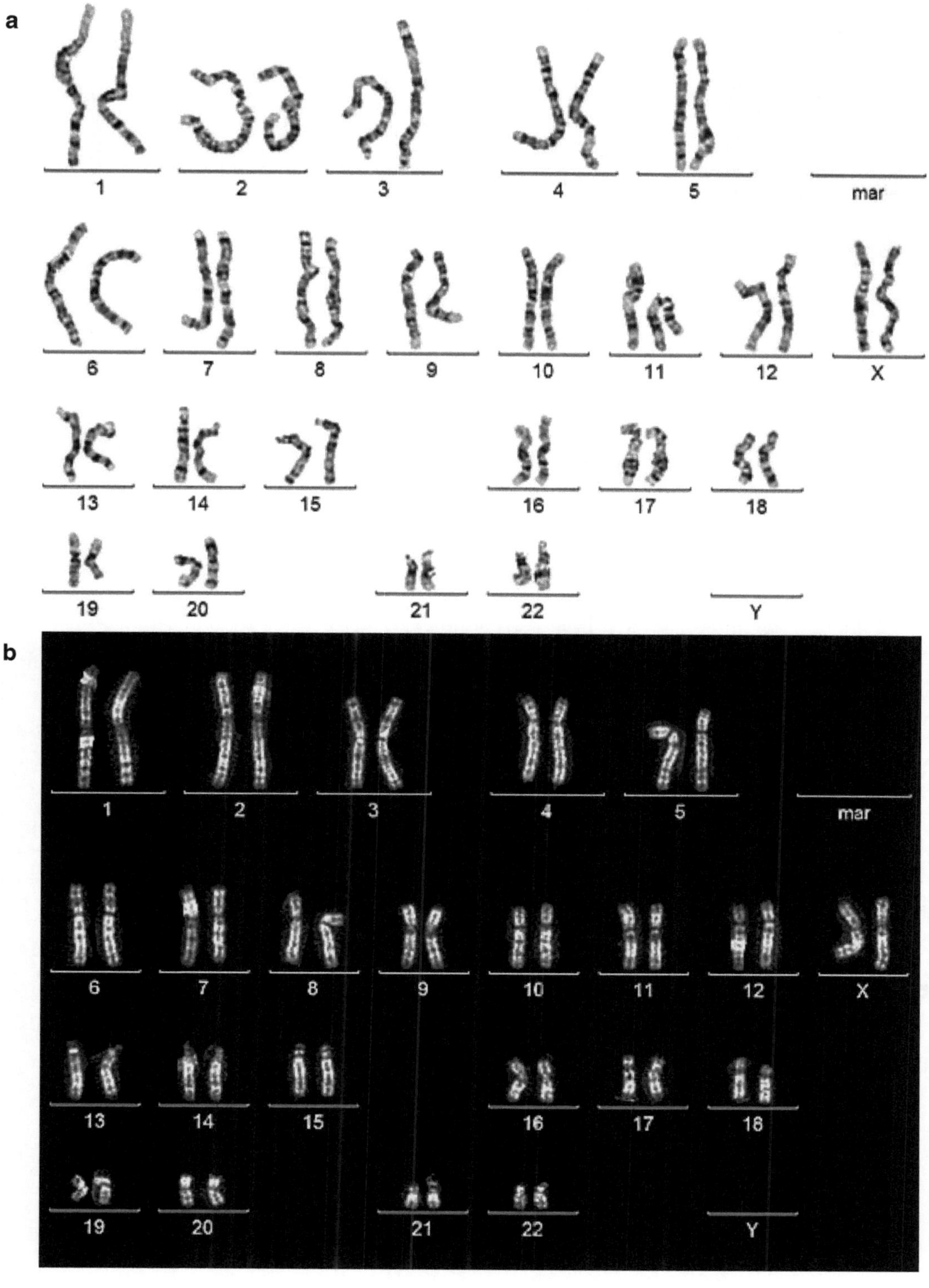

Fig. 6.26 Karyotype 46,XX,inv(6)(q25.1q25.3). **a** GTG-, **b** QFQ-banding, 600 and 400 bands per genome, **c** QFQ-, **d** GTG-banding, chromosomes from different mitoses, *left* normal, *right* inverted chromosome 6, **e** ideograms of chromosome 6, smallest inversion in the investigation group

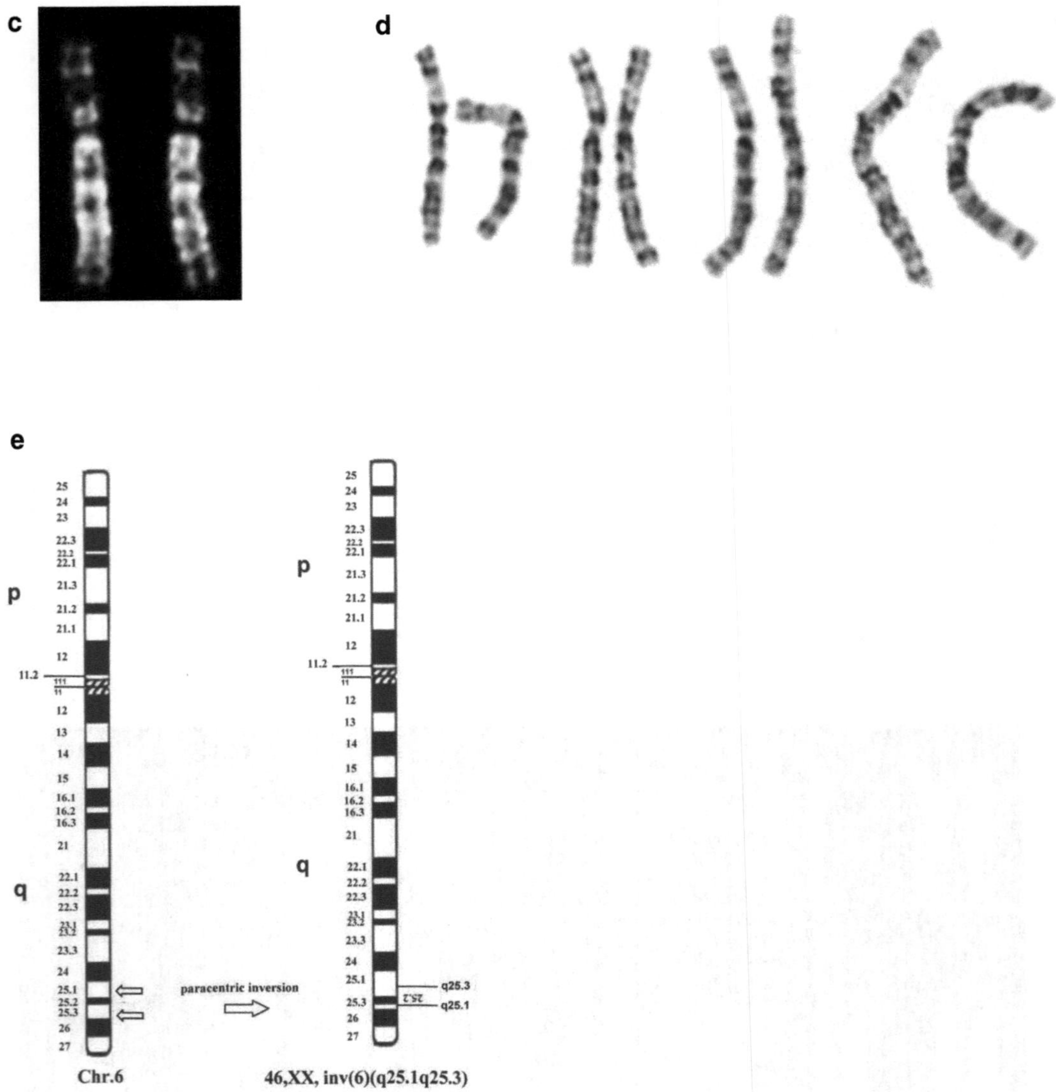

■ Fig. 6.26 (continued)

▪▪ Case 27: Karyotype: 46,XX,inv(7)(p14.2p21.2)

▪ Morphology
The paracentric inversion presented comprises about 50% of the length of the short arm and 18% of the total length of the chromosome. The diagnosis of the rearrangement is performed in a first step by a changed position of the band p21.2 to a localisation more proximal. For the analysis of the second breakpoint p14.2, a resolution of 550–700 bands per genome is necessary.

▪ Cave
When delineating the genetic risk of the carrier possible epigenetic alterations have also to be taken into account. In this case, it is recommended to contact colleagues whose work focusses on this field (◘ Fig. 6.27).

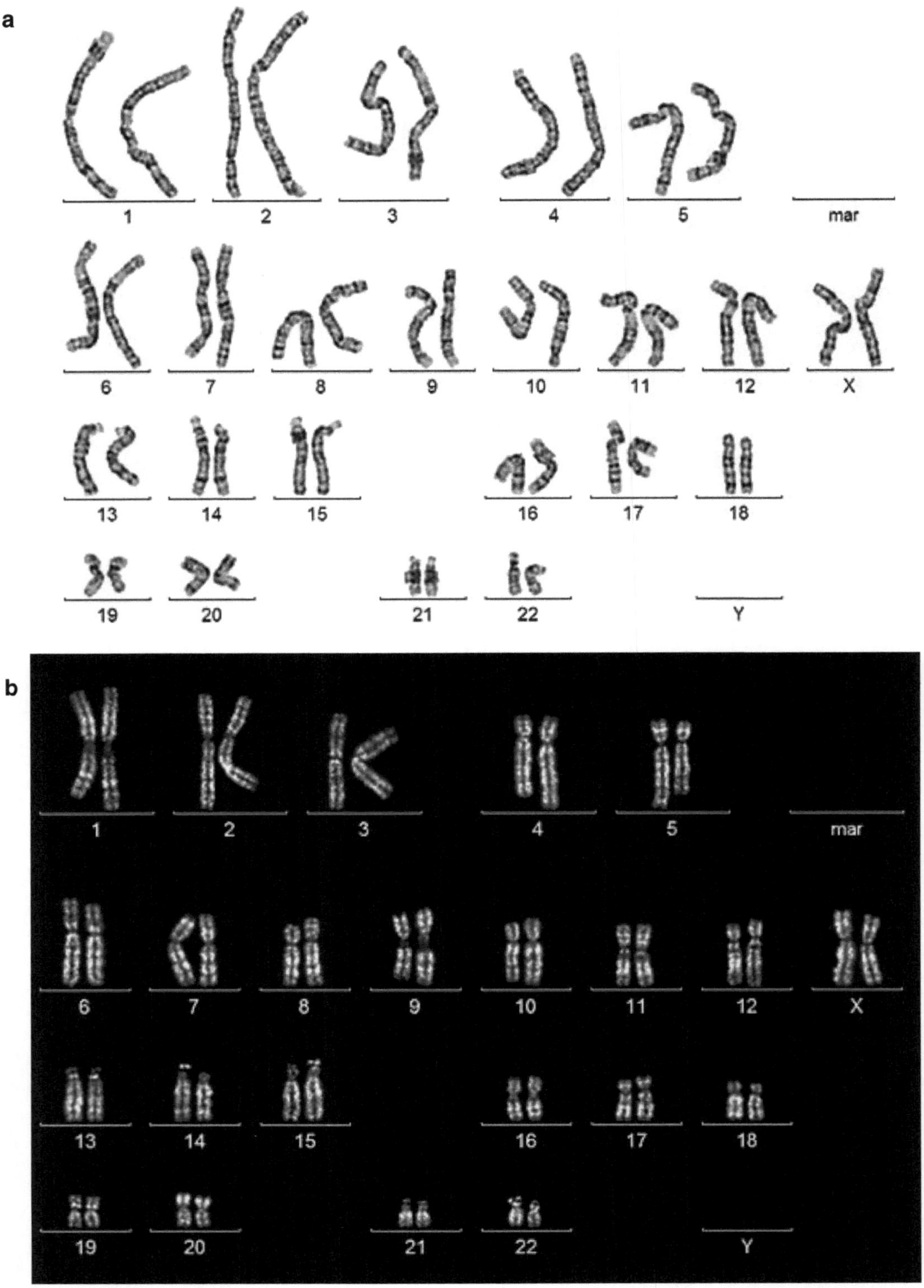

Fig. 6.27 Karyotype 46,XX,inv(7)(p14.2p21.2). **a** GTG-, **b** QFQ-banding, 700 and 400 bands per genome, **c** QFQ-, **d** GTG-banding, chromosomes from different mitoses, *left* normal, *right* inverted chromosome 7, **e** ideograms of chromosome 7, inverted segment of medium size

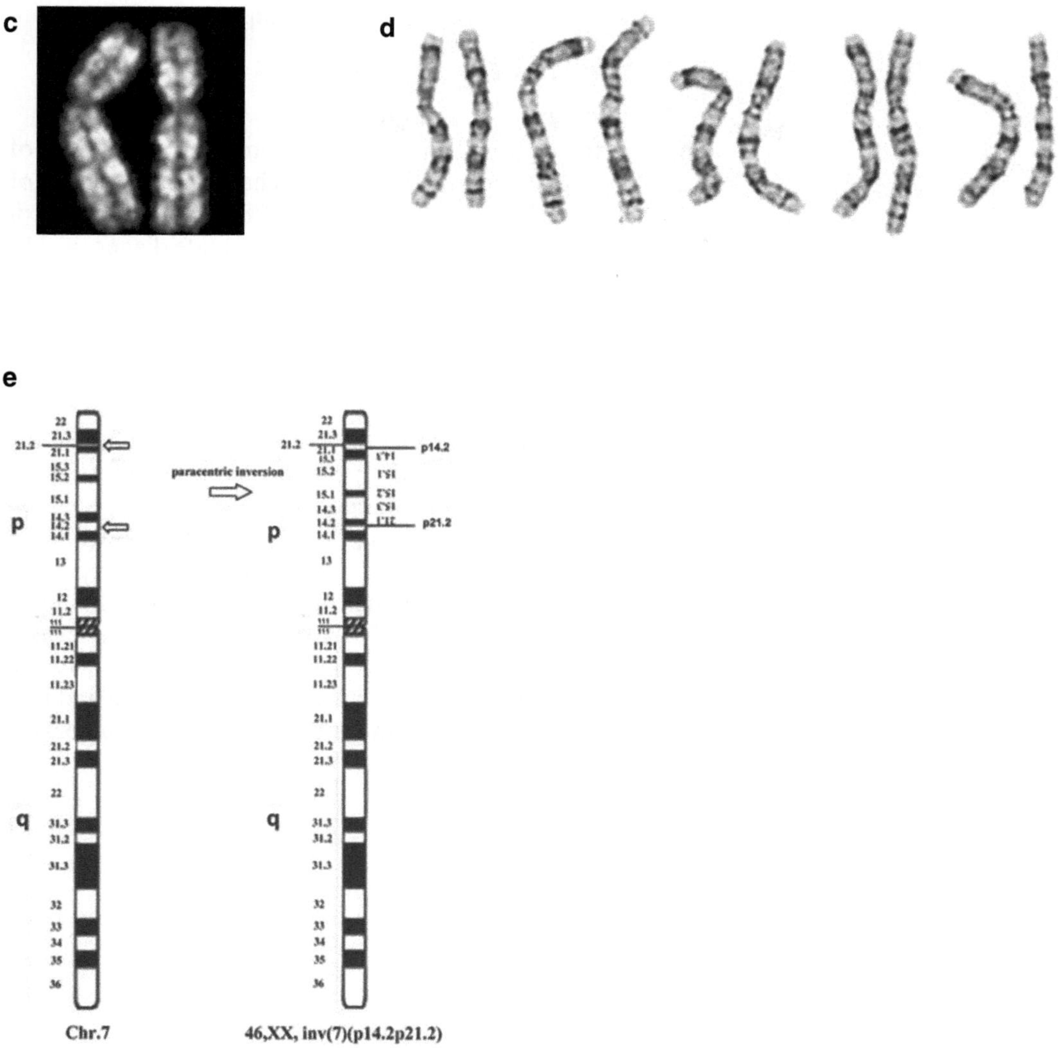

Fig. 6.27 (continued)

▪▪ Case 28: Karyotype: 46,XY,inv(8)(q13q24.22)

.

▪ Morphology

The inverted region in the long arm comprises about 40% of the total length of chromosome 8. A characteristic feature of this inversion is the changed position of 8q23 to a more proximal localisation, which enables a quick and safe diagnosis. The distal breakpoint seems to be a hot spot for aberration induction, as different types of rearrangements like deletions, rings and translocations have been described with this breakpoint.

▪ Cave

In the region 8q24.11q24.13 the mutation, mainly a deletion of the Langer-Giedion-syndrome, is localised. Therefore, when analysing an inversion including this region it is of special relevance to investigate if the position of a breakpoint is within or outside the critical region. Thus, when analysing a comparable inversion, the resolution of the chromosome bands should be 550–700 bands per genome. If the breakpoint can be located in the critical region, an additional microarray analysis has to be done. Genetic counselling has to be recommended and prenatal investigation should always be completed by microarray (◘ Fig. 6.28).

a

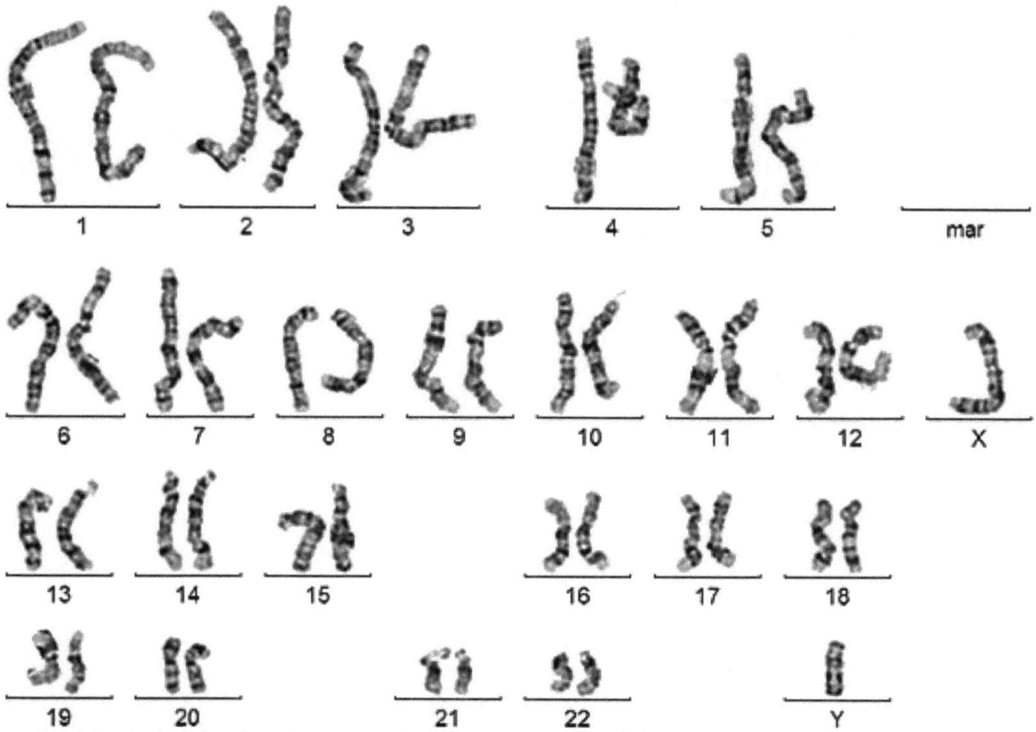

b

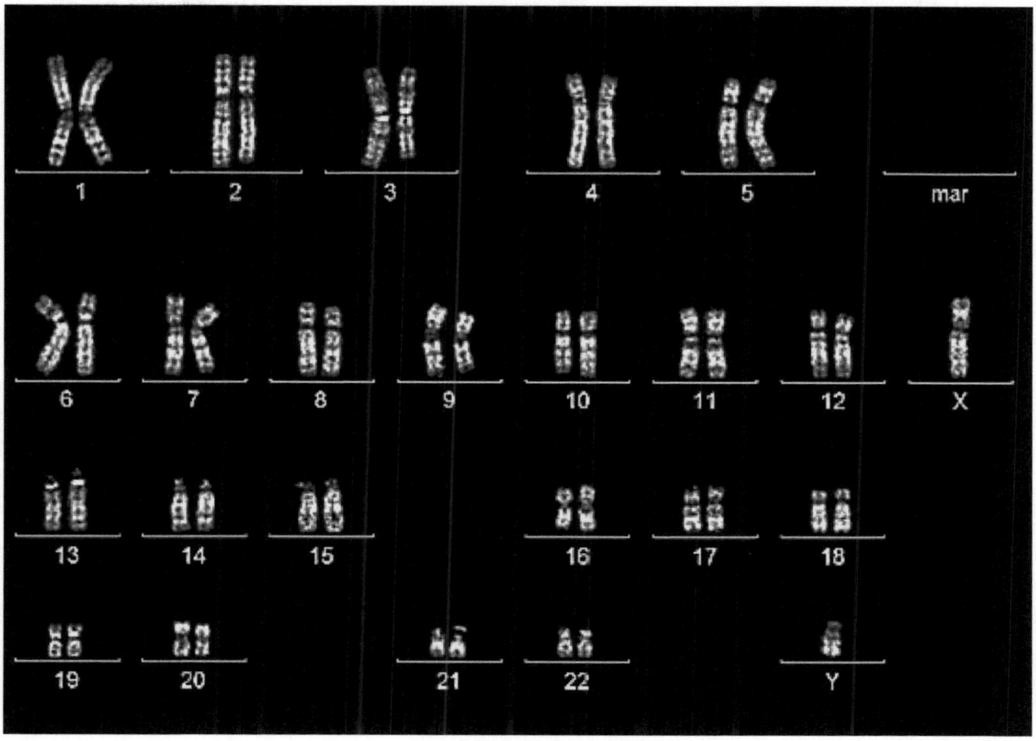

◘ Fig. 6.28 Karyotype 46,XY,inv(8)(q13q24.22). **a** GTG-, **b** QFQ-banding, 600 and 400 bands per genome, **c** QFQ-, **d** GTG-banding, chromosomes from different mitoses, *left* normal, *right* inverted chromosome 8, **e** ideograms of chromosome 8, medium to large size of the inverted segment

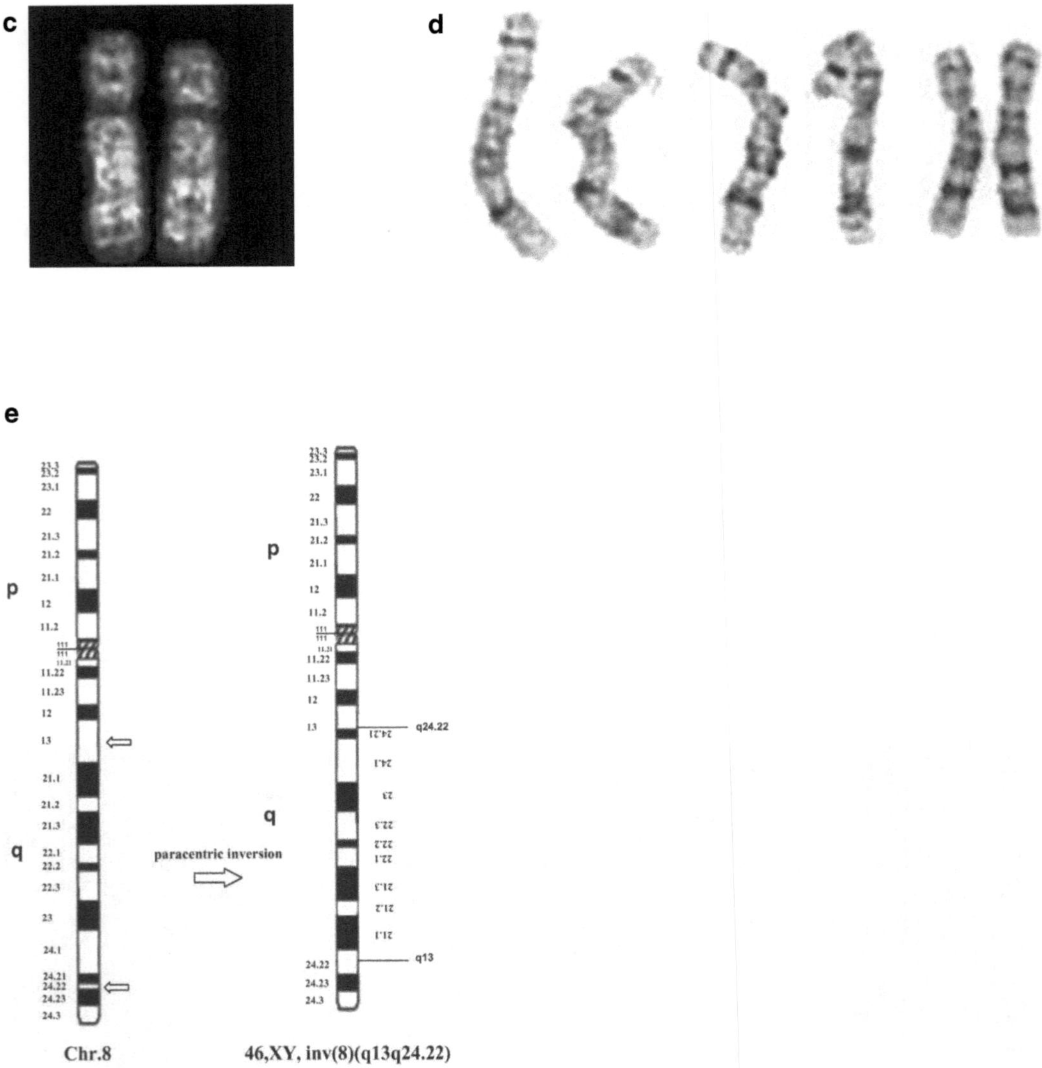

Fig. 6.28 (continued)

▪▪ Case 29: Karyotype: 46,XY,inv(9)(q32q34.3)

.

▪ Morphology

This distal localised paracentric inversion 9 comprises about 15% of the total length of the chromosome. Caused by the dislocation of the AT-rich band q33, this inversion can be easily recognised even by a resolution of less than 550 bands per genome. But the exact localisation of the breakpoints requires a resolution of 700 bands.

▪ Cave

The distal breakpoint q34.3 is an aberration hot spot. One of the three most frequent microdeletion syndromes is localised in this region. Furthermore, the band q34 is a frequent breakpoint in rearrangements occurring in different tumour diseases (◘ Fig. 6.29).

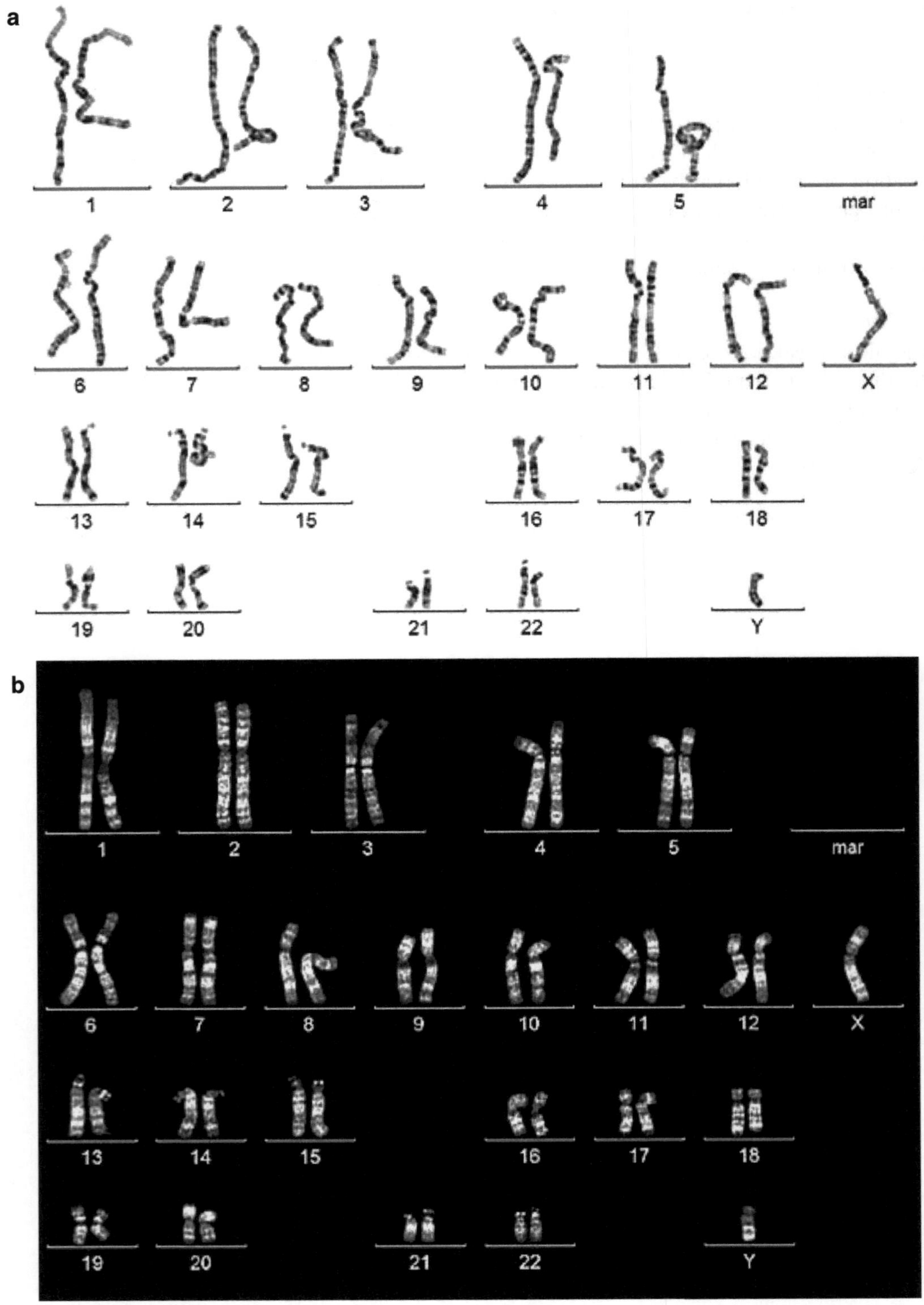

Fig. 6.29 Karyotype 46,XY,inv(9)(q32q34.3). **a** GTG-, **b** QFQ-banding, 700 and 400 bands per genome, **c** QFQ-, **d** GTG-banding, chromosomes from different mitoses, *left* normal, *right* inverted chromosome 9, **e** ideograms of chromosome 9, small distal inversion segment

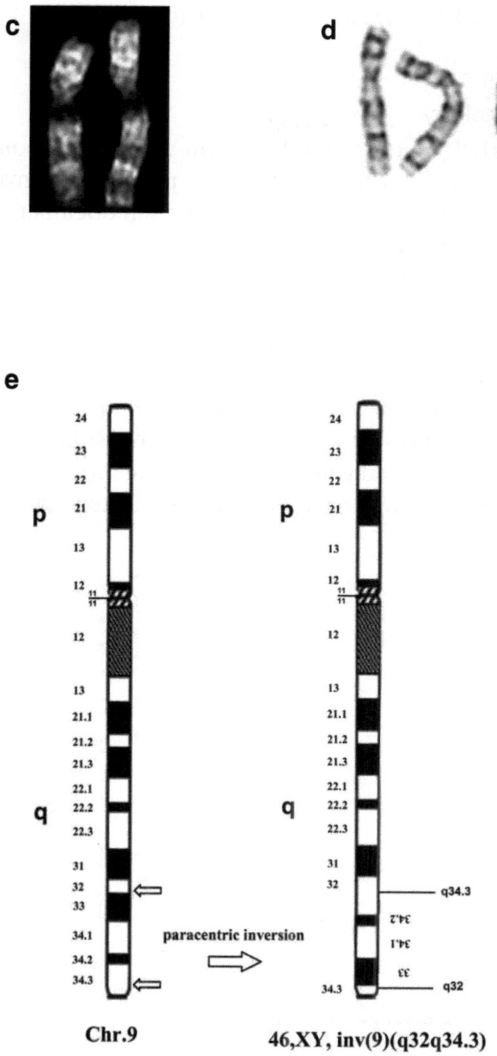

Fig. 6.29 (continued)

▪▪ Case 30: Karyotype: 46,XX,inv(11)(q21q23.3)

.

▪ Morphology

The inverted region in the long arm comprises about 15% of the total chromosome length. The paracentric inversion 11 with the breakpoints of our case has been documented worldwide in different ethnic populations. Besides, a Founder effect has been ascertained.

▪ Cave

In all family studies, this inversion was stated to be stable and to be always inherited as a balanced rearrangement. Therefore, relying on our present knowledge, this inversion does not present with an increased genetic risk. Investigations of relatives of a carrier or prenatal analyses are not indicated (◘ Fig. 6.30).

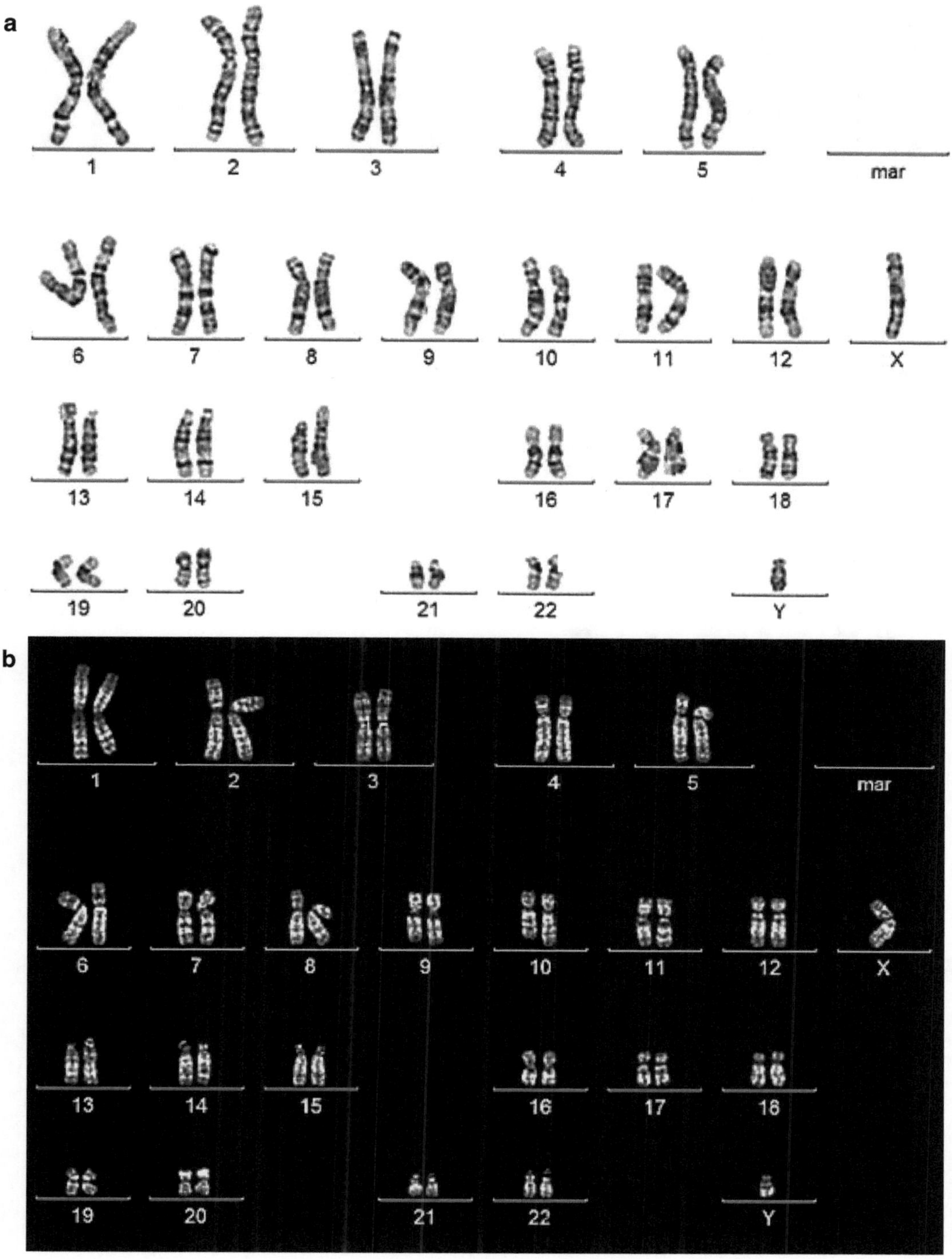

Fig. 6.30 Karyotype 46,XX,inv(11)(q21q23.3). **a** GTG-, **b** QFQ-banding, 550 and 400 bands per genome, **c** QFQ-, **d** GTG-banding, chromosomes from different mitoses, *left* normal, *right* inverted chromosome 11, **e** ideograms of chromosome 11, characteristic frequent breakpoints

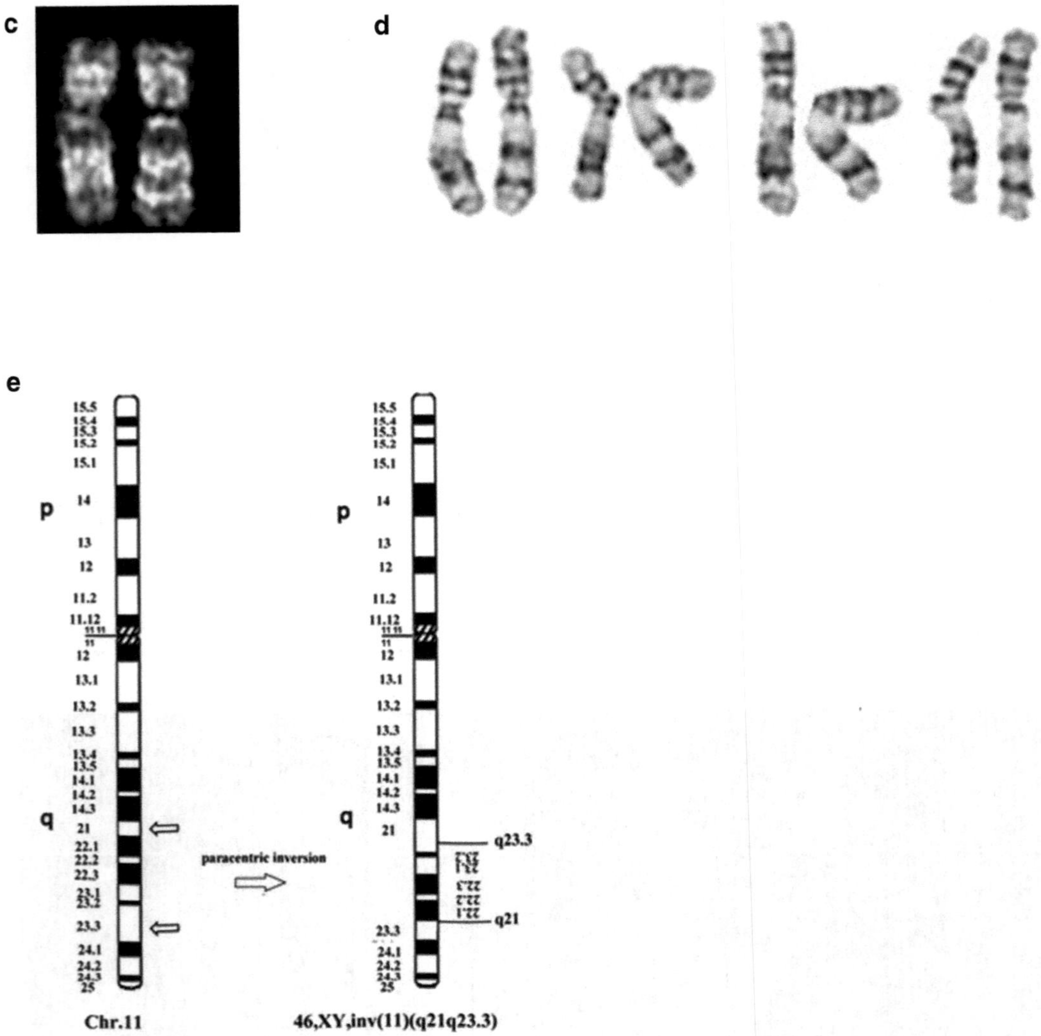

Fig. 6.30 (continued)

▪▪ Case 31: Karyotype: 46,XX,inv(12)(q21.1q24.1)

▪ Morphology

Presented here is a medium-sized, intermediate inversion in the long arm of chromosome 12, which comprises about 25% of the total length of the chromosome. The dislocation of band q23 leads to a striking altered banding pattern, already detectable at a resolution of 400–550 bands per genome, but the exact localisation of the breakpoints affords a higher resolution of at least 700 bands.

▪ Cave

According to the literature, the genetic risk for the carrier is low. This has to be taken into account when counselling the carrier and his family (◘ Fig. 6.31).

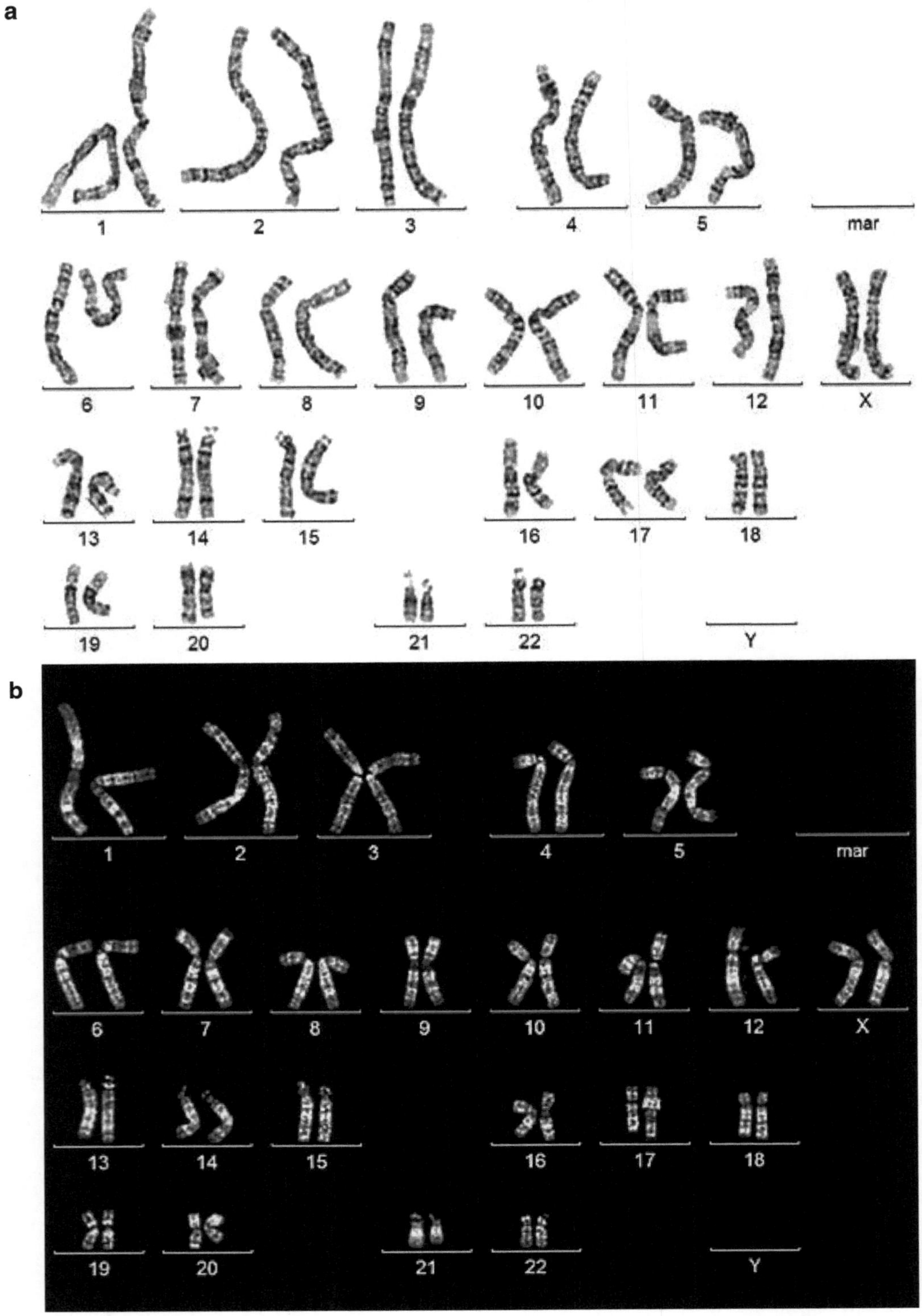

◘ Fig. 6.31 Karyotype 46,XX,inv(12)(q21.1q24.1). **a** GTG-, **b** QFQ-banding, 700 and 500 bands per genome, **c** QFQ-, **d** GTG-banding, chromosomes from different mitoses, *left* normal, *right* inverted chromosome 12, **e** ideograms of chromosome 12, intermediate breakpoints in the long arm

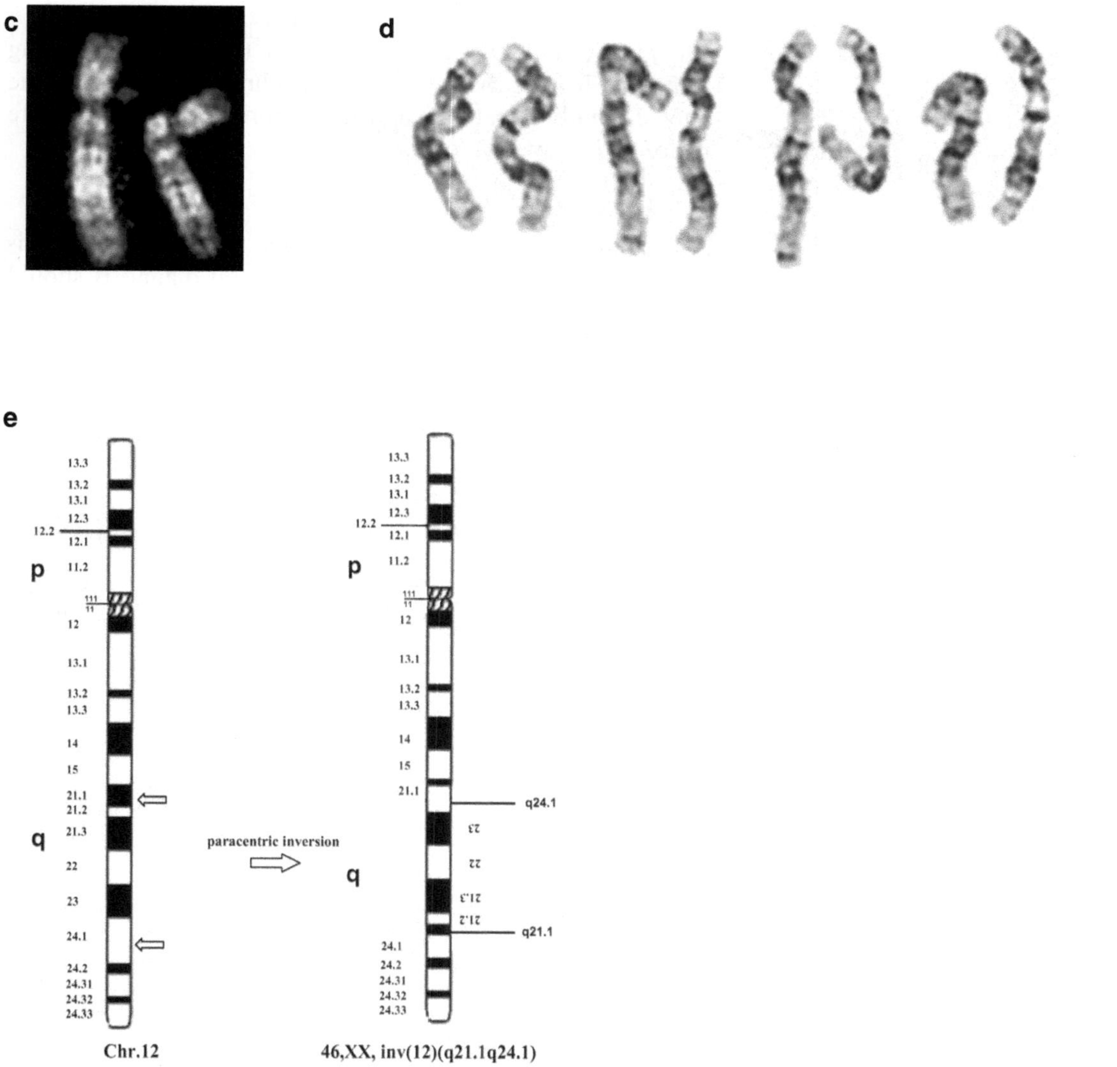

Fig. 6.31 (continued)

▪▪ Case 32: Karyotype: 46,XX,inv(13)(q12.1q22)

The inversion comprises about 50% of the euchromatic long arm, and 40% of the total length of chromosome 13. By the altered banding pattern in the proximal and intermediate region, it is easy to recognise.

▪ Cave

According to the literature, the genetic risk of the carrier is low. Therefore, the possibilities of prenatal diagnosis should be discussed, but there is no reason for a recommendation (◘ Fig. 6.32).

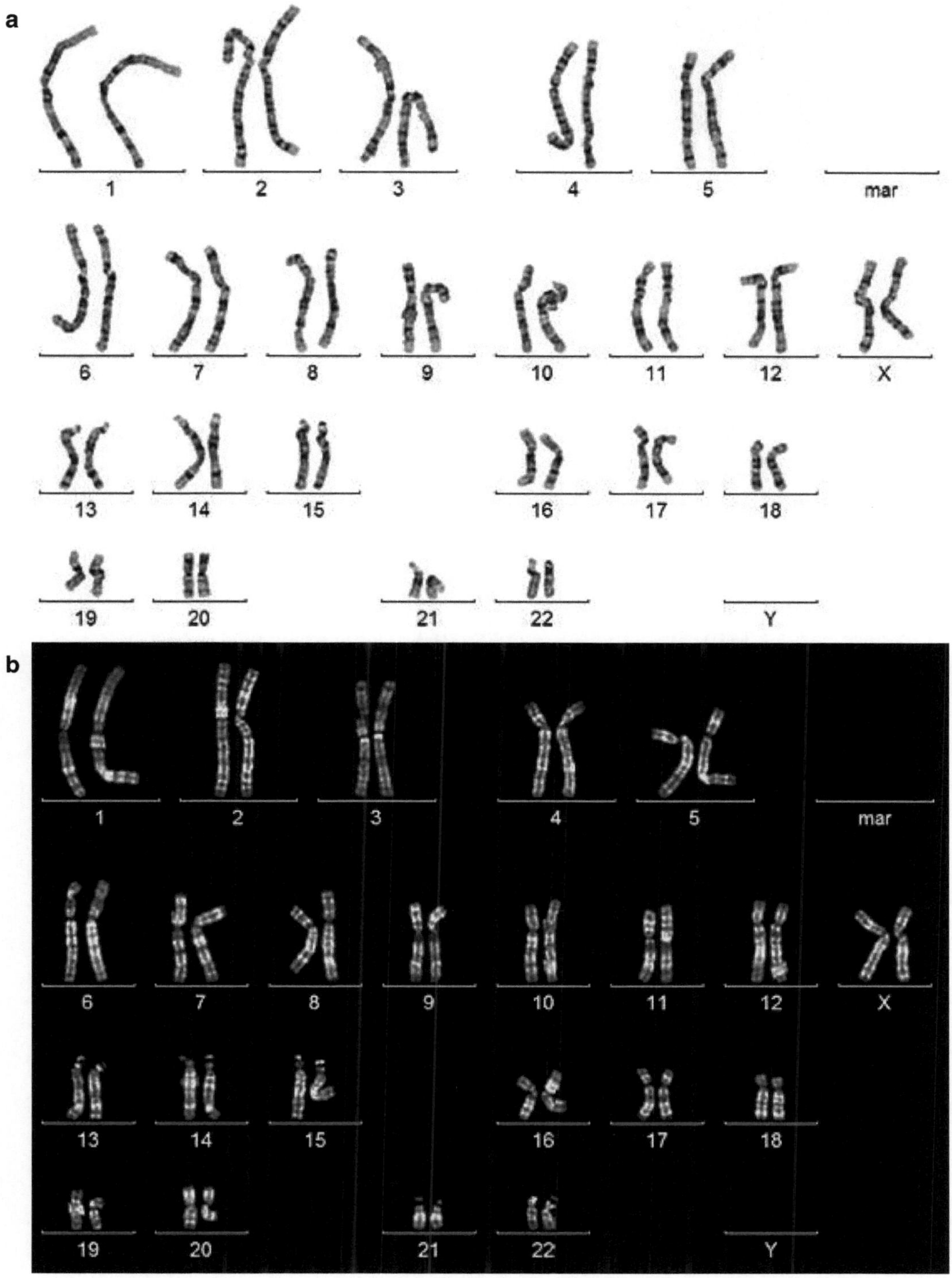

Fig. 6.32 Karyotype 46,XX,inv(13)(q12.1q22). **a** GTG-, **b** QFQ-banding, 700 and 550 bands per genome, **c** QFQ-, **d** GTG-banding, chromosomes from different mitoses, *left* normal, *right* inverted chromosome 13, **e** ideograms of chromosome 13, large inverted segment

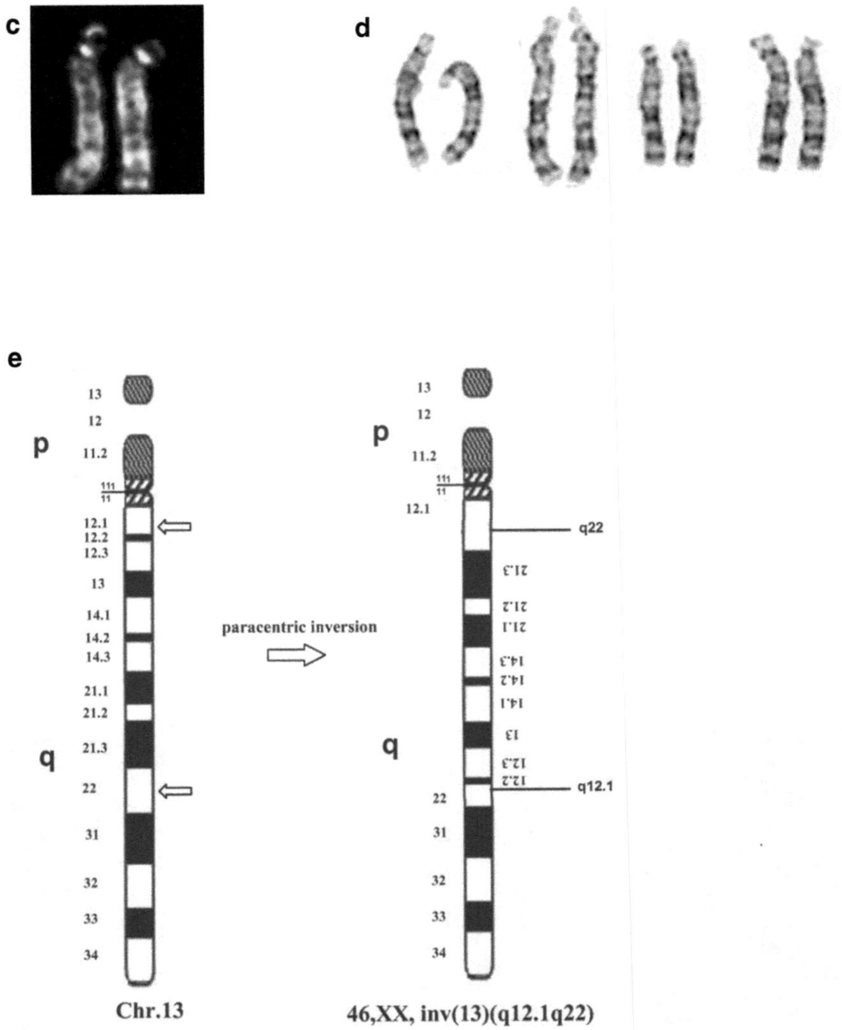

■ Fig. 6.32 (continued)

■■ Case 33: Karyotype: 46,XY,inv(13)(q13q32)

·

■ Morphology

This paracentric intermediate inversion 13 comprises about 60% of the length of the euchromatic long arm and 50% of the total length of the chromosome. The banding pattern is characteristically changed by shifting of the AT-rich bands q21 and q31.

■ Cave

The inverted region in this case is larger than in the previous one, thus it is attributed to cases with increased genetic risk. Genetic counselling should give the advice for the investigation of relatives and for prenatal diagnosis (◘ Fig. 6.33).

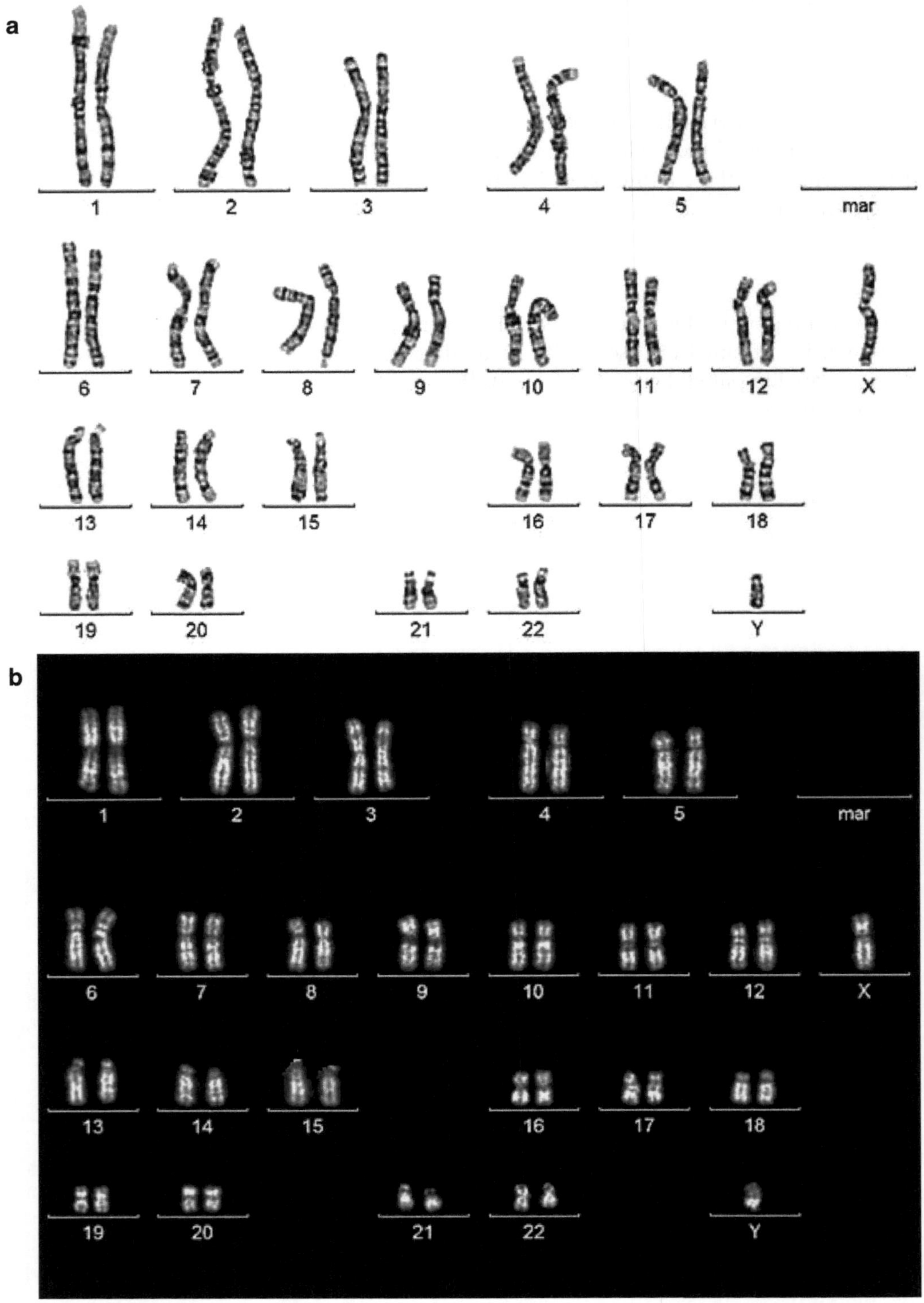

Fig. 6.33 Karyotype 46,XY,inv(13)(q13q32). **a** GTG-, **b** QFQ-banding, 600 and 400 bands per genome, **c** QFQ-, **d** GTG-banding, chromosomes from different mitoses, *left* normal, *right* inverted chromosome 13, **e** ideograms of chromosome 13, large inverted segment

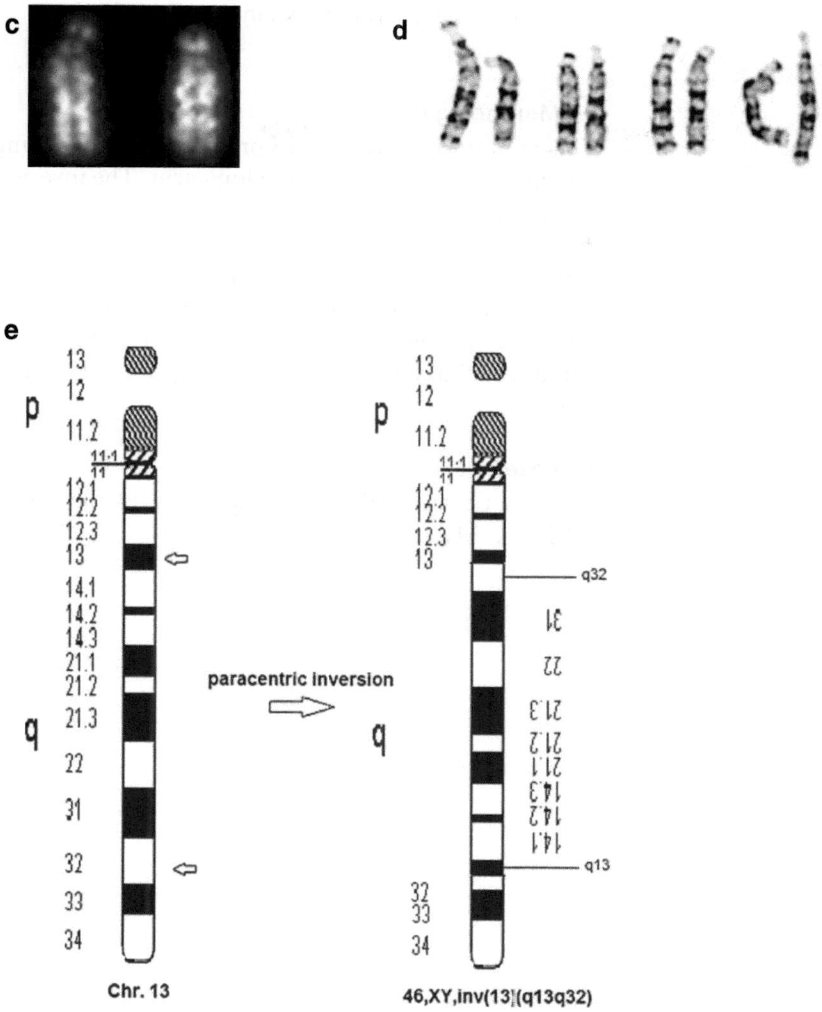

Fig. 6.33 (continued)

■■ Case 34: Karyotype: 46,XX,inv(13)(q21.2q31)

.

■ Morphology

This paracentric inversion 13 presents with a rearrangement in the intermediate region of the long arm. The inverted segment comprises about 25% of the euchromatic long arm and about 20% of the total length. The AT-rich bands q21 and q31 have changed their position and thus lead to an altered banding pattern. For the analysis of the inversion, a resolution of 550–700 bands per genome is necessary; otherwise, the relevant differentiation of q21 into the subbands q21.1, q21.2, q21.3 is not possible.

■ Cave

The genetic risk and the consequences for the carrier are low (◘ Fig. 6.34).

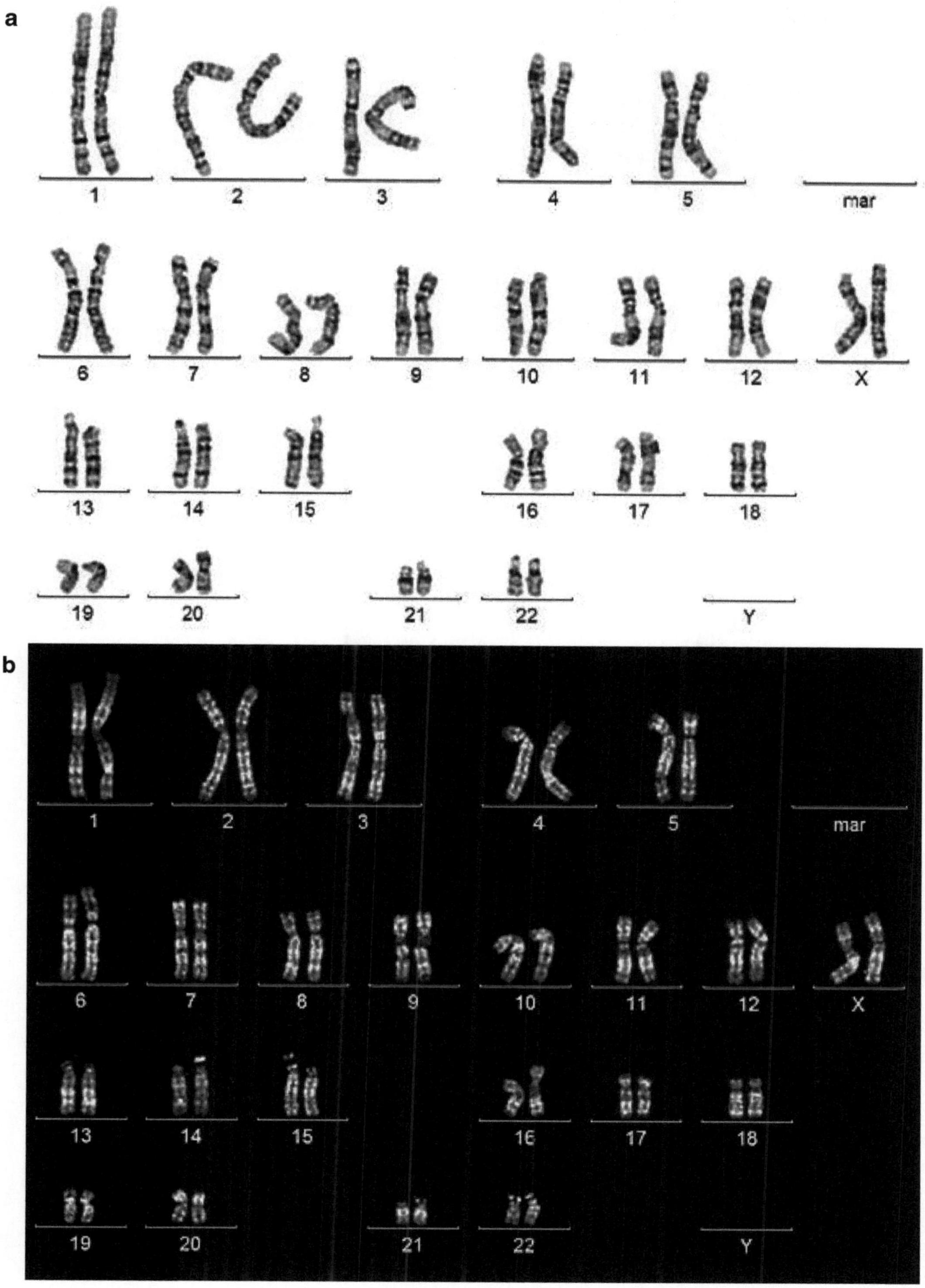

Fig. 6.34 Karyotype 46,XX,inv(13)(q21.2q31). **a** GTG-, **b** QFQ-banding, 500 and 400 bands per genome, **c** QFQ-, **d** GTG-banding, chromosomes from different mitoses, *left* normal, *right* inverted chromosome 13, **e** ideograms of chromosome 13, intermediate, medium-sized inverted segment

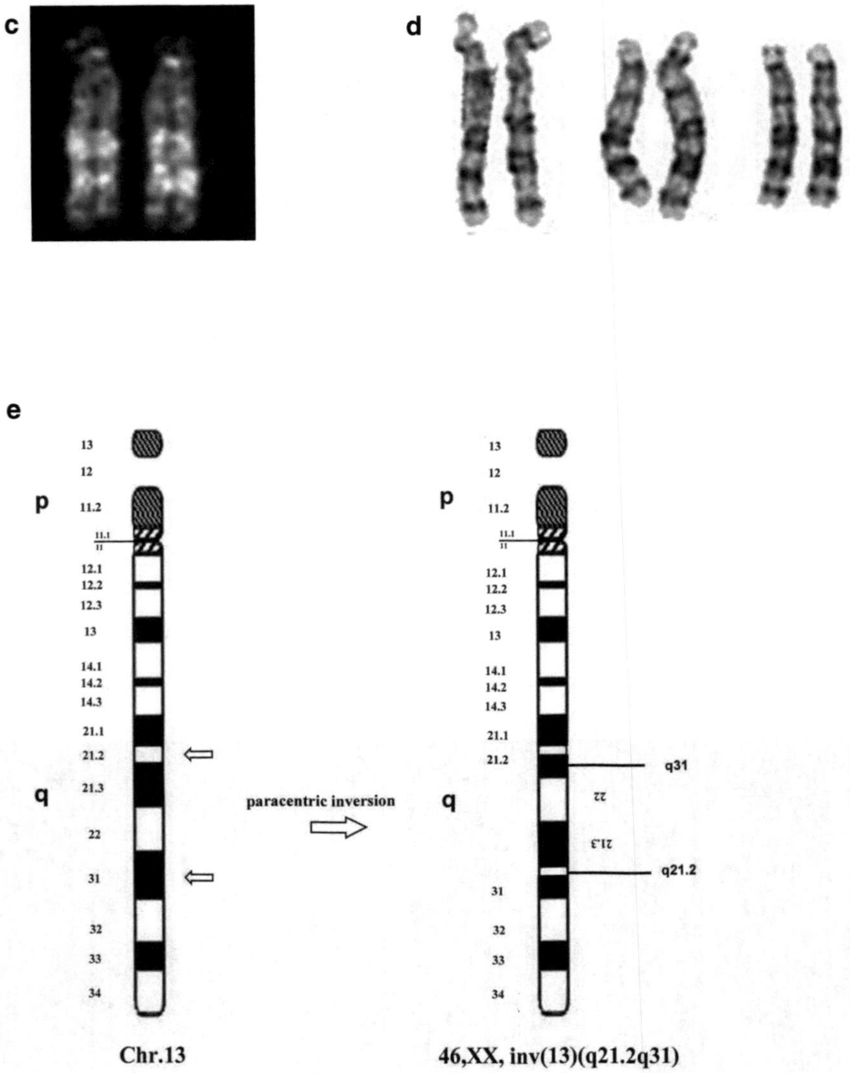

Fig. 6.34 (continued)

■■ **Case 35: Karyotype: 46,XX,inv(14)(q23.3q32.2),inv(3)(p11.2q12)(x2)**

The proband shows three inversions (two in heterochromatic and one in euchromatic regions).

■ **Morphology**

1. inv(3)(p11.2q12)(x2),
   This type of rearrangement is present in both chromosomes 3. Thus, the carrier is homozygous for the inversion. The presentation of the centromere adjacent heterochromatic block is usually performed by QFQ-banding, leading to a brilliant fluorescence of the region (fluorescence intensity i(5) according to ISCN). The inversion frequency in this genetic not relevant segment in the European population is about 2% (see: ► Sect. 15.6).
2. inv(14)(q23.3q32.2).
   The rearrangement is localised in the intermediate to distal region of the long arm. It comprises about 40% of the euchromatic long arm and 30% of the total length of the chromosome. The banding pattern has changed by the dislocation of the subbands of the AT-rich bands q21 and q31.

■ **Cave**

The analysis of a homozygosity of the inverted bands on chromosome 3 may be evidence that the carrier is descendant of parents which are related. This should be discussed in the course of genetic counselling. In the present case only, the low genetic risk of the paracentric inversion 14 has to be discussed in the course of genetic counselling (◘ Fig. 6.35).

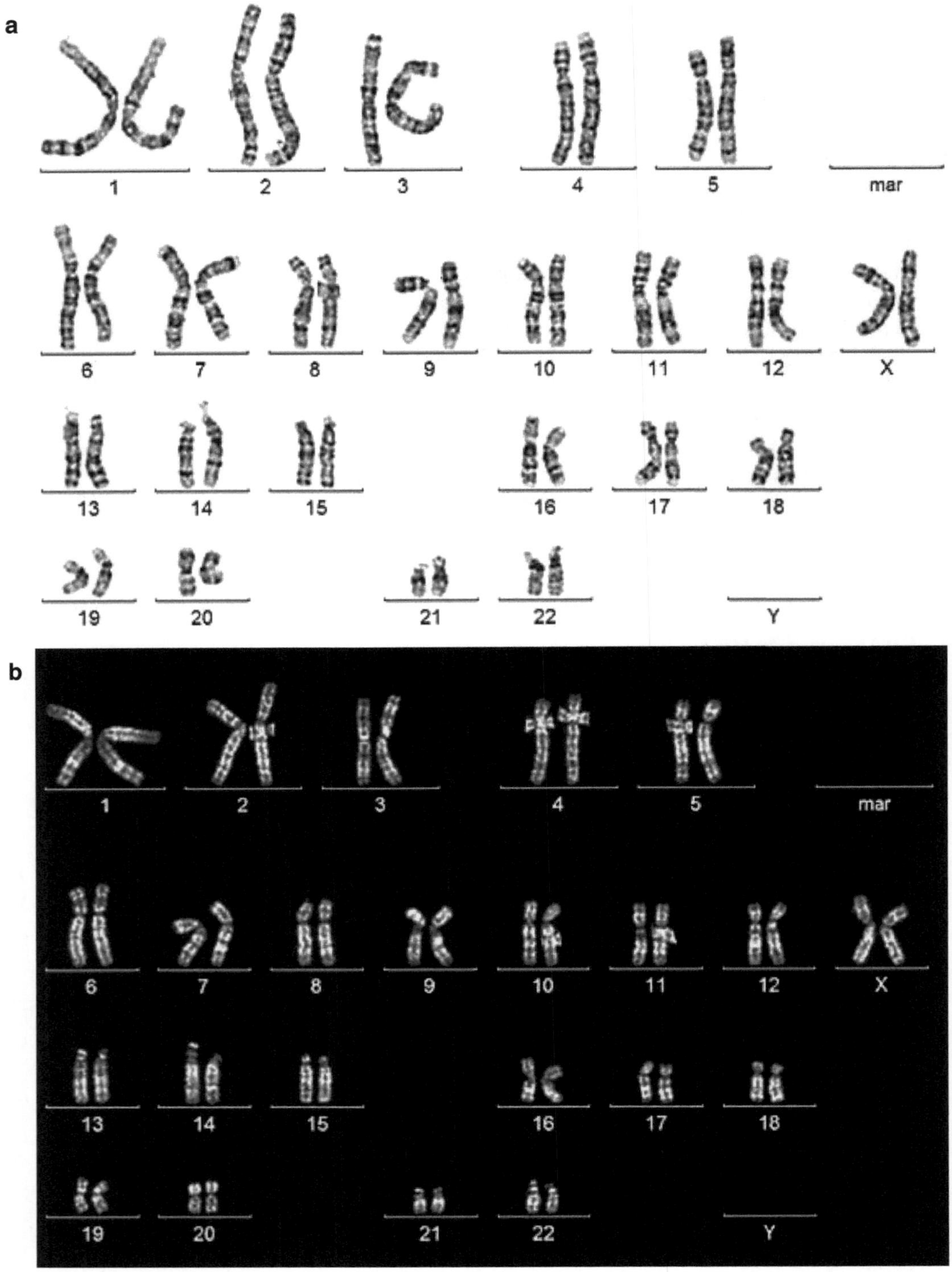

Fig. 6.35 Karyotype 46,XX,inv(14)(q23.3q32.2),inv(3)(p11.2q12)(x2). **a** GTG-, **b** QFQ-banding, 500 and 400 bands per genome, **c**, **d** QFQ-, **e** GTG-banding, chromosomes from different mitoses, **c** both chromosomes 3 show the pericentric inversion of the heterochromatin block; **d**, **e** chromosome 14: *left* normal, *right* inverted chromosome, **f** ideograms of chromosomes 3 and 14, inv(3) restricted to pericentromeric heterochromatin in both homologues, inv(14) medium sized in the intermediate to distal region

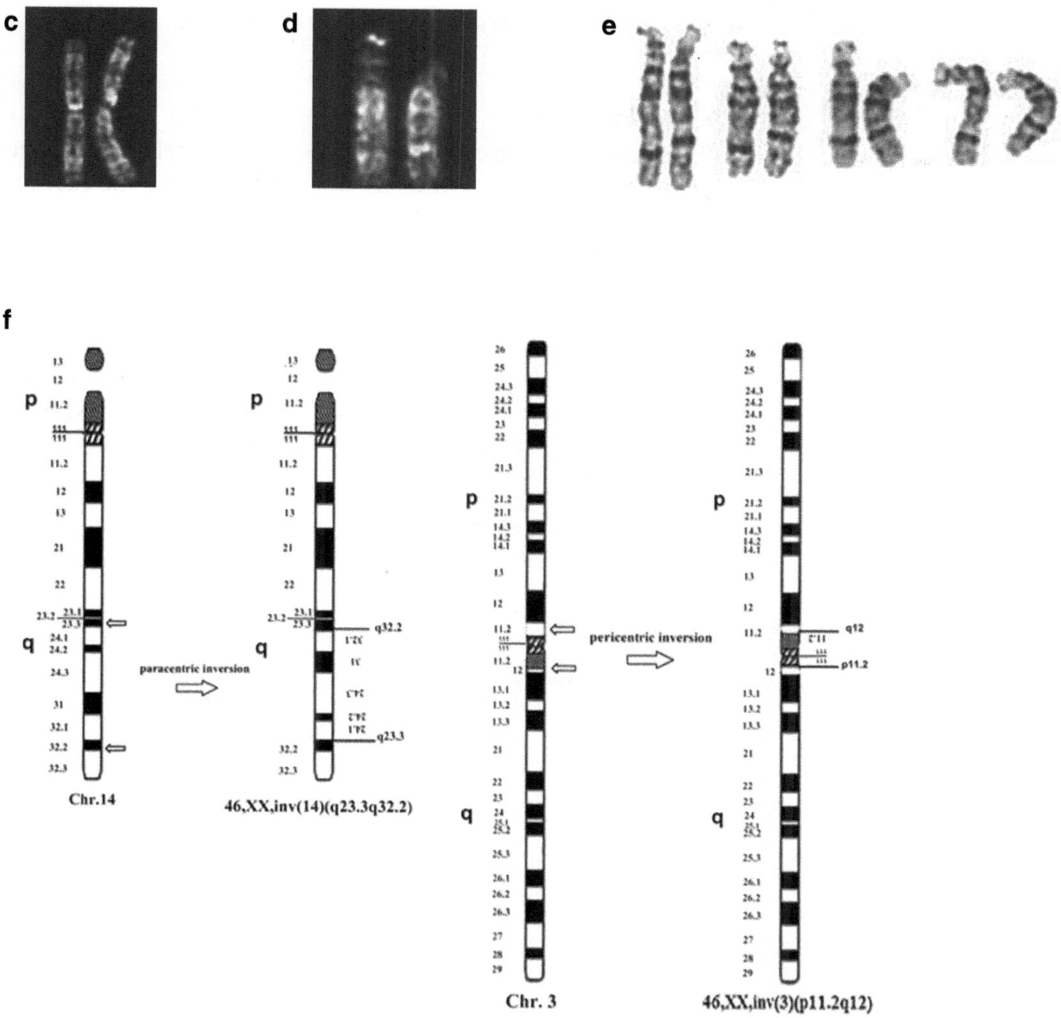

Fig. 6.35 (continued)

■■ Case 36: Karyotype: 46,XX,inv(15)(q24q26.3)

.

■ Morphology

This paracentric inversion is localised in the distal region of the long arm. The inversion comprises about 30% of the euchromatic long arm and 25% of the total length of the chromosome. Both breakpoints are localised in GC-rich regions, and q26.3 is a hot spot of aberrations for different types of rearrangement.

■ Cave

According to the literature, this type of inversion is inherited in balanced condition in about 99% of cases and can therefore be regarded as carrying no increased genetic risk for the carrier (◘ Fig. 6.36).

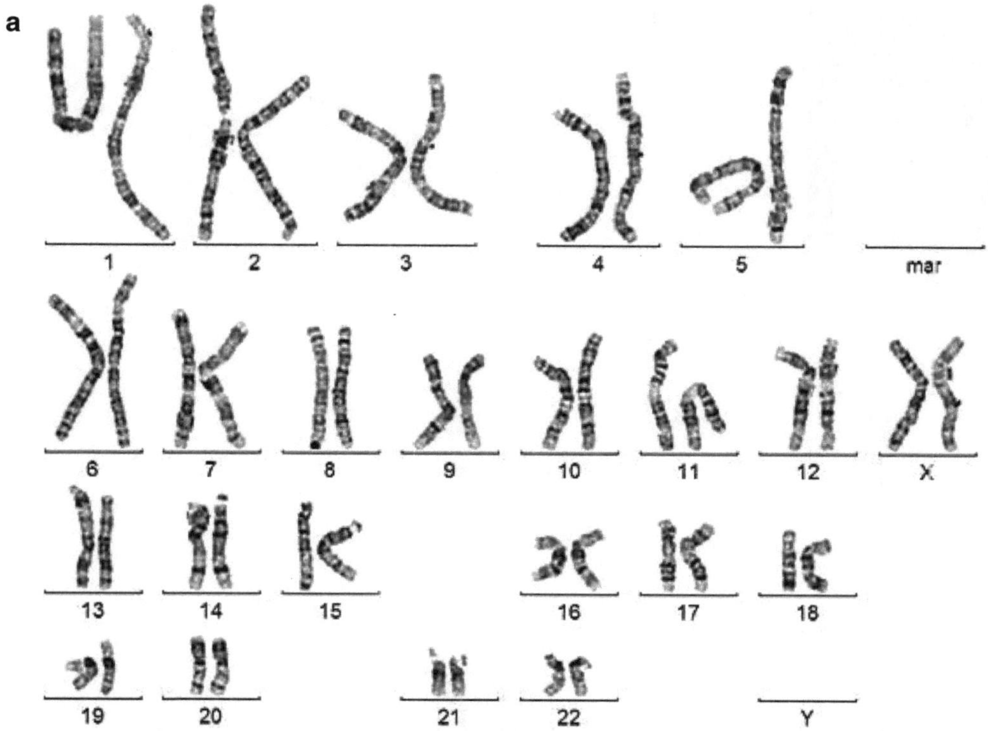

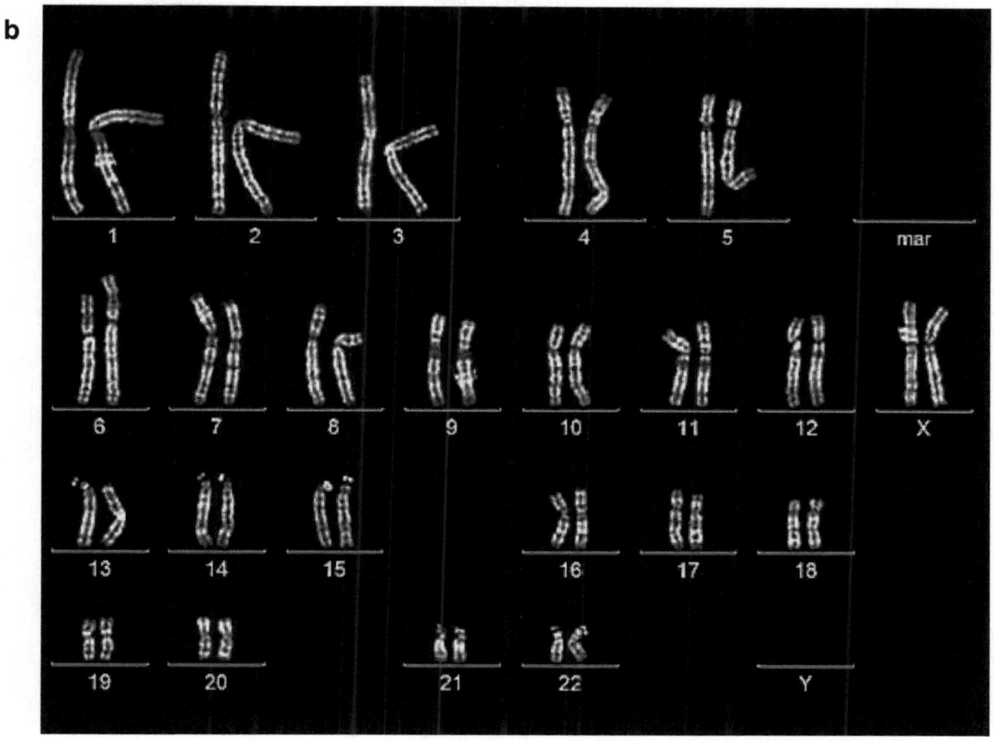

Fig. 6.36 Karyotype 46,XX,inv(15)(q24q26.3). **a** GTG-, **b** QFQ-banding, 700 and 550 bands per genome, **c** QFQ-, **d** GTG-banding, chromosomes from different mitoses, *left* normal, *right* inverted chromosome 15, **e** ideograms of chromosome 15, distal position of the inverted segment

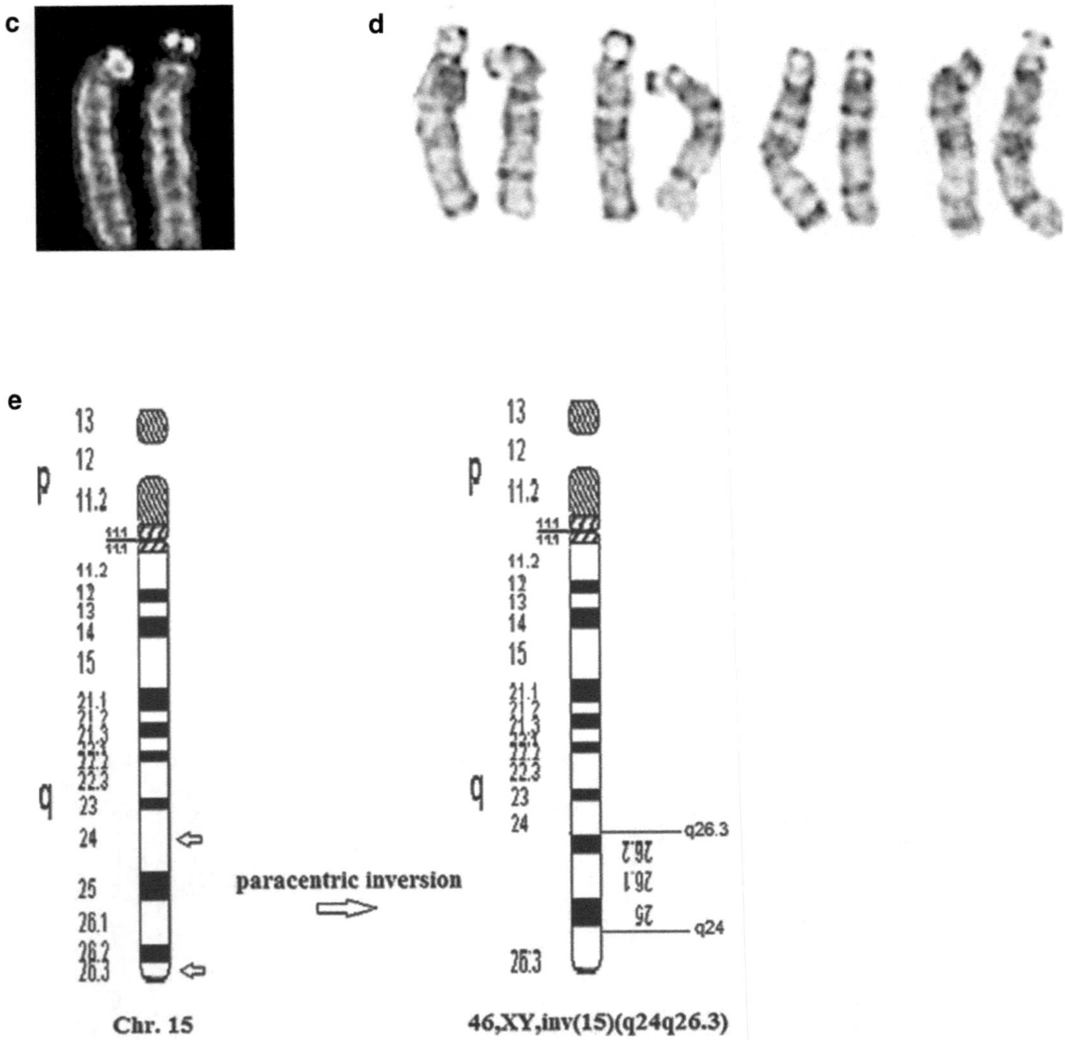

■ **Fig. 6.36** (continued)

▪▪ Case 37: Karyotype: 46,XY,inv(19)(p12p13.13)

▪ Morphology

The paracentric inversion in the short arm of chromosome 19 comprises about 15–20% of the total length of the chromosome. The proximal breakpoint is localised in the pericentromeric heterochromatin. The dislocation of this band into a more distal position leads to a striking alteration of the banding pattern. It received a position in an intermediate, GC-rich region. The analysis of this inversion affords a resolution of 700–850 bands per genome.

▪ Cave

The inverted region is relatively small and the low genetic risk is further reduced by the position of one breakpoint (p12) in a heterochromatic segment (◘ Fig. 6.37).

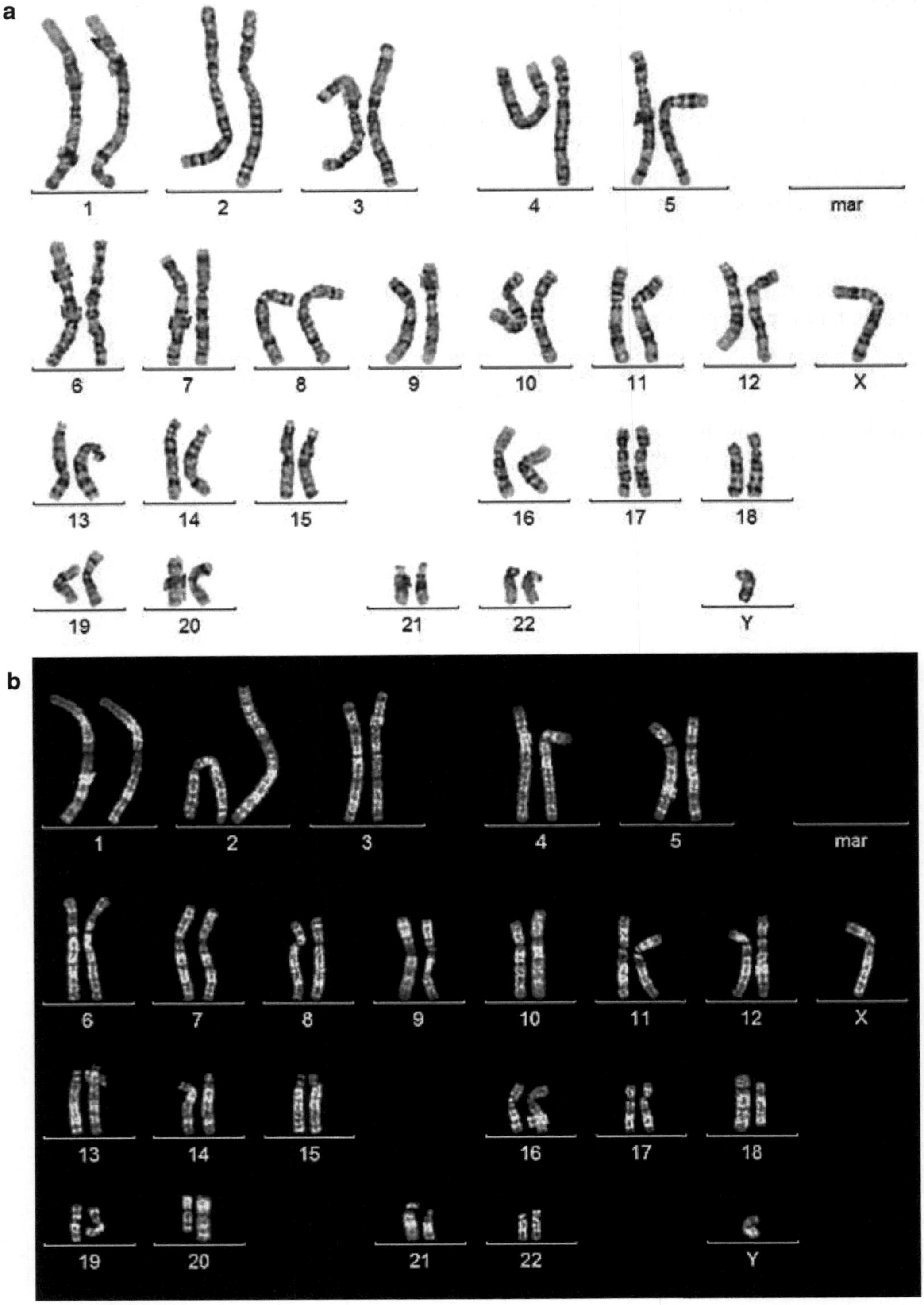

Fig. 6.37 Karyotype 46,XY,inv(19)(p12p13.13). **a** GTG-, **b** QFQ-banding, 700 and 500 bands per genome, **c** CBG-, **d** QFQ-, **e** GTG-banding, chromosomes from different mitoses, *left* normal, *right* inverted chromosome 19, **f** ideograms of chromosome 19, the inverted chromosome 19 is mainly characterised by the translocation of pericentromeric heterochromatin into the distal short arm

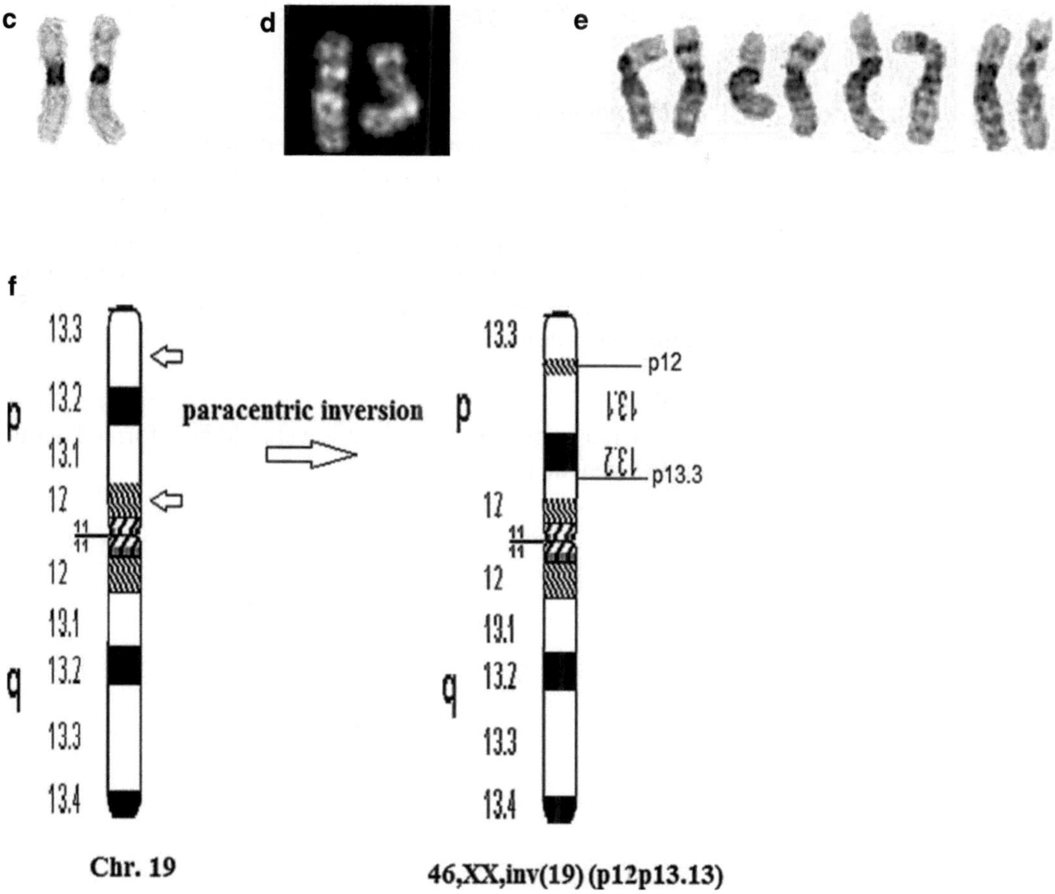

Fig. 6.37 (continued)

Case 38: Karyotype: 46,XX,inv(19)(q13.31q13.42)

.

Morphology

A paracentric inversion in the intermediate to distal region of the long arm of chromosome 19 is presented, which comprises about 25% of the total length of the chromosome. The analysis is done by the altered banding pattern, primary by the dislocation of the terminal AT-rich band q13.4 to the more proximal position. This investigation requires a resolution of 700–850 bands per genome.

Cave

The carrier of this type on inversion should be informed about his low genetic risk (■ Fig. 6.38).

a

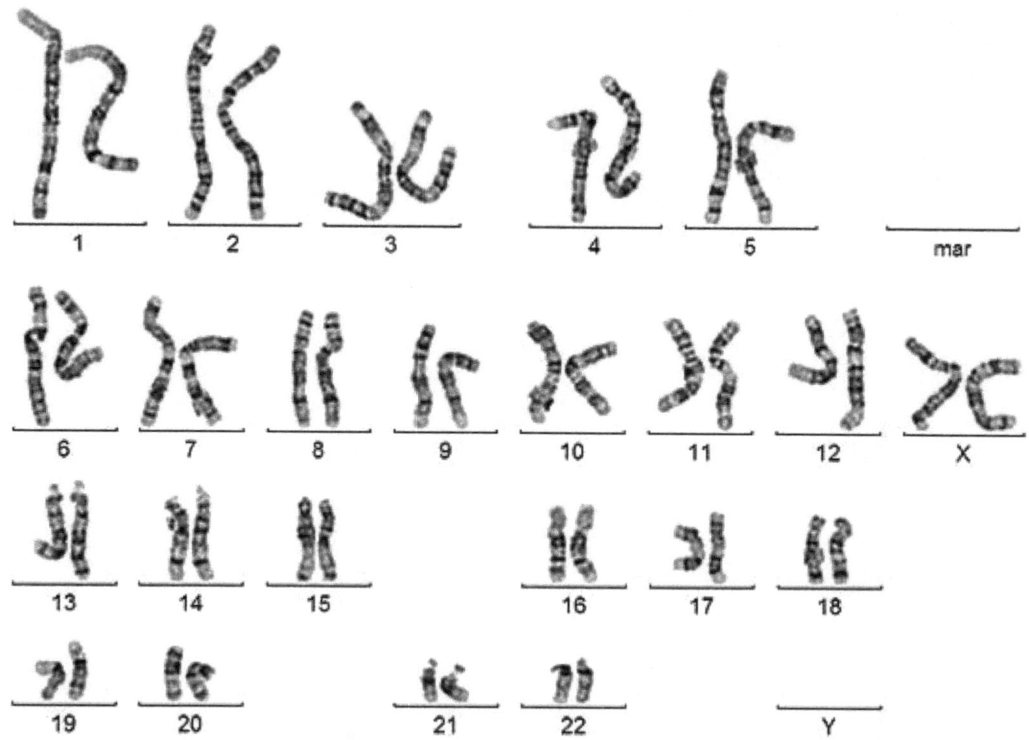

b

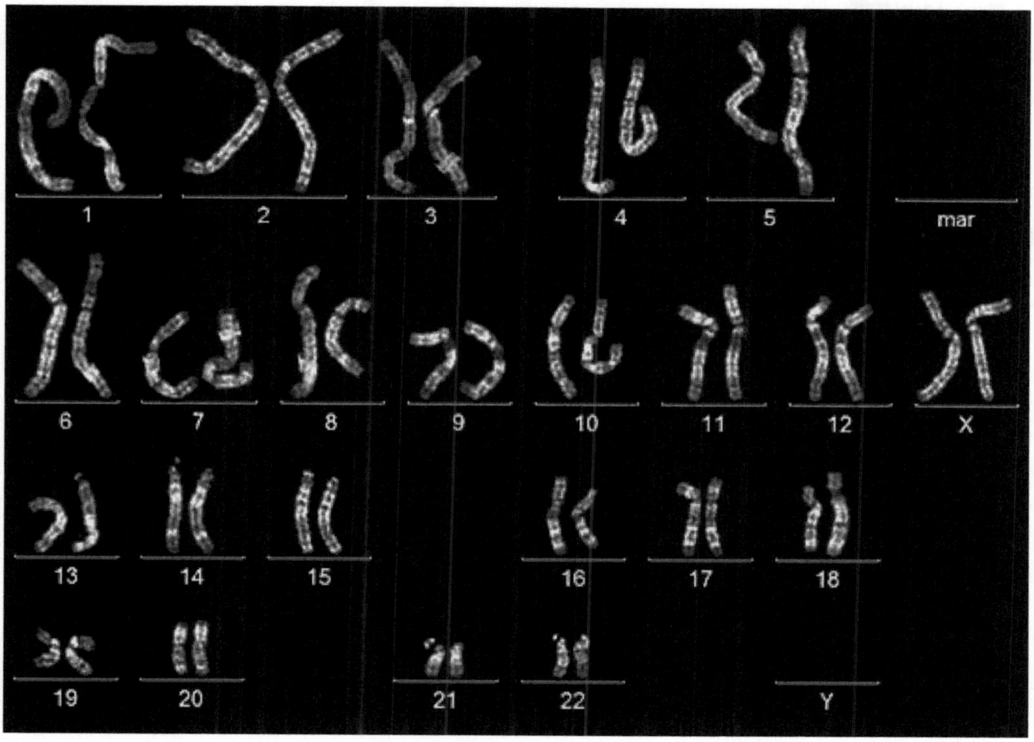

Fig. 6.38 Karyotype 46,XX,inv(19)(q13.31q13.42). **a** GTG-, **b** QFQ-banding, 800 and 700 bands per genome, **c** QFQ-, **d** GTG-banding, chromosomes from different mitoses, *left* normal, *right* inverted chromosome 19, **e** ideograms of chromosome 19, intermediate to distal inversion in the long arm

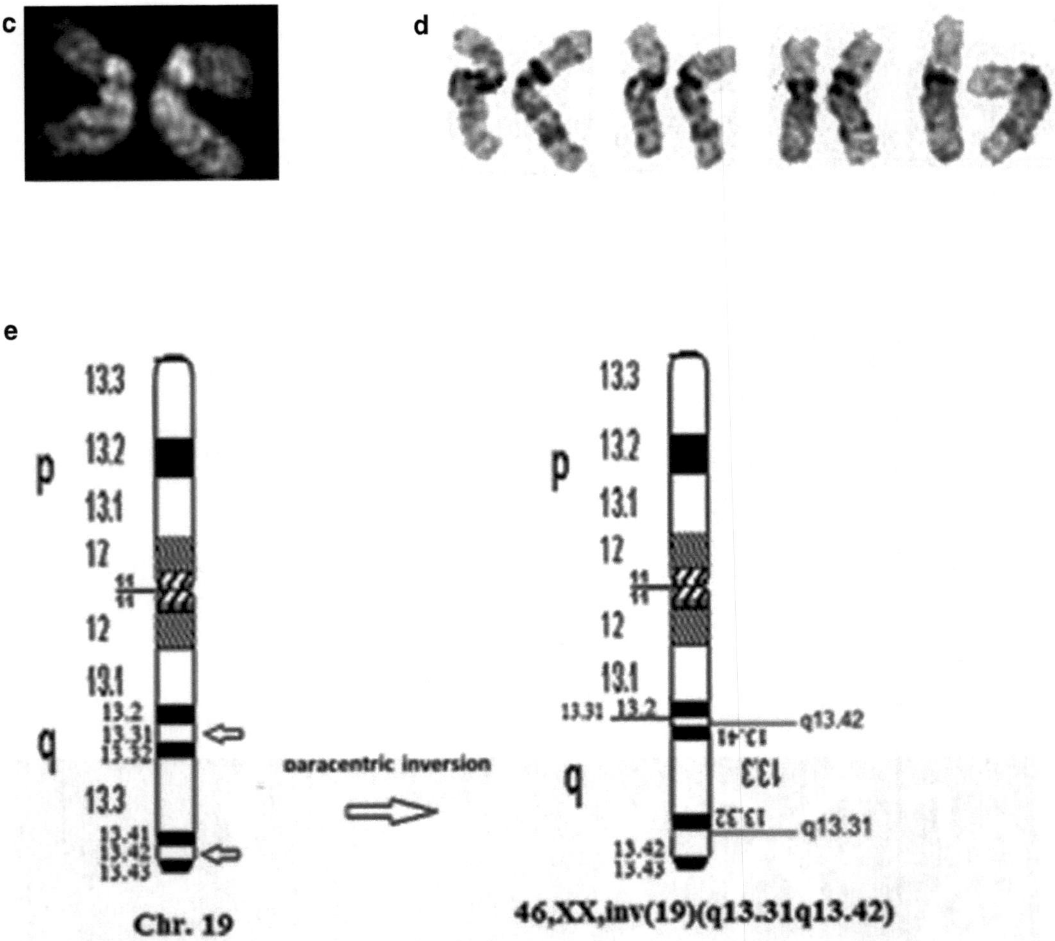

■ **Fig. 6.38** (continued)

■■ Case 39: Karyotype: 46,X,inv(X)(p11.3p22.32)

■ Morphology

This is a very characteristic and easy to diagnose inversion, as the main, AT-rich band in the short arm is now located in a more distal position. The length of the inverted segment is about 25% of the total length of the X-chromosome.

■ Cave

As the short arm of the X-chromosome is a gene-rich region, prenatal diagnosis is indicated for all inversion carriers, and additionally microarray should be performed, especially in male carriers to exclude unequal crossing over followed by deletion which can be especially relevant in the hemizygotic male (◘ Fig. 6.39).

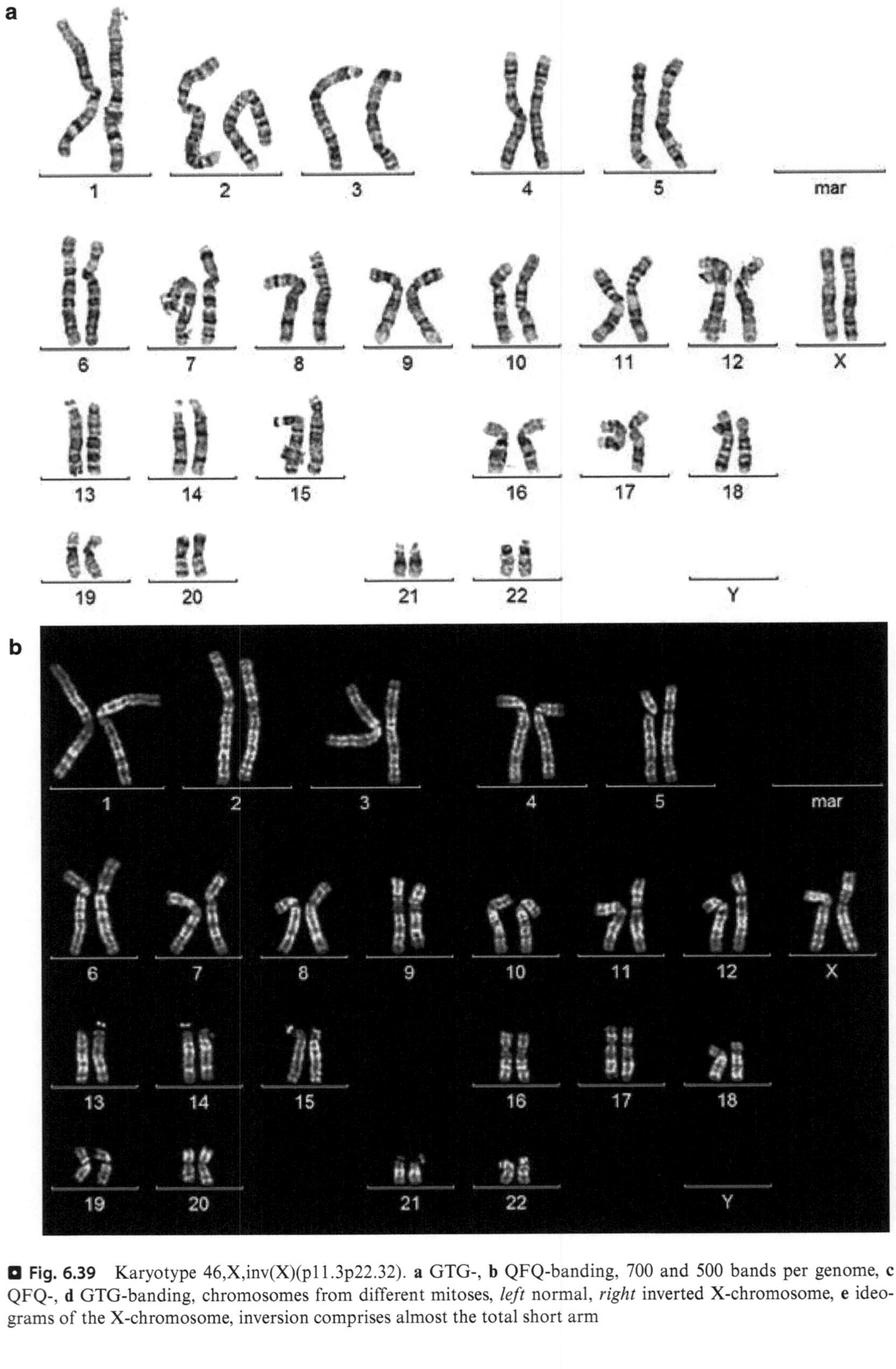

Fig. 6.39 Karyotype 46,X,inv(X)(p11.3p22.32). **a** GTG-, **b** QFQ-banding, 700 and 500 bands per genome, **c** QFQ-, **d** GTG-banding, chromosomes from different mitoses, *left* normal, *right* inverted X-chromosome, **e** ideograms of the X-chromosome, inversion comprises almost the total short arm

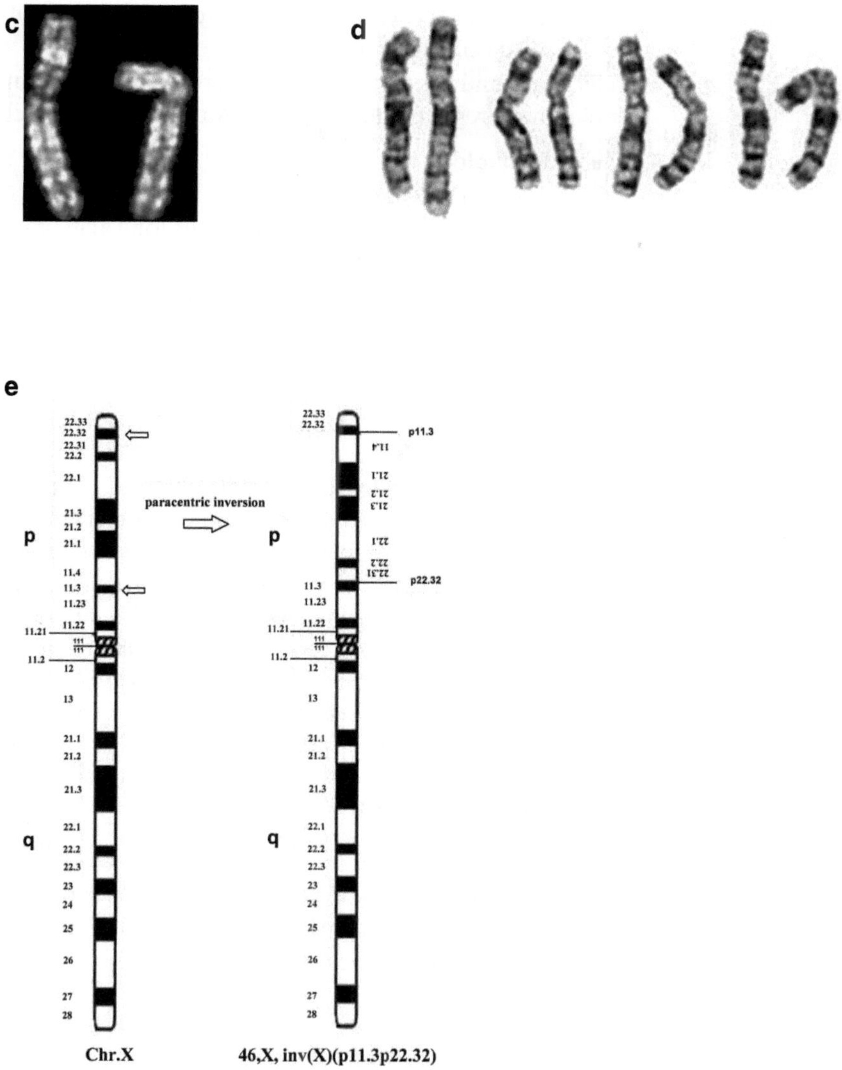

Fig. 6.39 (continued)

■■ Case 40: Karyotype: 46,X,inv(Y)(q11.21q12),inv(9)(p12q12),inv(9)(p12q13)

The proband has three different inversions, two of them in heterochromatic regions and one in euchromatin and heterochromatin.

■ Morphology

1. inv (Y)(q11.21q12),

   The paracentric inversion in the long arm has one breakpoint in the proximal AT-rich band q11.21 and the second one in the heterochromatic block q12. The segment comprises about 45% of the total length of the Y-chromosome, which is an individual percentage, as the length of q12 is very variable. The analysis of the inversion is performed by a combination of GTG, QFQ and CBG-banding, as the distal heterochromatic block is dislocated and the banding pattern in the euchromatin has changed. The aberration is easy to recognise, but the breakpoint determination affords a resolution of at least 550 bands per genome.
2. Type I: inv(9)(p12q13) and Type II: inv(9)(p12q12).

   Both chromosomes 9 show, classified as heteromorphisms or variants, pericentric inversions of the heterochromatin. Whereas one of them presents with a complete inversion of the heterochromatic block q12 (Type I, frequency in the European population 1%) the second one shows a partial inversion (Type II, frequency in the European population 10%) (see: ► Sect. 15.6).

■ Cave

The heterochromatin inversions of chromosome 9 present with no increased genetic risk. The paracentric inversion Y can cause male fertility problems (◘ Fig. 6.40).

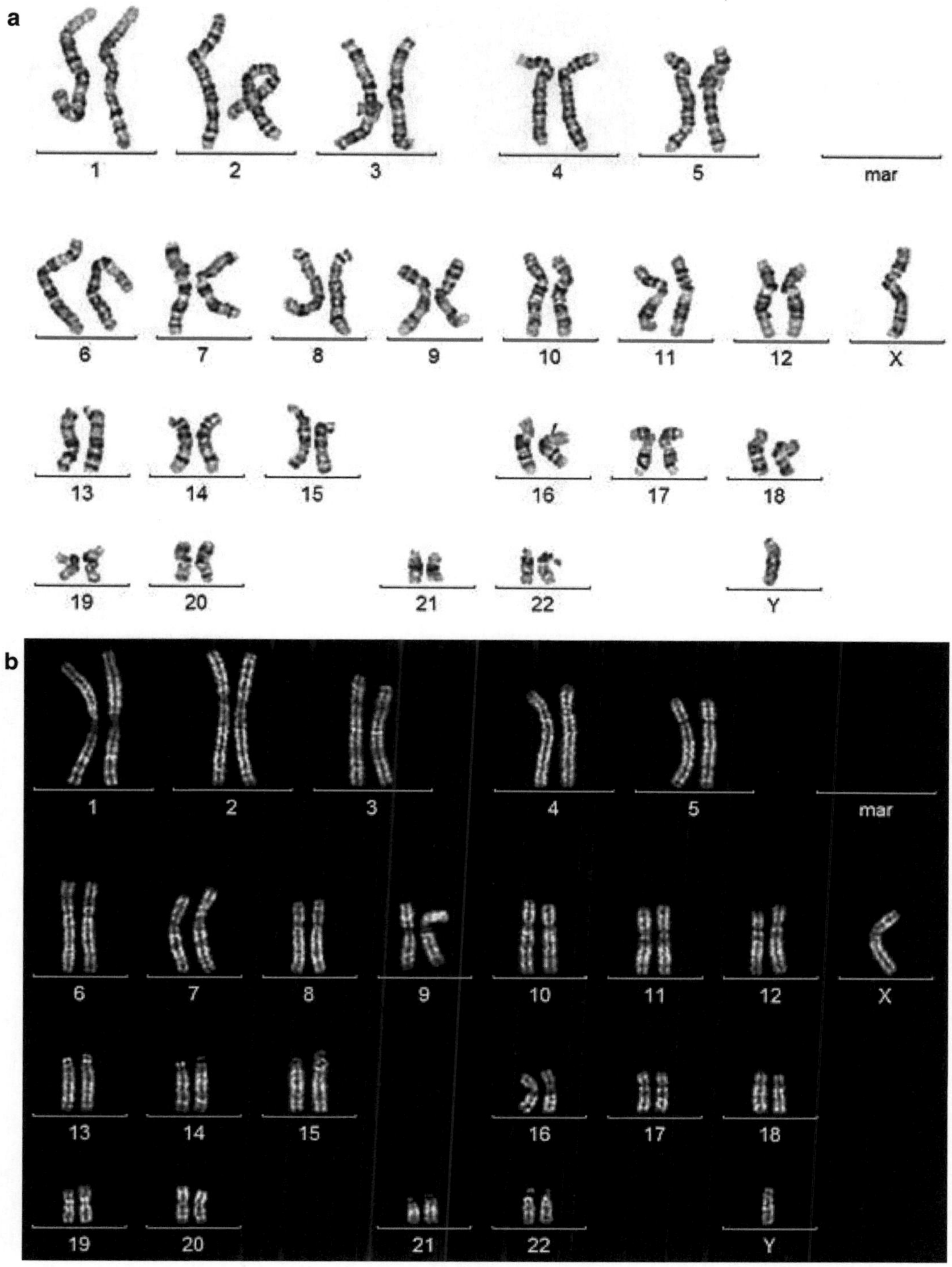

Fig. 6.40 Karyotype 46,X,inv(Y)(q11.21q12),inv(9)(p12q12),inv(9)(p12q13). **a** GTG-, **b** QFQ-banding, 600 and 400 bands per genome, **c** CBG-, **d** QFQ-, **e** GTG-banding, chromosomes from different mitoses, **d/1**: Y-chromosome from a normal control person, **f** ideograms of the Y-chromosome, one breakpoint in the constitutive heterochromatin, **g** ideograms of a normal (control person) and two inverted chromosomes 9, all breakpoints in the constitutive heterochromatin

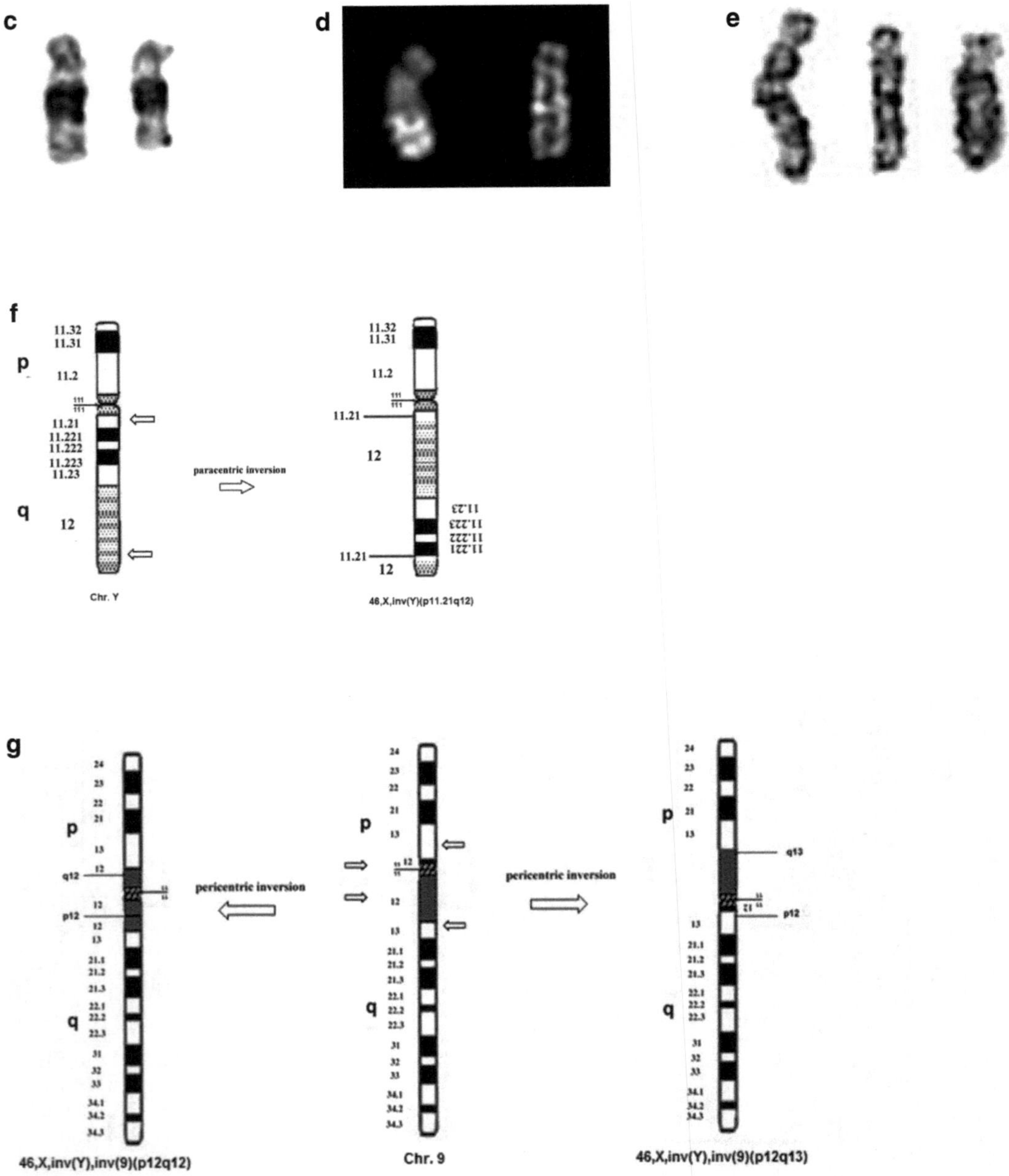

Fig. 6.40 (continued)

6.3 Summary

The collection presented here includes 47 inversions. For reasons of space, only 40 were described in detail. The total investigation group comprises paracentric and pericentric inversions in frequencies of 58% and 42%, which means that paracentric inversions were more frequent than pericentric ones in the investigation group. Furthermore, an unequal interchromosomal distribution of the chromosomes involved was noted: the large chromosomes 1–12 and the X-chromosome showed 73% of the inversions, chromosomes 13–22 and Y only 27% and in the latter group chromosomes 17, 18, 21 and 22 showed no inversions at all.

In pericentric inversions, the breakpoints were approximately uniformly distributed in proximal, intermediate and distal chromosome regions, whereas in paracentric inversions about 52% were located intermediate, 33% distal and only 15% proximal.

The size of the inverted regions ranges from 4% to a maximum of 80% of the total length of the chromosome involved.

All inversions shown could be safely distinguished by the altered banding pattern from insertions.

Deletions

Contents

7.1 Deletions – 166

7.2 Summary – 227

C. Behrend et al., *Human Chromosome Atlas*, https://doi.org/10.1007/978-3-031-10588-3_7

7.1 Deletions

41. 46,XY,del(1)(p36.3)
42. 46,XY,del(2)(q31q32.1),inv(4)(p12q11.2)
43. 46,XY,del(4)(q31.3)
44. 46,XX,del(5)(p14.2)
45. 46,XX,del(5)(p15.1)
46. 46,XY,del(8)(p23) [40]/46,XY[1]
47. 46,XX,del(9)(p22p24),inv(4)(p12q11.2)
48. 46,XX,del(10)(p13p15.1)
49. 46,XY,del(13)(q21.2q31)
50. 46,XY,del(15)(q11.2q13),inv(4)(p12q11.2)
51. 46,XY,del(17)(p11.2),inv(4)(p12q11.2)
52. 46,XX,del(17)(q23.1q23.3),inv(3)(p11.1q11.2)2x
53. 46,XX,del(18)(p11.2),dup(7)(q36)t der(18),t(7;18)(q36;p11.2) (D18S552-,7QTEL20 +),inv(9)(p12q13)
54. 46,XX,del(18)(q21.3),inv(4)(p12q11.2)
55. 46,X,del(X)(p11.23)
56. 46,X,del(X)(p22.1)
57. 46,X,del(X)(p22.1p22.2),inv(4)(p12q11.2)
58. 46,X,del(X)(q21.2)
59. 46,X,del(X)(q27.2)
60. 46,X,del(Y)(q11.22).

■■ Case 41: Karyotype: 46,XY,del(1)(p36.3)

Indication for chromosome analysis: Characteristic symptoms of the deletion 1p36.

■ Morphology

The deleted chromosome ends with the AT-band p36.2. A resolution of 550–700 bands per genome was necessary, to get a precise diagnosis in the case presented here.

■ Cave

Chromosome investigations in the parents should be performed to exclude the presence of a balanced reciprocal translocation in one of them. After these additional analyses, a summarising genetic counselling should be offered to them (◘ Fig. 7.1).

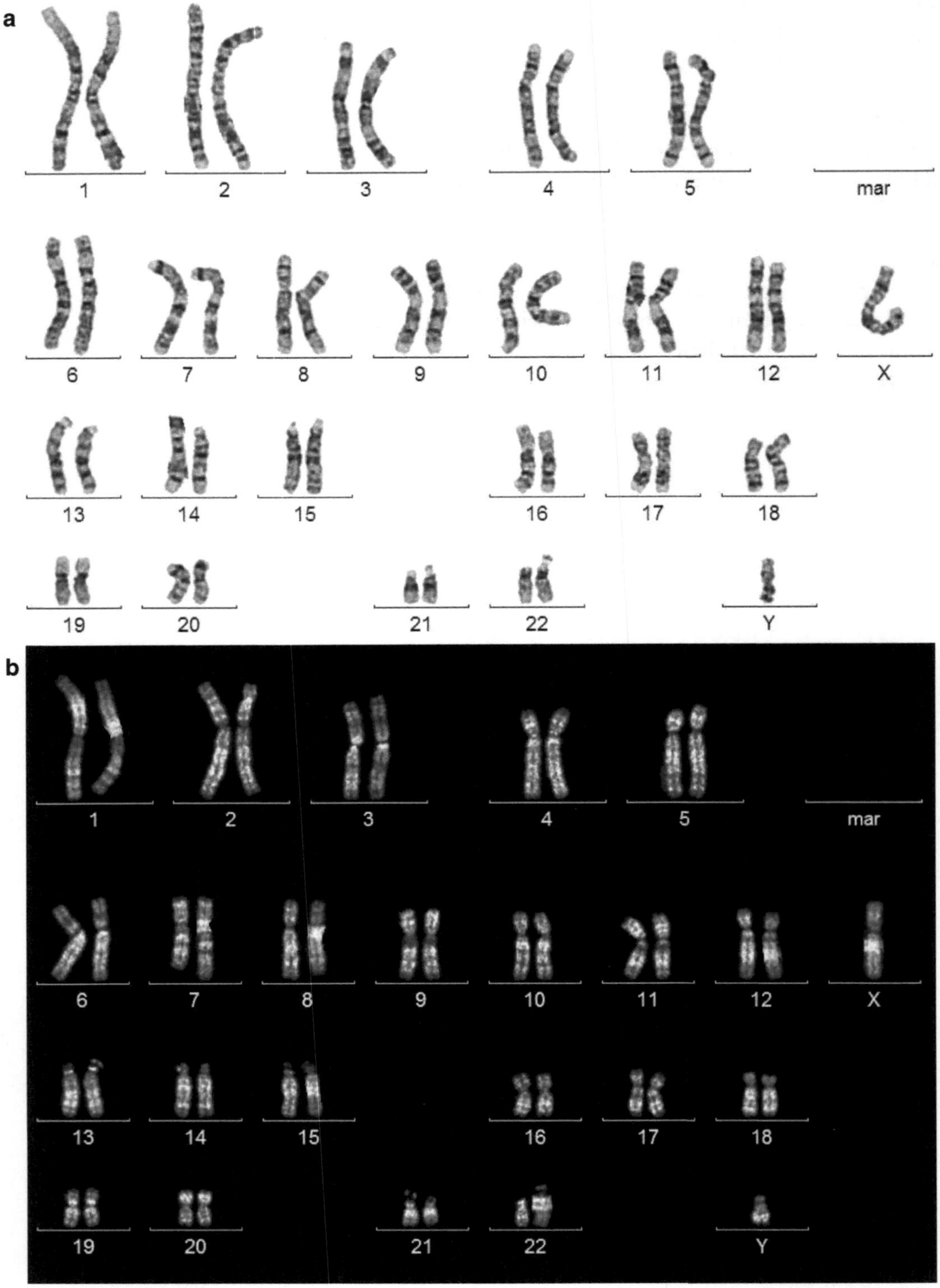

■ **Fig. 7.1** Karyotype 46,XY,del(1)(p36.3). **a** GTG-, **b** QFQ-banding, 600 and 400 bands per genome, **c** QFQ-, **d** GTG-banding, chromosomes from different mitoses, *left* normal, *right* deleted chromosome 1, **e** ideograms of chromosome 1, small distal deletion in the short arm

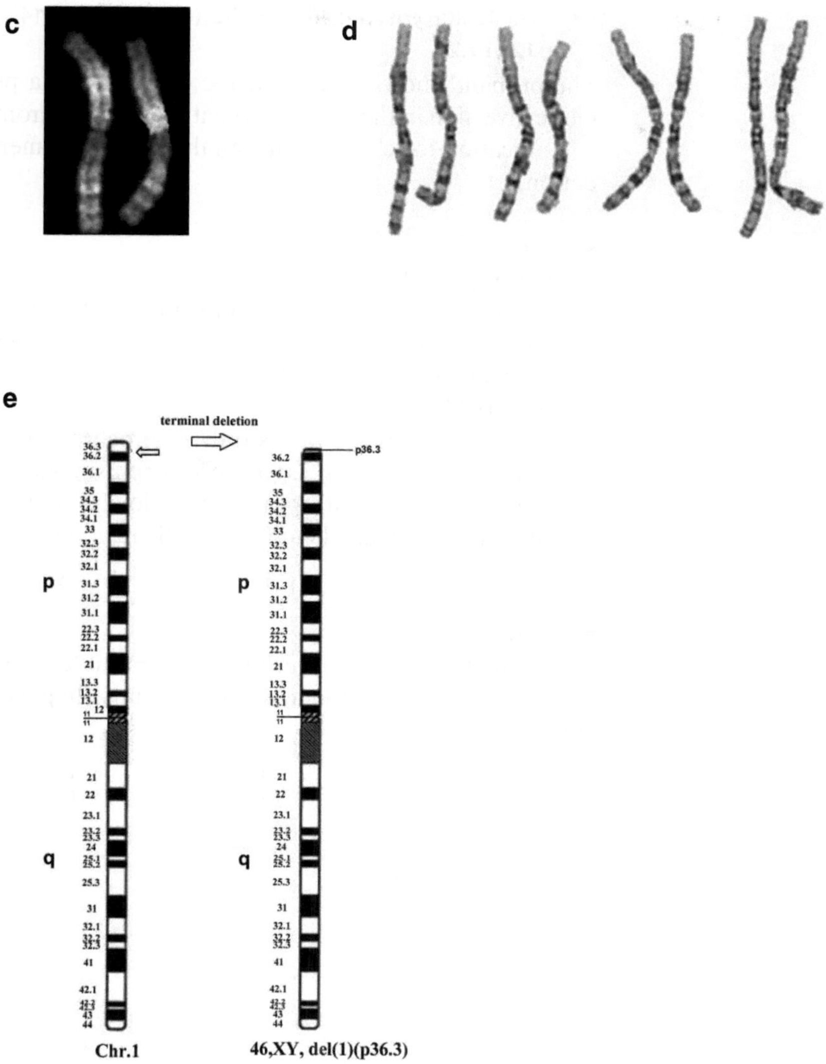

Fig. 7.1 (continued)

▪▪ Case 42: Karyotype: 46,XY,del(2)(q31q32.1),inv(4)(p12q11.2)

The proband shows a euchromatic deletion and a pericentromeric inversion in the heterochromatin block of chromosome 4.

Indication for chromosome analysis: Developmental retardation.

▪ Morphology

1. del(2)(q31q32.1)

   The interstitial deletion in the long arm of chromosome 2 leads to an altered banding pattern and to a change in the index of long and short arm.
2. inv(4)(p12q11.2)

   This inversion involves only heterochromatic segments and is therefore classified as polymorphism without genetic relevance. In the European population, it is the most frequent inversion in a heterochromatic region with 7.6% (see: ► Sect. 15.6).

▪ Cave

The parents must be investigated to exclude a paracentric inversion in one of them. Genetic counselling is indicated (◘ Fig. 7.2).

a

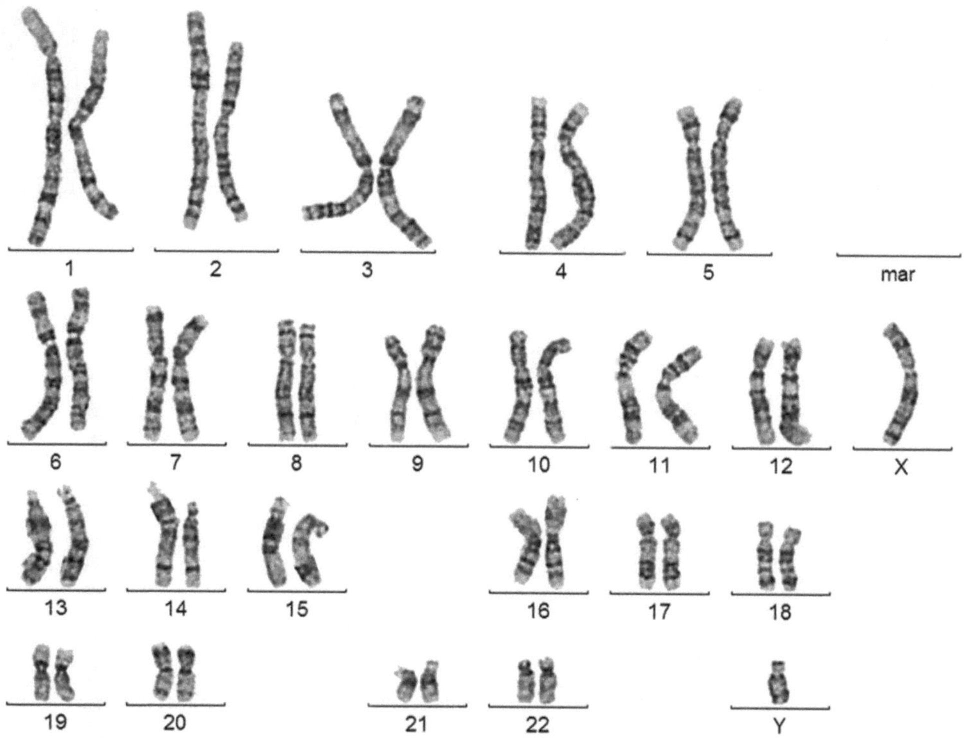

b

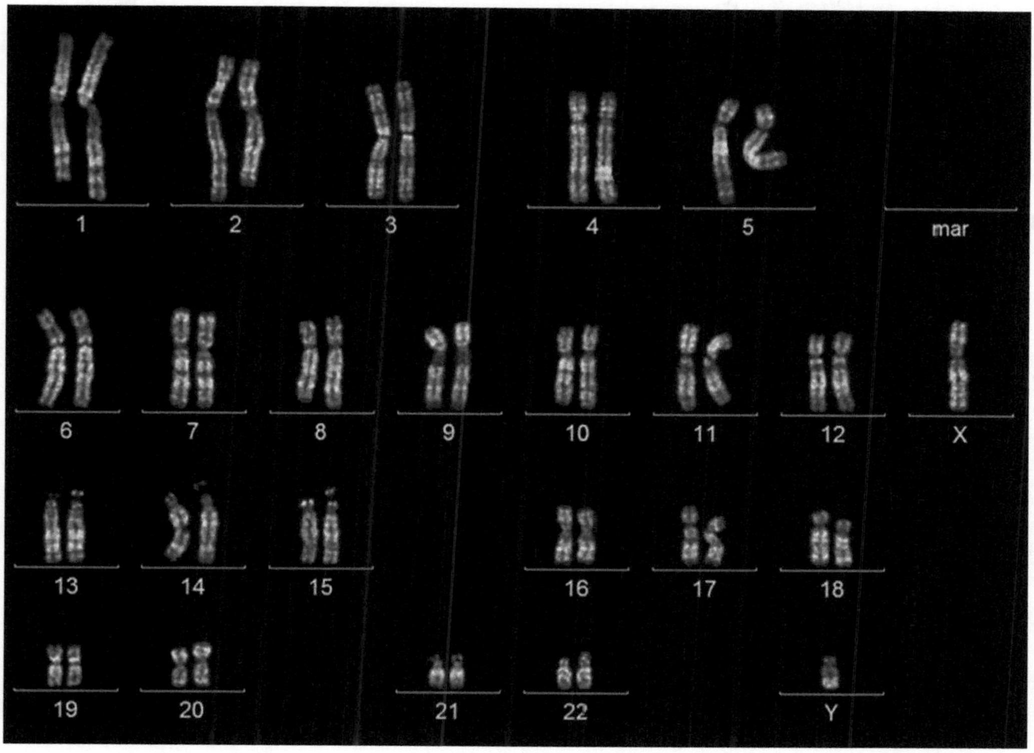

Fig. 7.2 Karyotype 46,XY,del(2)(q31q32.1),inv(4)(p12q11.2). **a** GTG-, **b** QFQ-banding, 700 and 400 bands per genome, **c** QFQ-, **d** GTG-banding, chromosomes from different mitoses, *left* normal, *right* deleted chromosome 2, **e** *left* normal, *right* inverted chromosome 4, **f** ideograms of chromosome 2, short interstitial deletion in the long arm, **g** ideograms of chromosome 4, pericentric inversion restricted to the heterochromatin

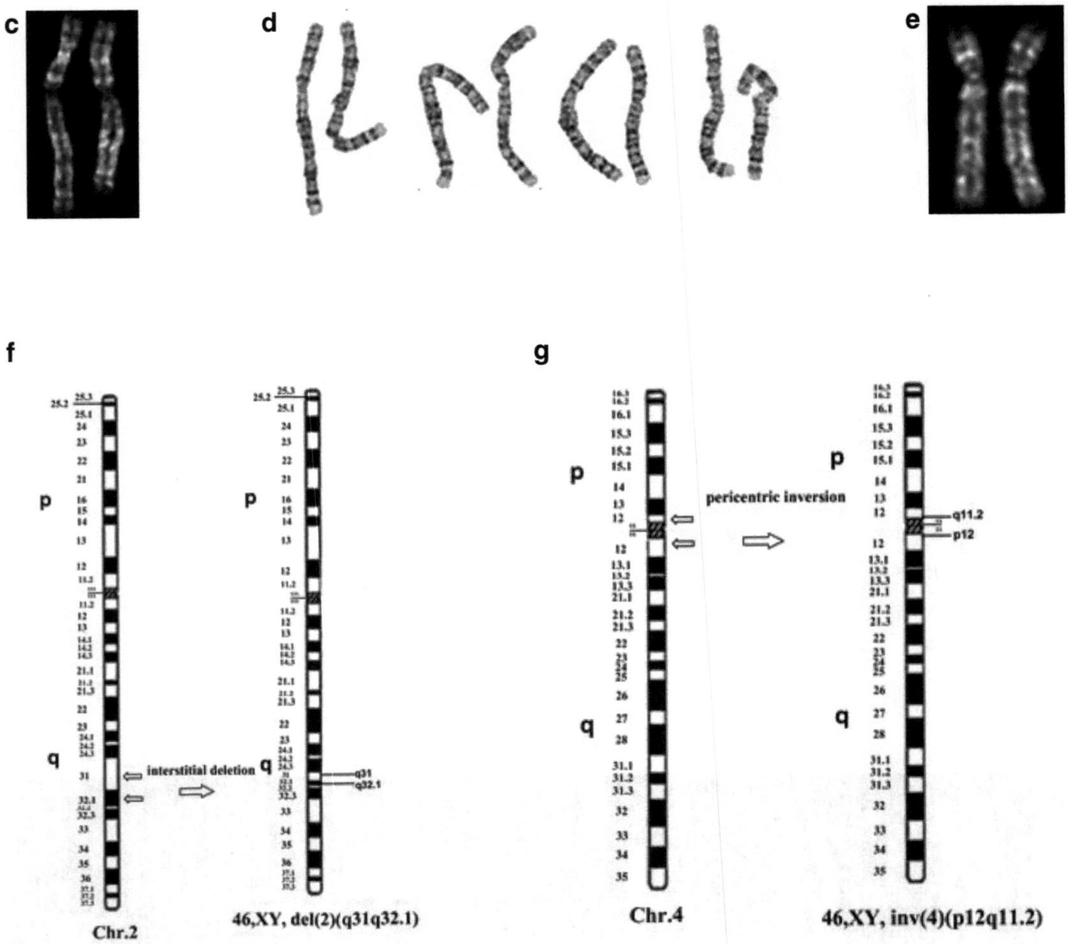

Fig. 7.2 (continued)

▪▪ Case 43: Karyotype: 46,XY,del(4)(q31.3)

Indication for chromosome analysis: Characteristic clinical symptoms of the deletion syndrome 4q31.

▪ Morphology

The long arm of the deleted chromosome has only about 70% of its original length left.

▪ Cave

The chromosomes of the parents should be investigated to exclude a reciprocal translocation in one of them and thereafter genetic counselling should be offered (◘ Fig. 7.3).

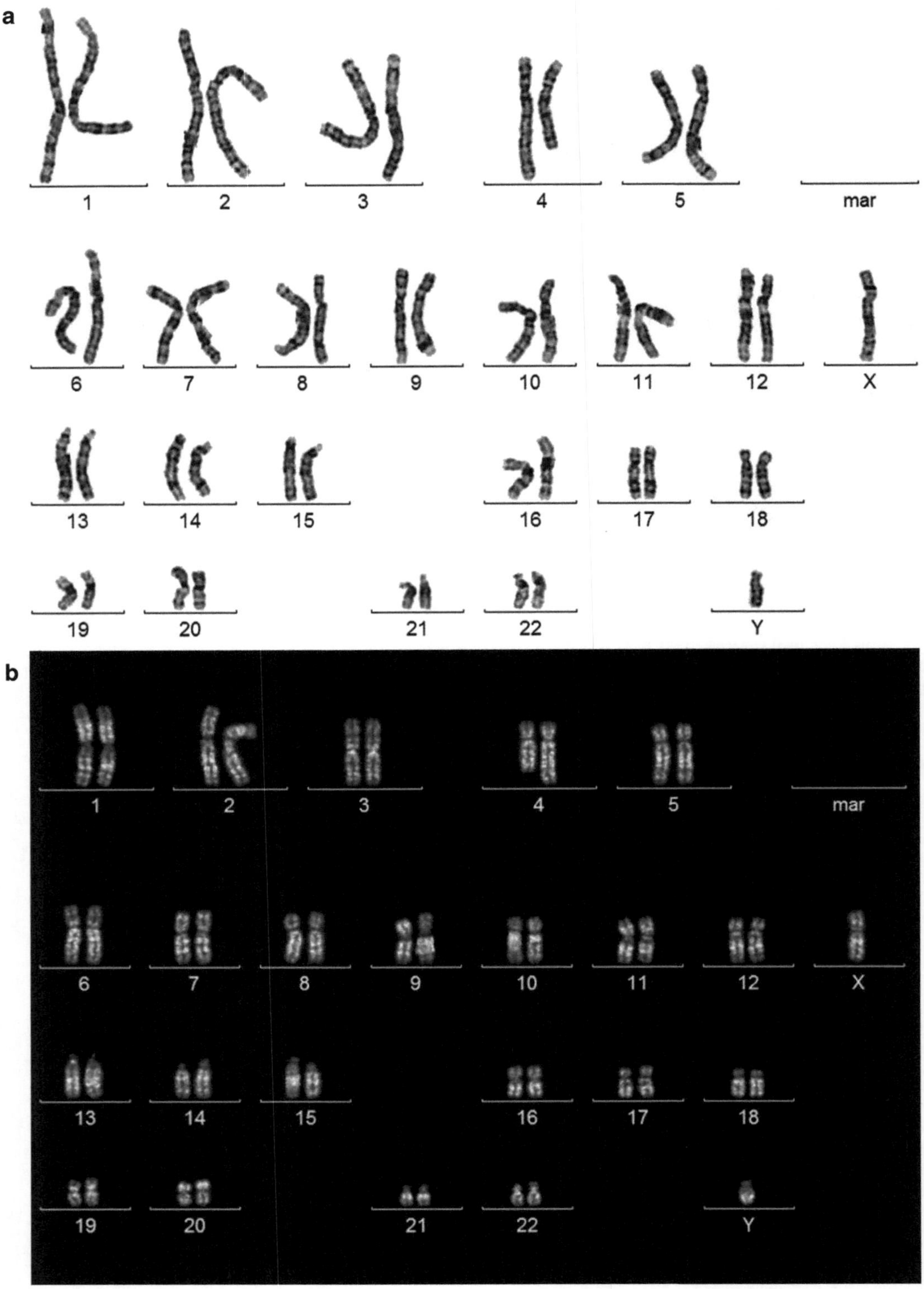

Fig. 7.3 Karyotype 46,XY,del(4)(q31.3). **a** GTG-, **b** QFQ-banding, 600 and 400 bands per genome, **c** QFQ-, **d** GTG-banding, chromosomes from different mitoses, *left* normal, *right* deleted chromosome 4, **e** ideograms of chromosome 4, deletion distal in the long arm

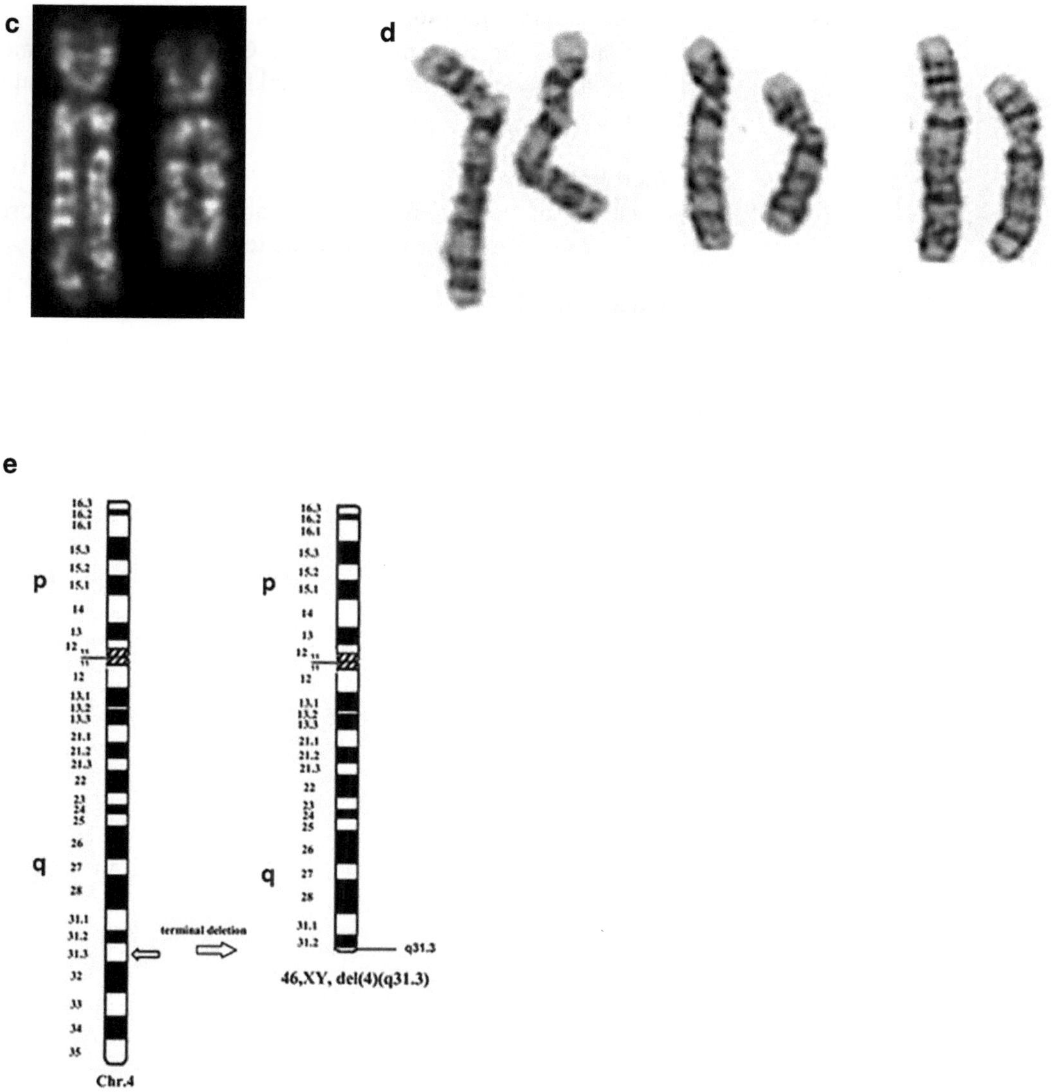

Fig. 7.3 (continued)

Case 44: Karyotype: 46,XX,del(5)(p14.2)

Indication for chromosome analysis: Cri-du-chat syndrome with typical crying, characteristic facial features and moderate retardation.

Morphology

The deletion of the patient comprises about 50% of the short arm of chromosome 5 and 15% of the total length.

Cave

In about 8–15% of the cases, the deletion 5p originated from the inheritance of an unbalanced translocation of parental origin. Therefore, the chromosomes of the parents should be investigated to enable a realistic risk estimation (Fig. 7.4).

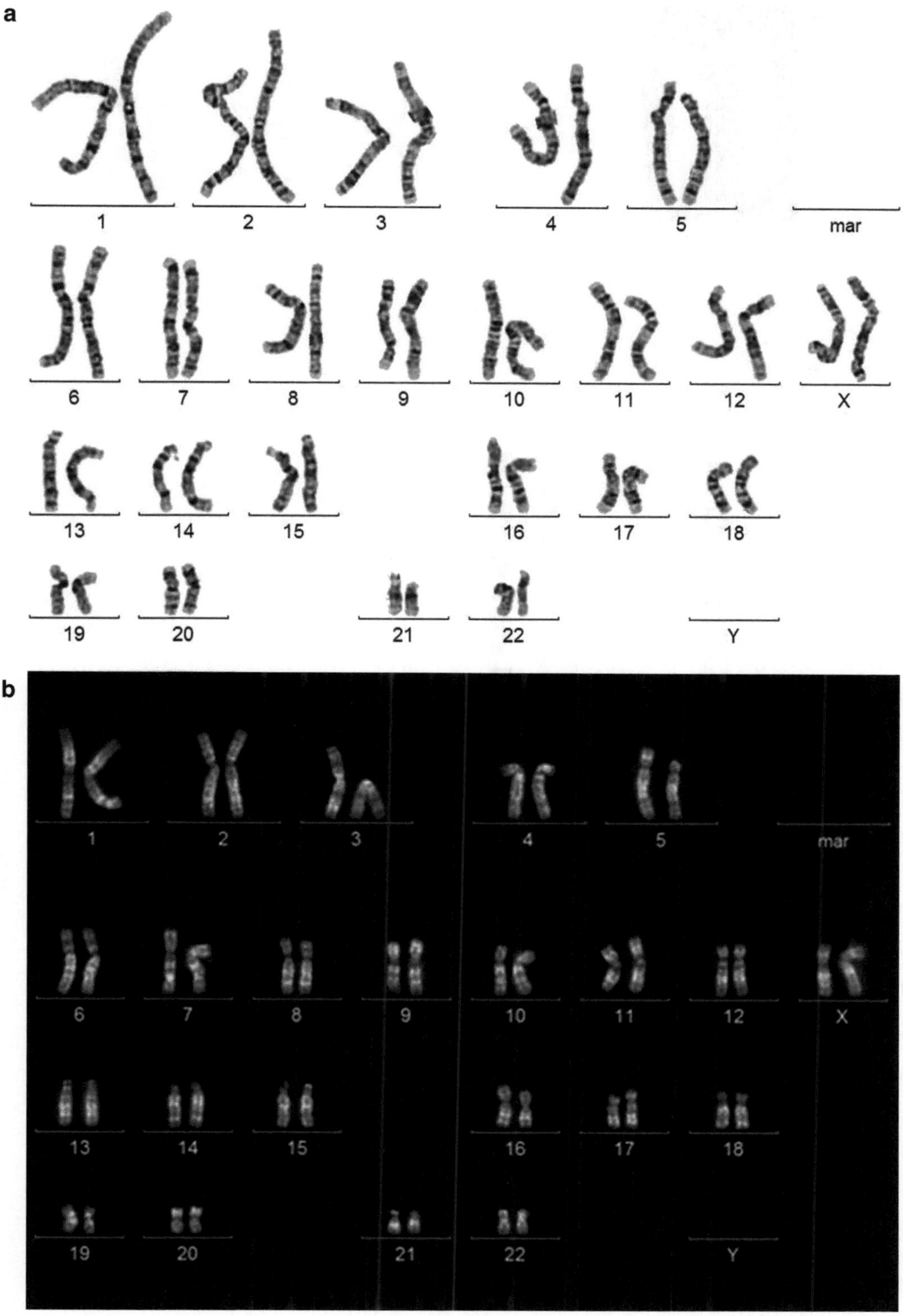

Fig. 7.4 Karyotype 46,XX,del(5)(p14.2). **a** GTG-, **b** QFQ-banding, 600 and 400 bands per genome, **c** QFQ-, **d** GTG-banding, chromosomes from different mitoses, *left* normal, *right* deleted chromosome 5, **e** ideograms of chromosome 5. Terminal deletion of about 30% of the short arm

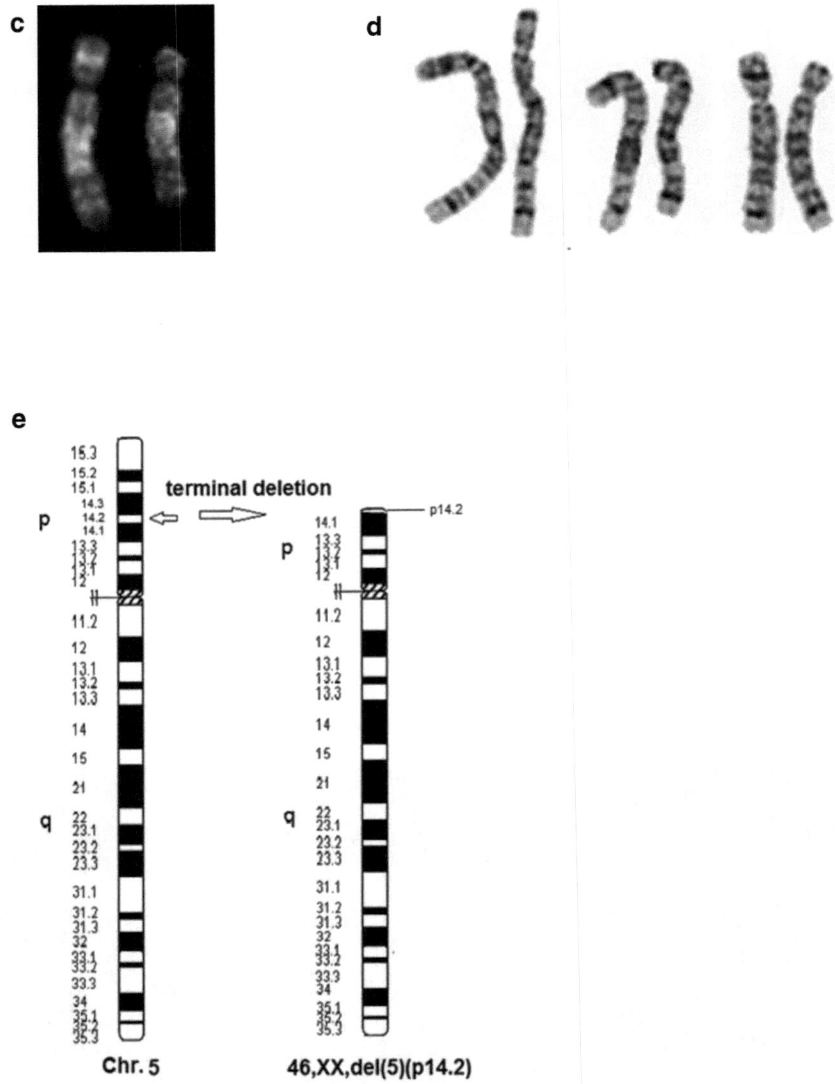

Fig. 7.4 (continued)

▪▪ Case 45: Karyotype: 46,XX,del(5)(p15.1)

Indication for chromosome analysis: Typical crying of children with Cri-du-chat syndrome, but not typical facial features, mild to moderate retardation.

▪ Morphology

The deletion shows a more distal breakpoint than in the previous case which is the reason for the less severe phenotype in the patient. In both cases (44 and 45), the banding resolution has to be at least 550 bands per genome. Otherwise, the breakpoints cannot be determined safely. But this is relevant for a prognosis of development.

▪ Cave

The same criteria have to be taken into account as in Case 44 (◘ Fig. 7.5).

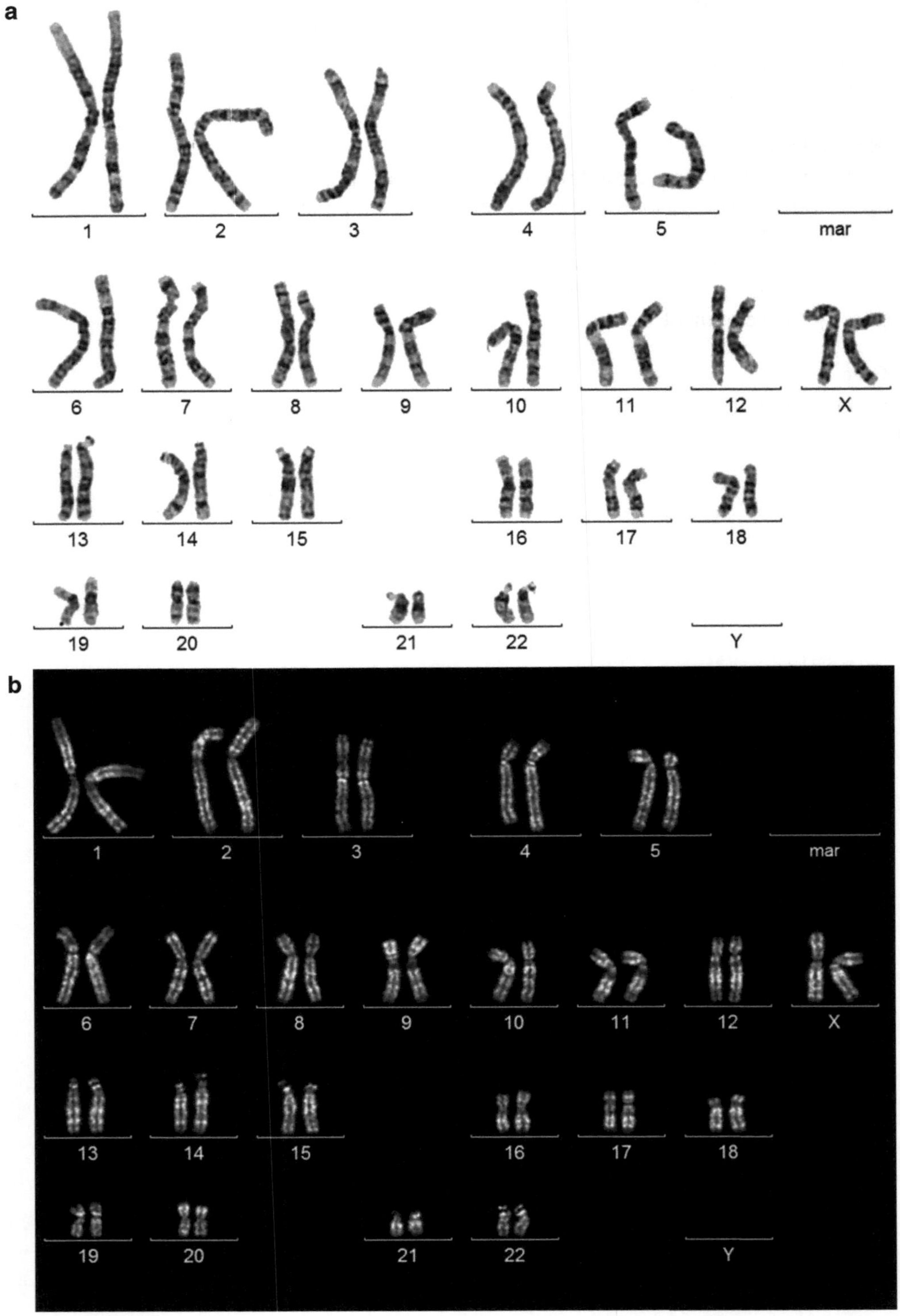

Fig. 7.5 Karyotype 46,XX,del(5)(p15.1). **a** GTG-, **b** QFQ-banding, 600 and 400 bands per genome, **c** QFQ-, **d** GTG-banding, chromosomes from different mitoses, c *left* normal, *right* deleted chromosome 5, **e** ideograms of chromosome 5. Terminal deletion of about 30% of the short arm, and 9% of the total length of chromosome 5

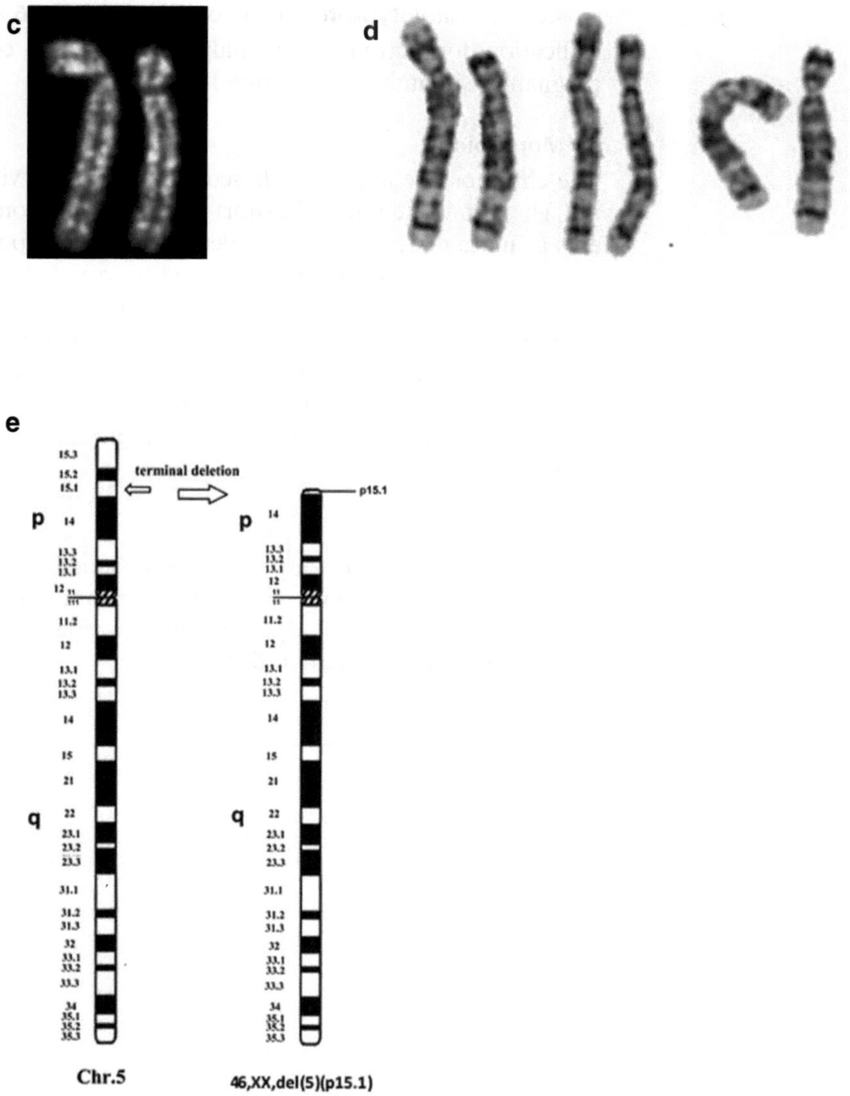

Fig. 7.5 (continued)

▪▪ Case 46: Karyotype: mos46,XY,del(8)(p23) [40]/46,XY[1]

Indication for chromosome analysis: Pathologic course of pregnancy, spontaneous abortion first trimenon.

▪ Morphology

The chromosome analysis is based on a chorionic villus sample. The distal deletion in the short arm of chromosome 8 presents as mosaic. Therefore, the deletion must be of postzygotic origin caused by an abnormality in the course of the cell cycle and it originated in the pathologic conceptus de novo. As the analysis was performed on a biopsy of the abortion material, quantitative analyses of the two karyotypes were not possible as well as their distribution in different cell systems.

The terminal deletion 8p comprises about 15% of the short arm.

▪ Cave

This case makes it obvious that chromosome investigations always afford the analysis of a sufficient number of mitoses. Additional microarray analyses will not provide the investigator with further information (◘ Fig. 7.6).

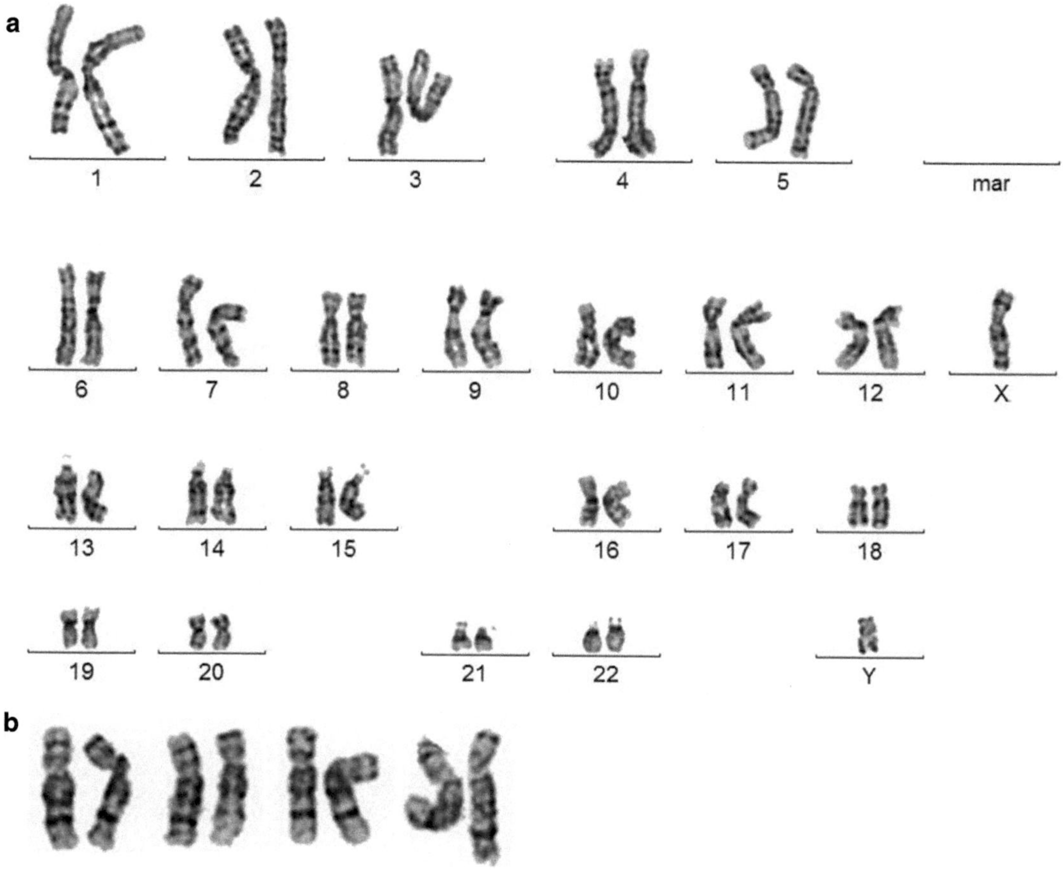

Fig. 7.6 Karyotype mos 46,XY,del(8)(p23) [40]/46,XY[1]. **a** GTG-banding: 550 bands per genome, **b** GTG-banding, chromosomes from different mitoses, *left* normal, *right* deleted chromosome 8, **c** ideograms of chromosome 8, small terminal deletion of the short arm

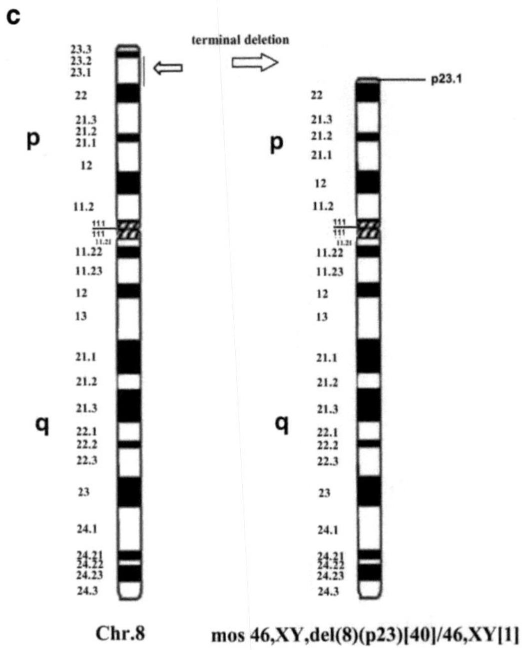

Fig. 7.6 (continued)

▪▪ Case 47: Karyotype: 46,XX,del(9)(p22p24),inv(4)(p12q11.2)

Two rearrangements are diagnosed, one in the euchromatin and the second in the constitutive heterochromatin.

Indication for chromosome analysis: Developmental retardation.

▪ Morphology

1. del(9)(p22p24)
   The interstitial deletion in the short arm comprises about 10% of the total length of chromosome 9 and about 40% of the short arm. The centromere shows a more acrocentric position and the banding pattern in the short arm has changed; thus, the analysis is easy to perform. But the breakpoint localisation requires a resolution of about 550 bands per genome.
2. inv(4)(p12q11.2)
   (see: ▶ Sect. 15.6).

▪ Cave

The chromosomes of the parents must be analysed before genetic counselling to exclude a parental paracentric inversion (◘ Fig. 7.7).

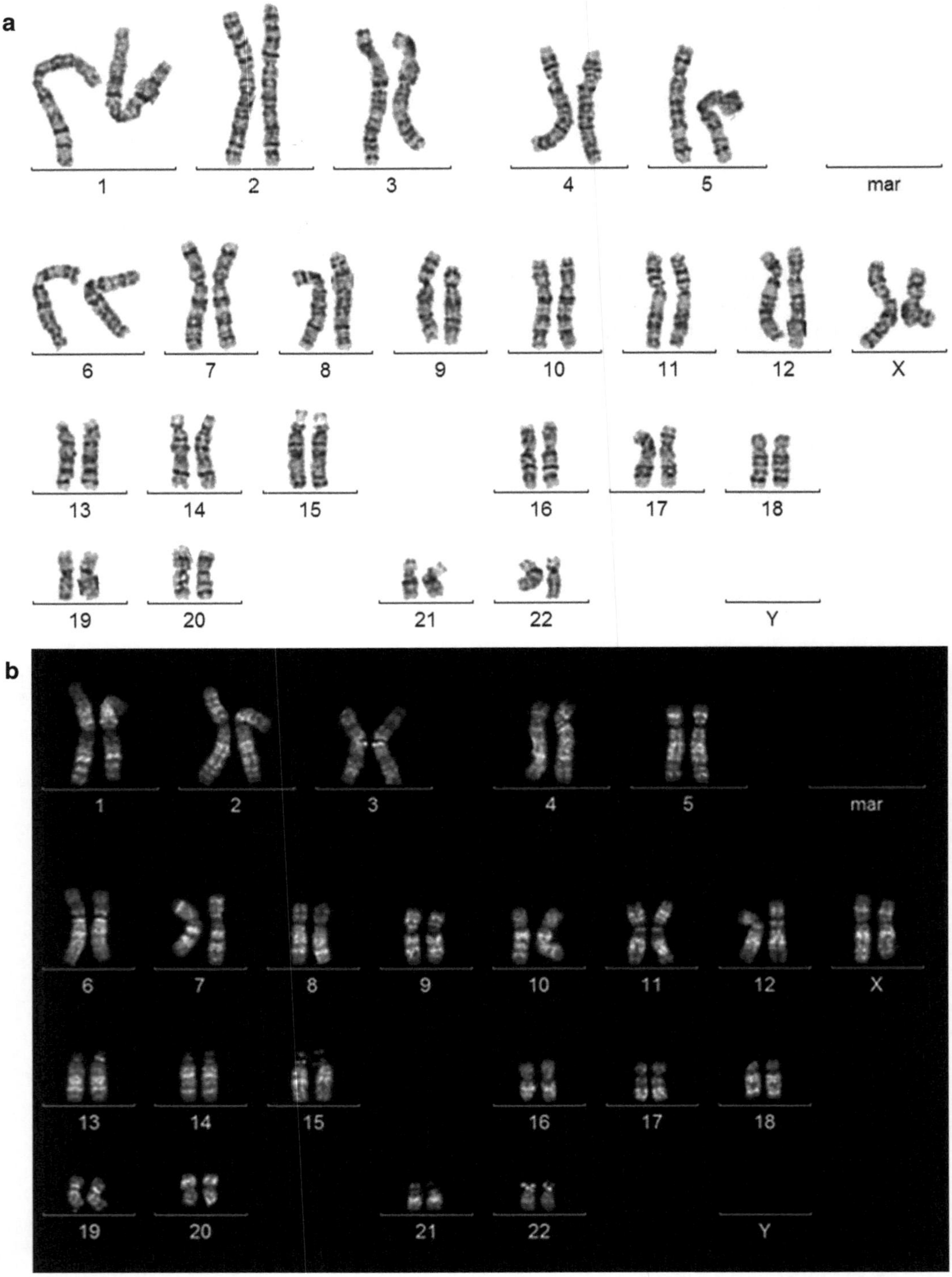

Fig. 7.7 Karyotype 46,XX,del(9)(p22p24),inv(4)(p12q11.2). **a** GTG-, **b** QFQ-banding, 550 and 400 bands per genome, **c** GTG-banding, chromosomes from different mitoses, *left* normal, *right* deleted chromosome 9, inv(4) see Case 4, **d** ideograms of chromosome 9 with terminal deletion in the short arm

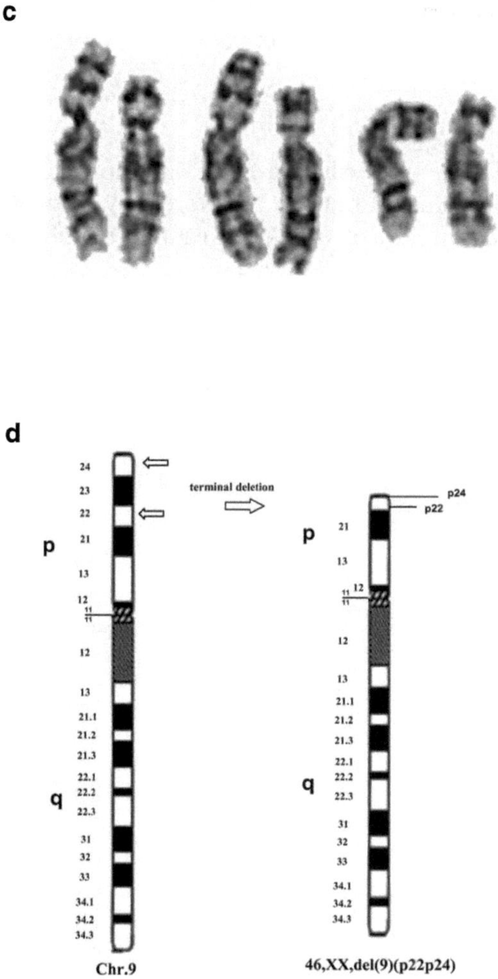

Fig. 7.7 (continued)

▪▪ Case 48: Karyotype: 46,XX,del(10)(p13p15.1)

Indication for chromosome analysis: Developmental retardation.

▪ Morphology

The interstitial deletion in the short arm of chromosome 10 includes the AT-rich band p14 and is therefore easy to diagnose, it comprises about 30% of the short arm and 8% of the total chromosome length.

▪ Cave

Family investigations should be performed before the genetic counselling to exclude a parental inversion (◘ Fig. 7.8).

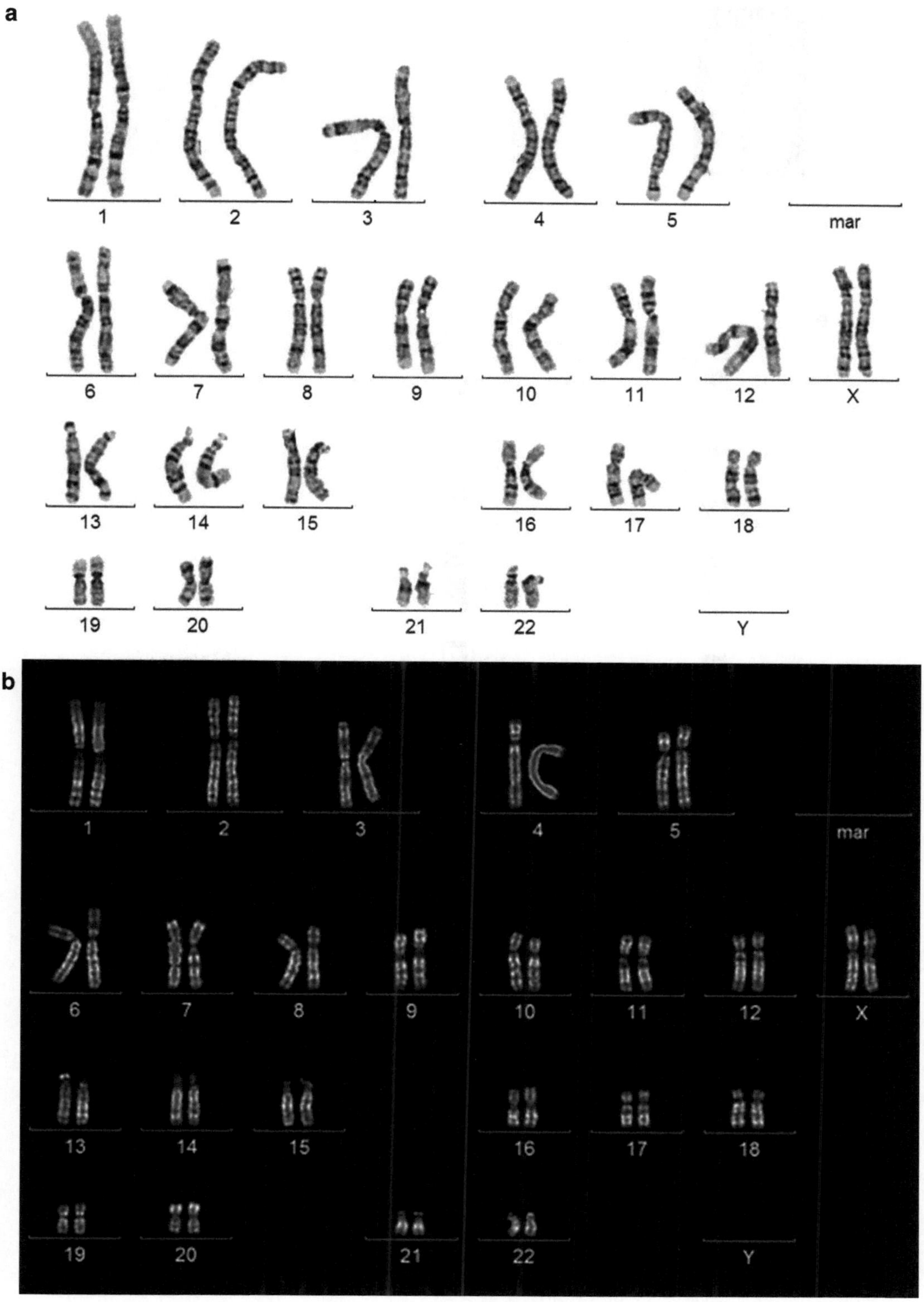

Fig. 7.8 Karyotype 46,XX,del(10)(p13p15.1). **a** GTG-, **b** QFQ-banding, 600 and 400 bands per genome, **c** QFQ-, **d** GTG-banding, chromosomes from different mitoses, *left* normal, *right* deleted chromosome 10, **e** ideograms of the chromosome 10 with the interstitial deletion in the short arm

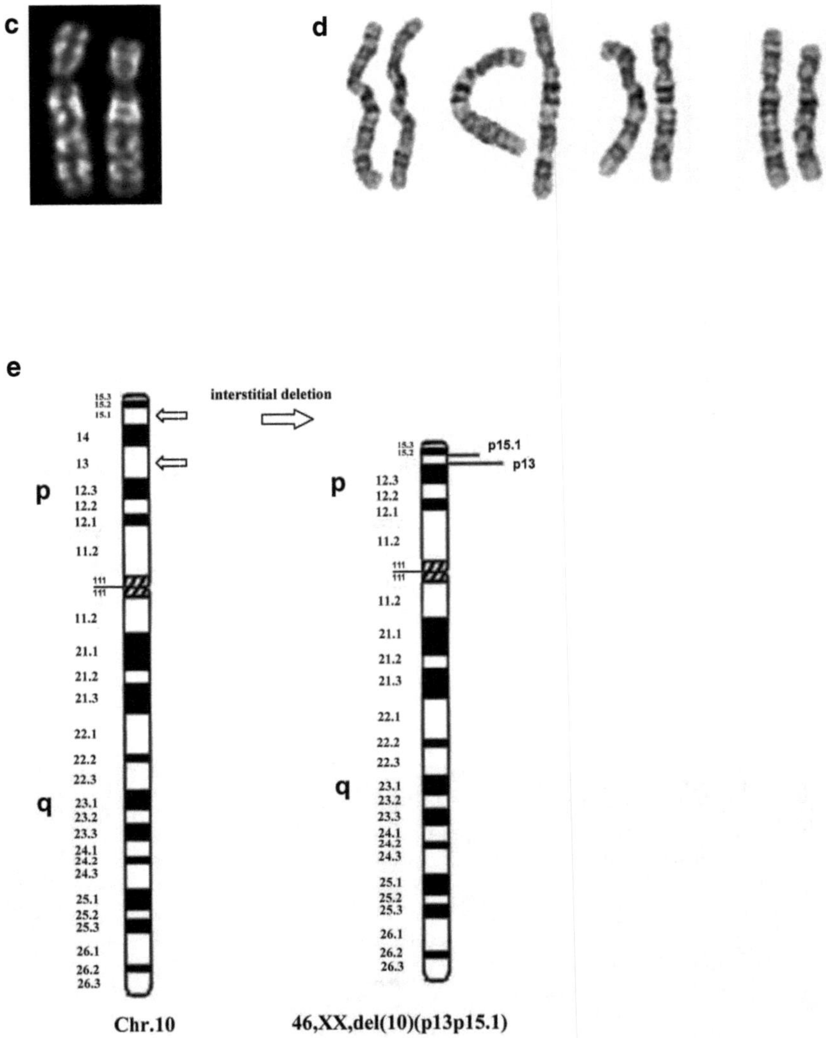

Fig. 7.8 (continued)

■■ Case 49: Karyotype: 46,XY,del(13)(q21.2q31)

Indication for chromosome analysis: Multiple malformations. The pattern of clinical symptoms is in accordance with findings from the literature in cases of similar deletions. The prognosis is unfavourable.

■ Morphology

The interstitial deletion 13q comprises about 25% of the euchromatic regions of the long arm.

■ Cave

The aberration was confirmed to be de novo therefore the genetic risk for the parents is less than 1%, this is caused by a possible gonadal mosaic in one of them (◘ Fig. 7.9).

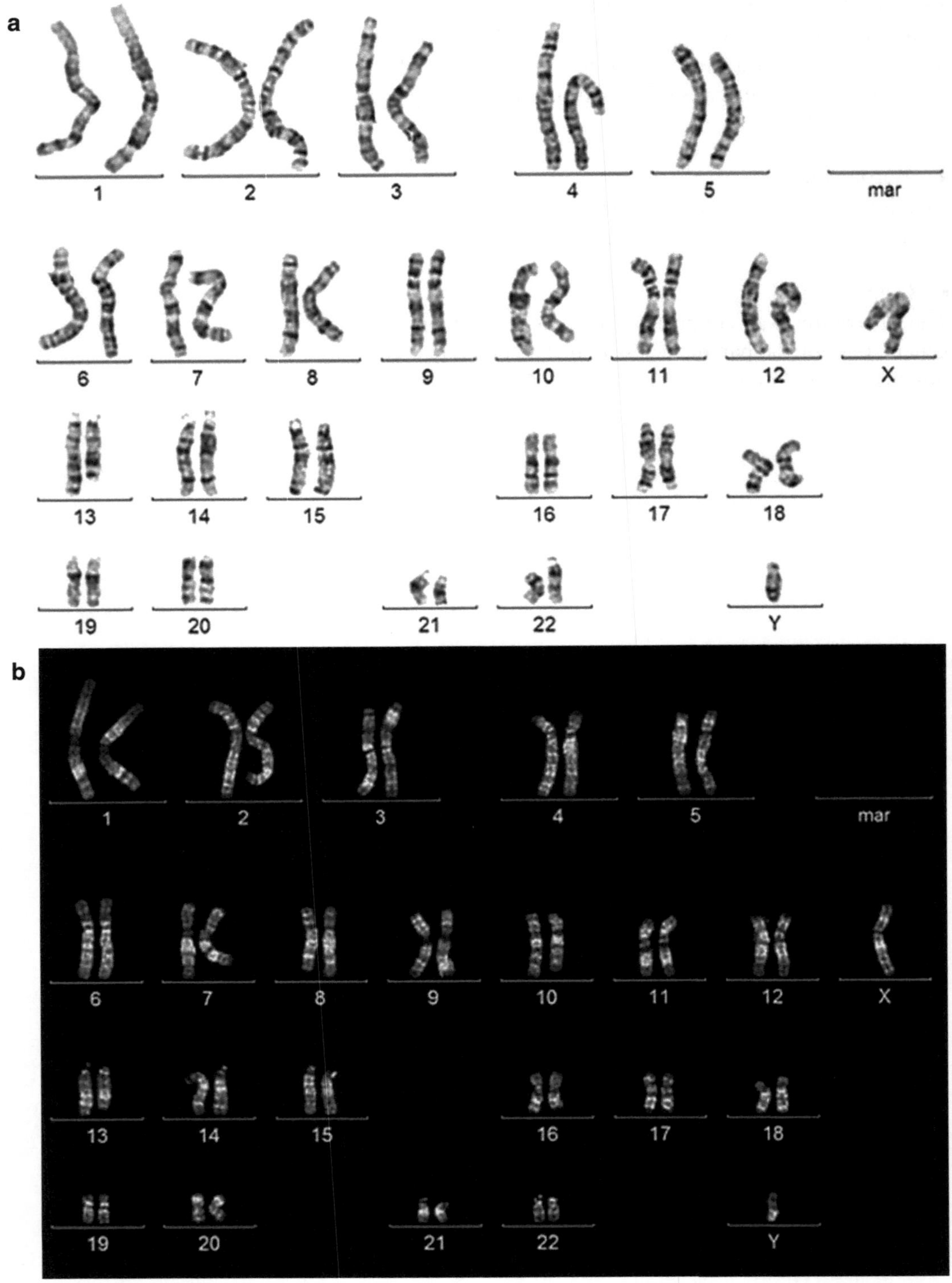

Fig. 7.9 Karyotype 46,XY,del(13)(q21.2q31). **a** GTG-, **b** QFQ-banding, 550 and 400 bands per genome, **c** QFQ-, **d** GTG-banding, chromosomes from different mitoses, *left* normal, *right* deleted chromosome 13, **e** ideograms of chromosome 13, interstitial deletion intermediate in the long arm

c

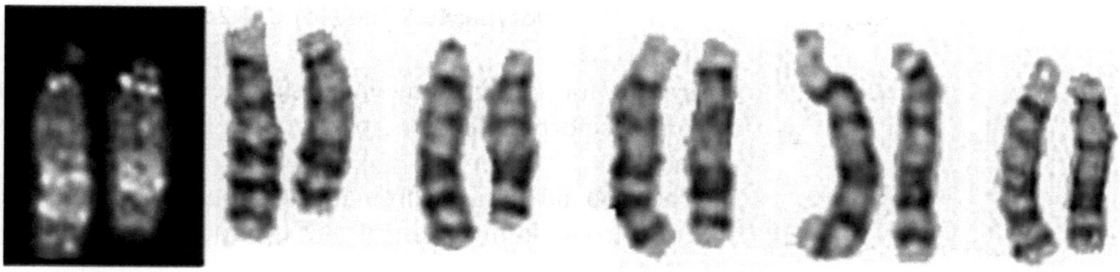

d

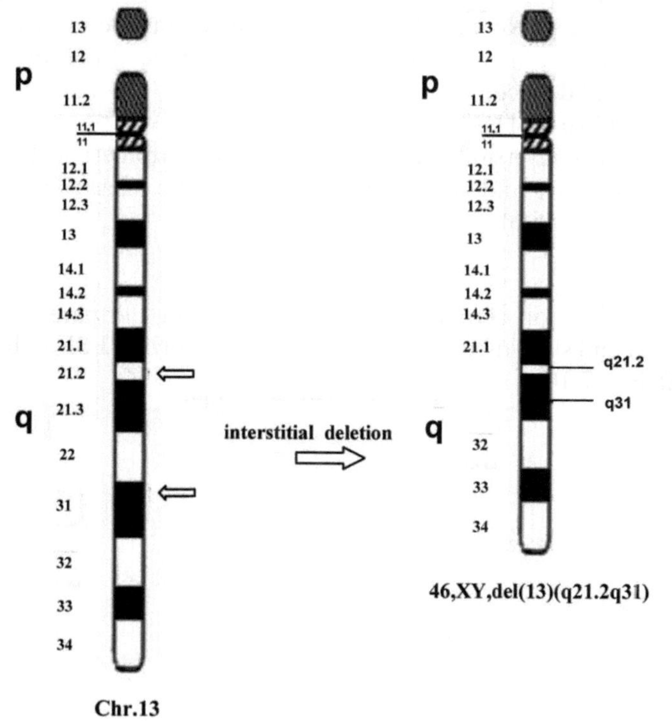

Fig. 7.9 (continued)

▪▪ Case 50: Karyotype: 46,XY,del(15)(q11.2q13),inv(4)(p12q11.2)

Indication for chromosome analysis: Developmental retardation, dysmorphic features, phenotype of Prader Willi syndrome.

The proband has a euchromatic deletion 15, combined with a pericentromeric inversion of the heterochromatic block in chromosome 4.

▪ Morphology

The deletion 15q comprises about 4% of the euchromatic long arm.

1. del(15)(q11.2q13)

   The proximal deletion in the long arm is the most frequent one in chromosome 15. The phenotype of the carrier is well defined.
2. inv(4)(p12q11.2)

   The inversion includes only the pericentromeric heterochromatin and is therefore without genetic relevance (see: ► Sect. 15.6).

▪ Cave

Indication for chromosome investigations in the parents is given and should be combined with FISH applying LSI probes (◘ Fig. 7.10).

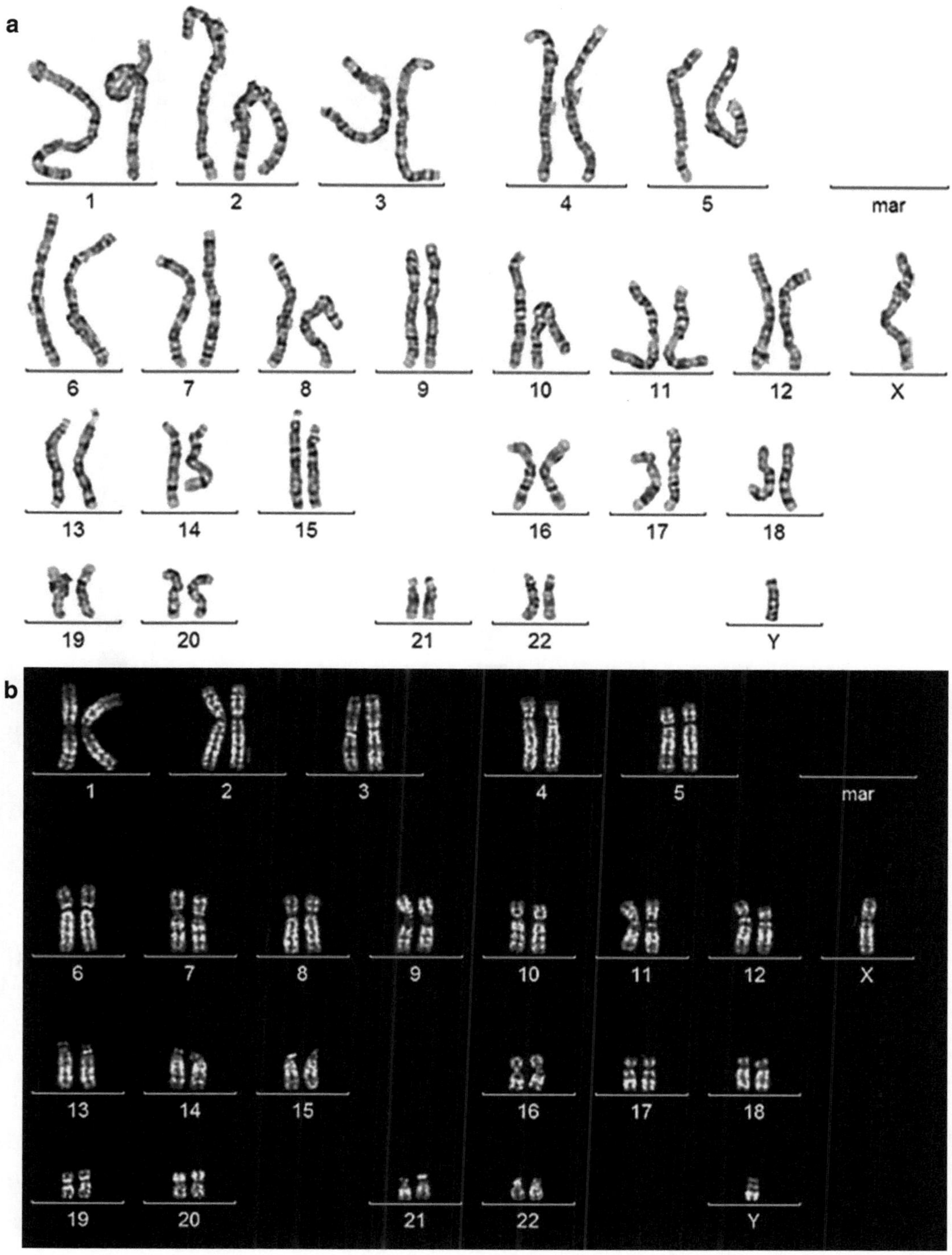

Fig. 7.10 Karyotype 46,XY,del(15)(q11.2q13),inv(4)(p12q11.2). **a** GTG-, **b** QFQ-banding, 700 and 400 bands per genome, **c** GTG-banding, chromosomes from different mitoses, *left* normal, *right* deleted chromosome 15, inv(4) see Case 4, **d** ideograms of chromosome 15, small proximal deletion

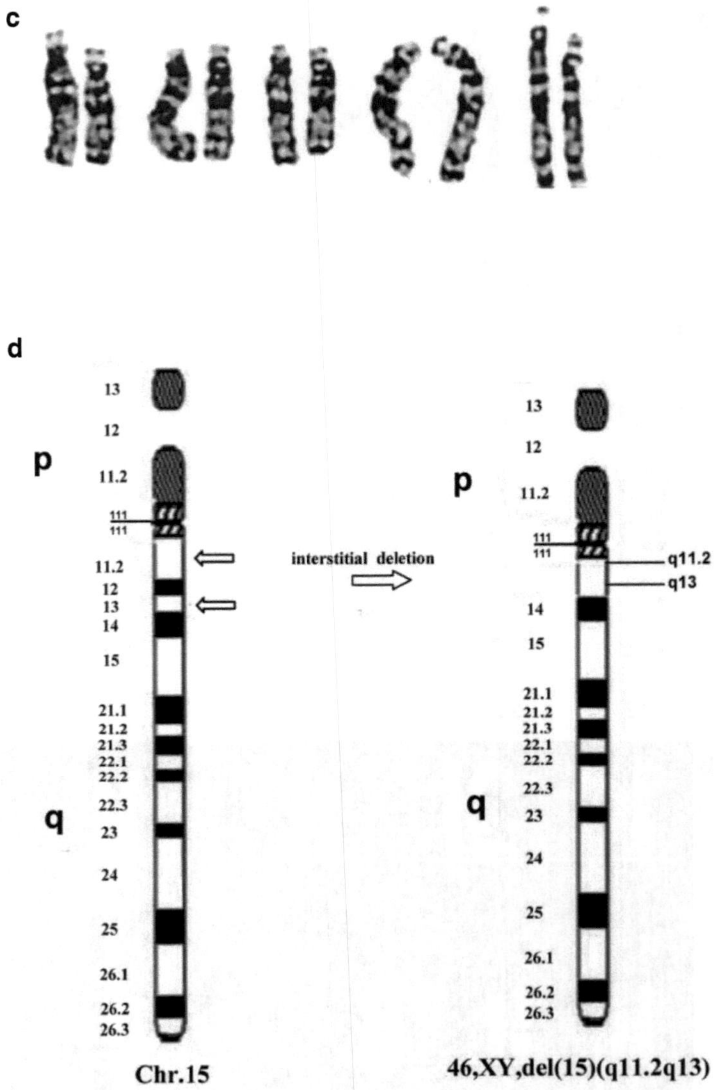

Fig. 7.10 (continued)

■■ Case 51: Karyotype: 46,XY,del(17)(p11.2),inv(4)(p12q11.2)

Indication for chromosome analysis: Smith-Magenis syndrome.

■ Morphology

1. The proximal, GC-rich band p11.2 in the short arm of chromosome 17 is deleted, it comprises about 15% of its length. Thus, the following AT-band p12 has changed its position from the middle of the short arm to one neighbouring the pericentromeric heterochromatin. The phenotype of the short arm changed from a tulip-shaped appearance to a U-shape.
2. inv(4)(p12q11.2) (see: ► Sect. 15.6).

■ Cave

The de novo origin of the aberration 17p has to be confirmed before genetic counselling (◘ Fig. 7.11).

a

b

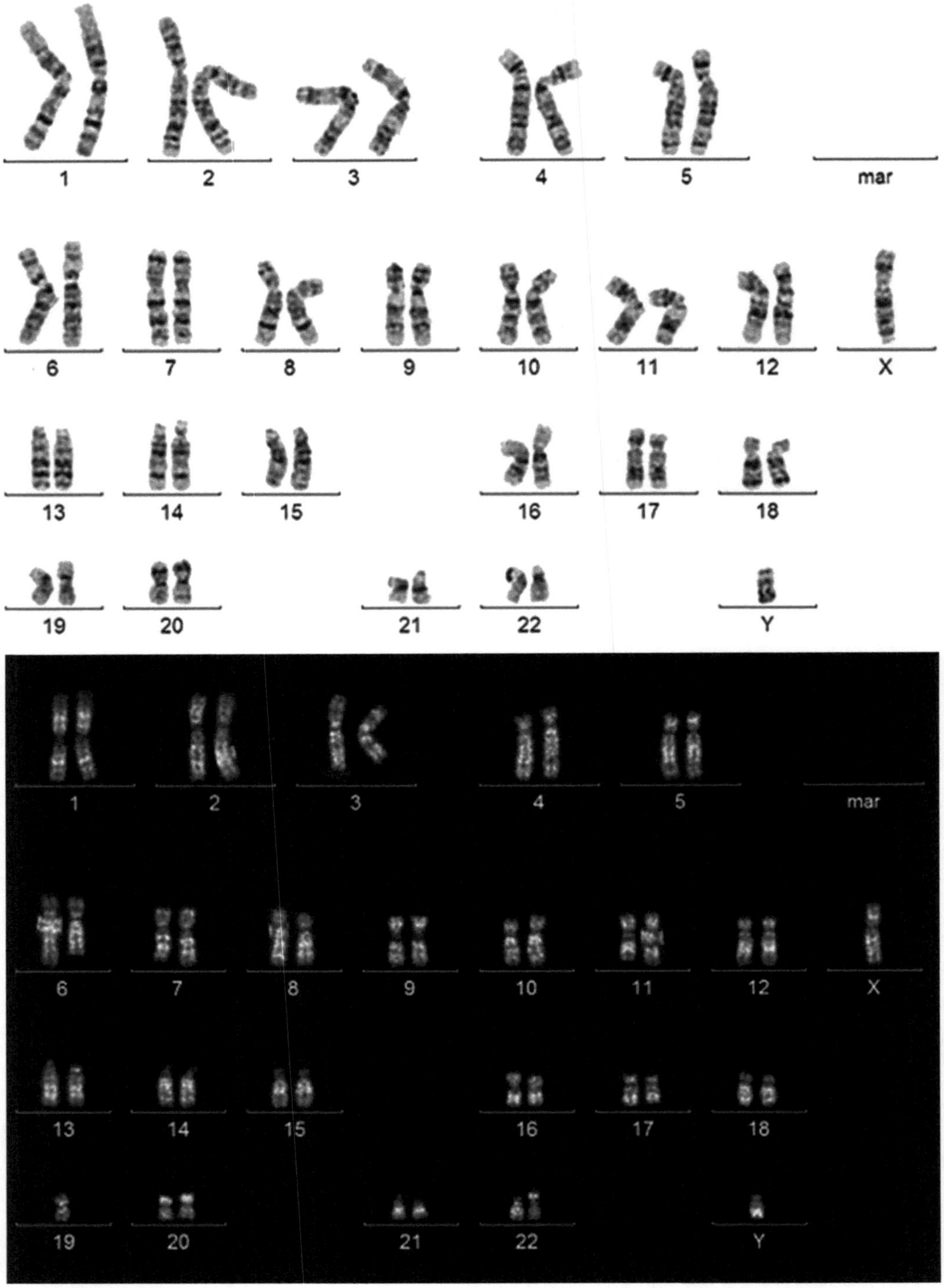

◘ Fig. 7.11 Karyotype 46,XY,del(17)(p11.2),inv(4)(p12q11.2). **a** GTG-, **b** QFQ-banding, 500 and 400 bands per genome, **c** GTG-banding, chromosomes from different mitoses, *left* normal, *right* deleted chromosome 17, inv(4) see Case 4, **d** ideograms of chromosome 17, small interstitial short arm deletion

c
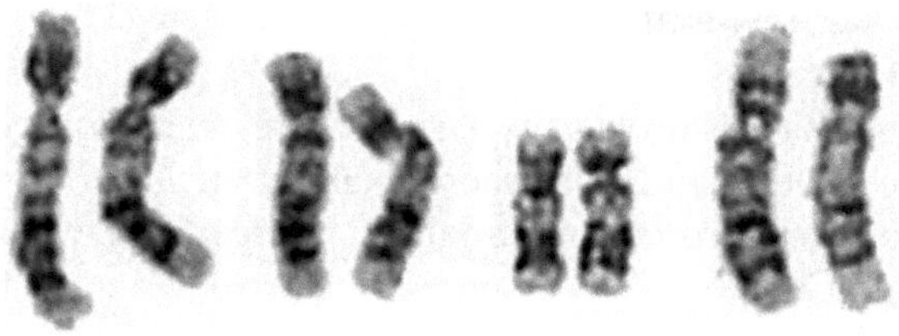

d
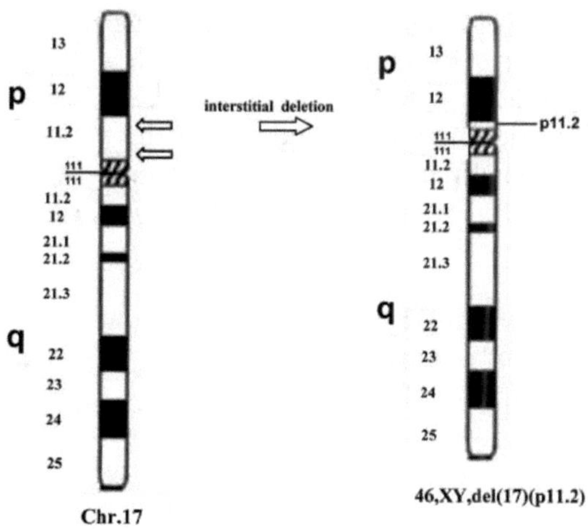

Fig. 7.11 (continued)

▪▪ Case 52: Karyotype: 46,XX,del(17)(q23.1q23.3),inv(3)(p11.2q11.2)

Chromosome analysis reveals two structural rearrangements, a euchromatic deletion and a heterochromatic inversion.

Indication for chromosome analysis: Unspecific developmental retardation.

▪ Morphology

1. del(17)(q23.1q23.3)

   The patient presents with an interstitial deletion of subband 17q23.2 in the distal region of the long arm. This and other comparable small structural aberrations are only detectable at a resolution of at least 700 bands per genome. The deletion comprises about 2.8% of the long arm and 2% of the total length of the chromosome.

2. inv(3)2x(p11.1q11.2)

   This pericentric inversion includes only the pericentromeric heterochromatin and is therefore without genetic relevance (see: ► Sect. 15.6).

▪ Cave

Before genetic counselling an inversion in chromosome 17 should be excluded in the parents (◘ Fig. 7.12).

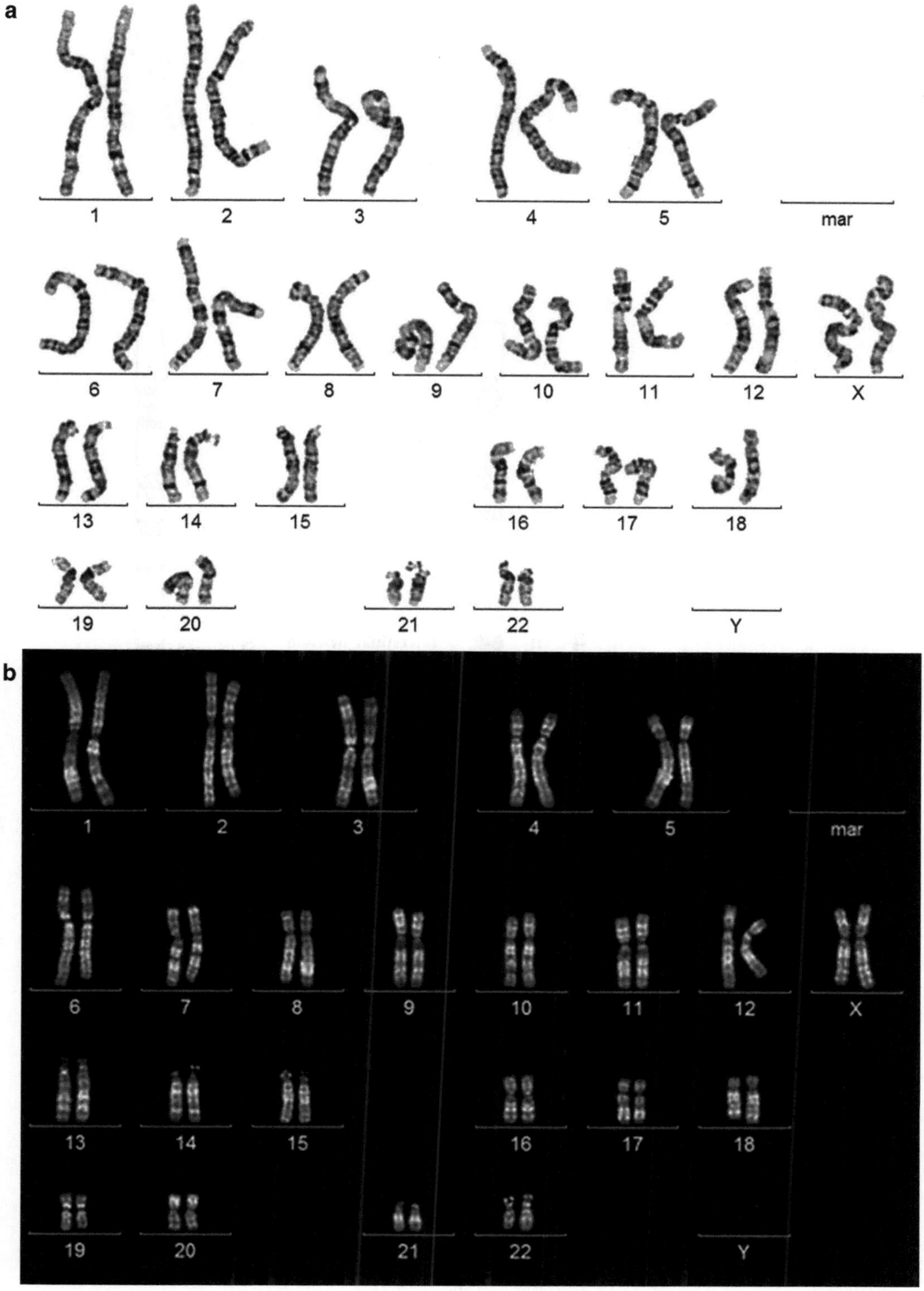

Fig. 7.12 Karyotype 46,XX,del(17)(q23.1q23.3),inv(3)(p11.1q11.2)2x. **a** GTG-, **b** QFQ-banding, 850 and 550 bands per genome, **c** GTG-banding, chromosomes from different mitoses, *left* normal, *right* deleted chromosome 17, **d** ideograms of chromosome 17, interstitial long arm deletion

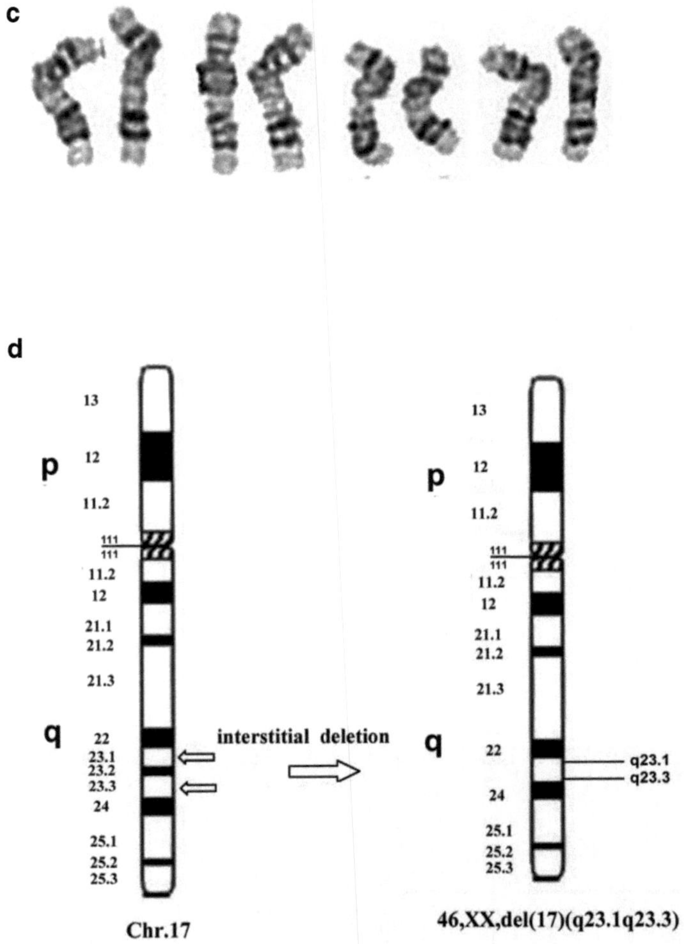

■ **Fig. 7.12** (continued)

▪▪ Case 53: Karyotype: 46,XX,del(18)(p11.2)dup(7)(q36)ish der(18),t(7;18)(q36;p11.2)(D18S552-,7QTEL20+),inv(9)(p12q13)

Indication for chromosome analysis: Developmental retardation, phenotype of de Grouchy syndrome I.

The patient presents with one pericentric inversion 9 and a deletion derived from a parental translocation 7/18.

▪ Morphology

1. del(18)(p11.2)

   The terminal deletion 18p is known as a frequent structural aberration and is called de Grouchy syndrome I. The phenotype of the patient is in good accordance with the clinical findings in the syndrome. The duplication 7q36 also present in the patient does not reveal any specific clinical symptoms. In the majority of cases known from the literature, the deletion 18p originated de novo in the patients. But the case presented here demonstrates that other types of aberration such as inversions or translocations in the parents can lead to an unbalanced rearrangement in the child. In the case presented here, by microscopy only one of the translocation chromosomes (18) was detectable, the other one (7) was only analysable by subtelomere FISH.
2. inv(9)(p12q13) Type I

   The pericentric inversion of chromosome 9 can be defined as polymorphism without genetic relevance. Both breakpoints are localised in the constitutive heterochromatin (see Case 40 of inversions).

▪ Cave

Only by additional FISH investigations the reciprocal translocation 7/18 could be analysed in the patient and the carrier mother. These findings are an impressive example for the relevance of additional investigations in structural chromosome aberrations (▪ Fig. 7.13).

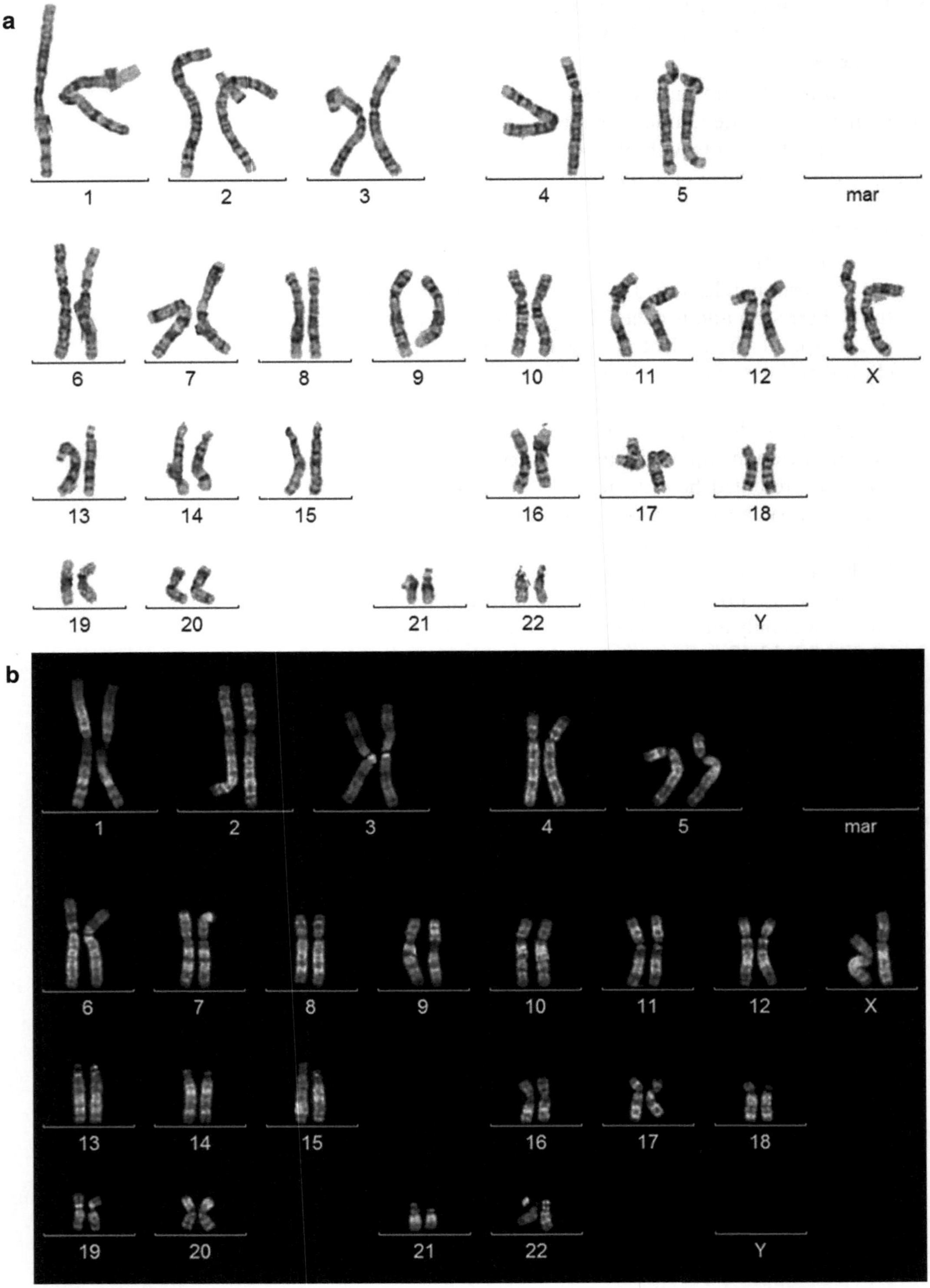

Fig. 7.13 Karyotype 46,XX,del(18)(p11.2),dup(7)(q36),t(7;18)(q36;p11.2),inv(9)(p12q13). **a** GTG-, **b** QFQ-banding, 550 and 400 bands per genome, **c** GTG-banding, chromosomes from different mitoses, *left* normal, *right* deleted chromosome 18, inv(9) see Case 40 of inversions, **d** ideograms of breakpoints and rearrangements of unbalanced reciprocal translocation 7;18

c

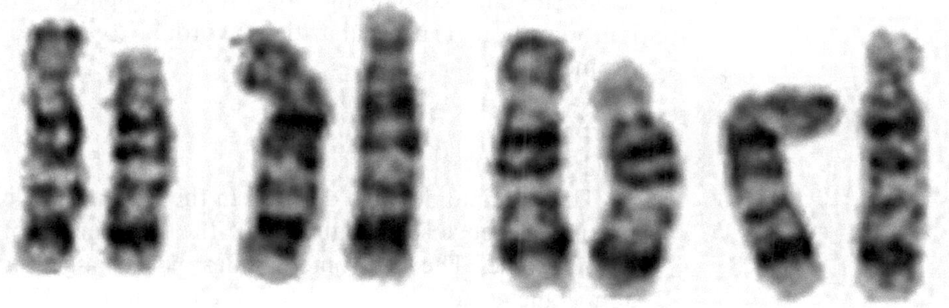

d

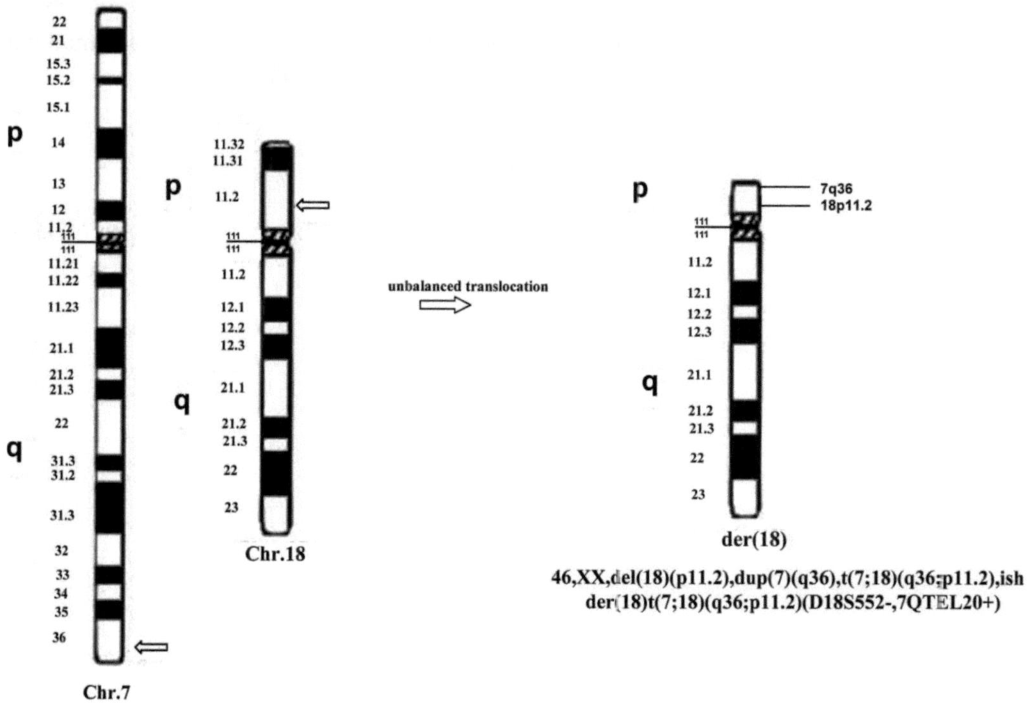

Fig. 7.13 (continued)

▪▪ Case 54: Karyotype: 46,XX,del(18)(q21.3),inv(4)(p12q11.2)

Indication for chromosome analysis: Mild developmental retardation, morphological peculiarities according to de Grouchy syndrome II.

▪ Morphology

1. del(18)(q21.3)
   The terminal de novo deletion in the long arm of chromosome 18 encloses about 25% of the total length of the chromosome. The centromere index has changed to an almost metacentric position and the characteristic AT-band q22 is deleted. The breakpoint q21.3 is a frequent region of rearrangements.
2. inv(4)(p12q11.2)
   (see: ► Sect. 15.6)

▪ Cave

The de novo origin of the deletion has to be confirmed before genetic counselling of the family (◘ Fig. 7.14).

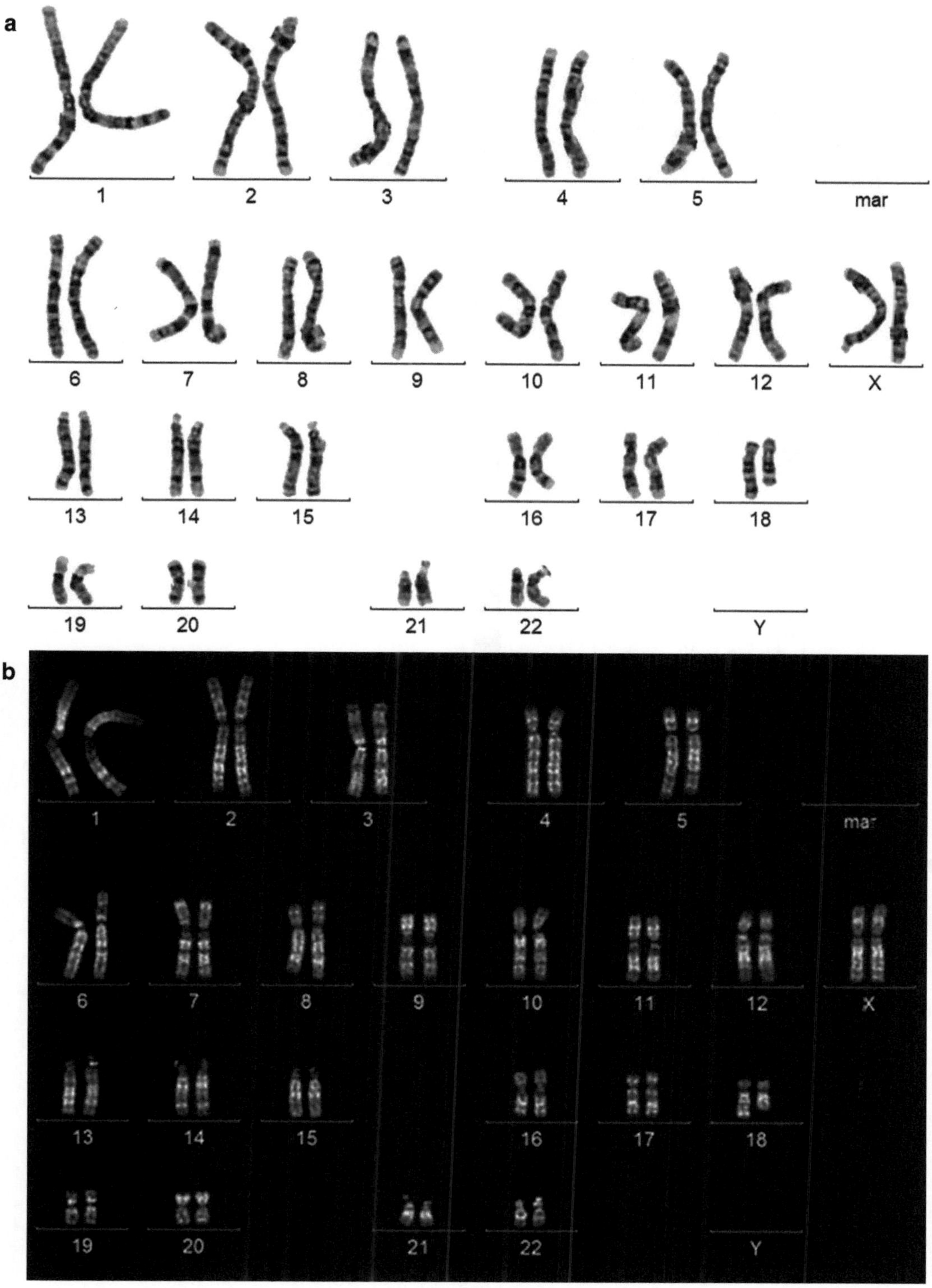

Fig. 7.14 Karyotype 46,XX,del(18)(q21.3),inv(4)(p12q11.2). **a** GTG-, **b** QFQ-banding, 550 and 400 bands per genome, **c** GTG-banding, chromosomes from different mitoses, *left* normal, *right* deleted chromosome 18, inv(4) see Case 2 of deletion, **d** ideograms of chromosome 18, terminal deletion in the long arm, inv(4) (see: ► Sect. 15.6)

c

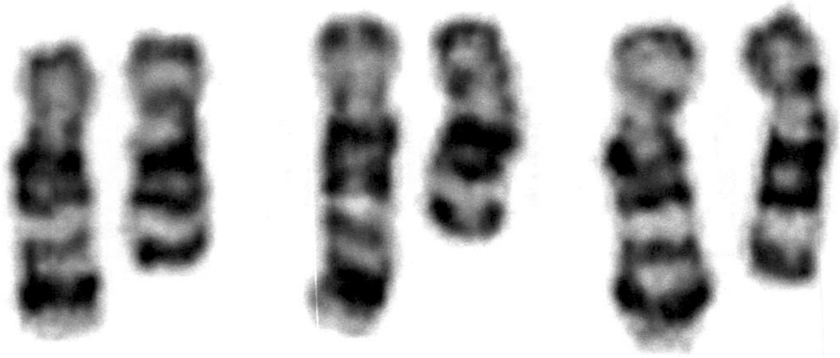

d

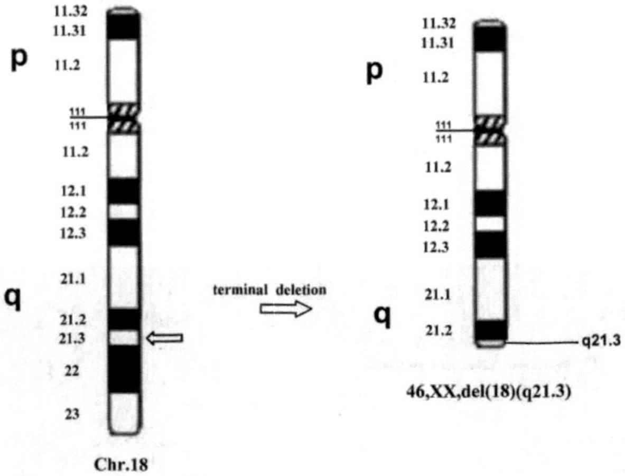

■ **Fig. 7.14** (continued)

Case 55: Karyotype: 46,X,del(X)(p11.23)

Indication for chromosome analysis: Clinical symptoms of Turner syndrome.

Morphology

The terminal deletion of the short arm of the X-chromosome comprises about 30% of the total length. The deleted chromosome is stable in its inheritance in the course of cell divisions. A second cell line was not detectable.

Cave

The chromosomes of the mother should be investigated before genetic counselling to exclude a balanced translocation or an inversion (Fig. 7.15).

7

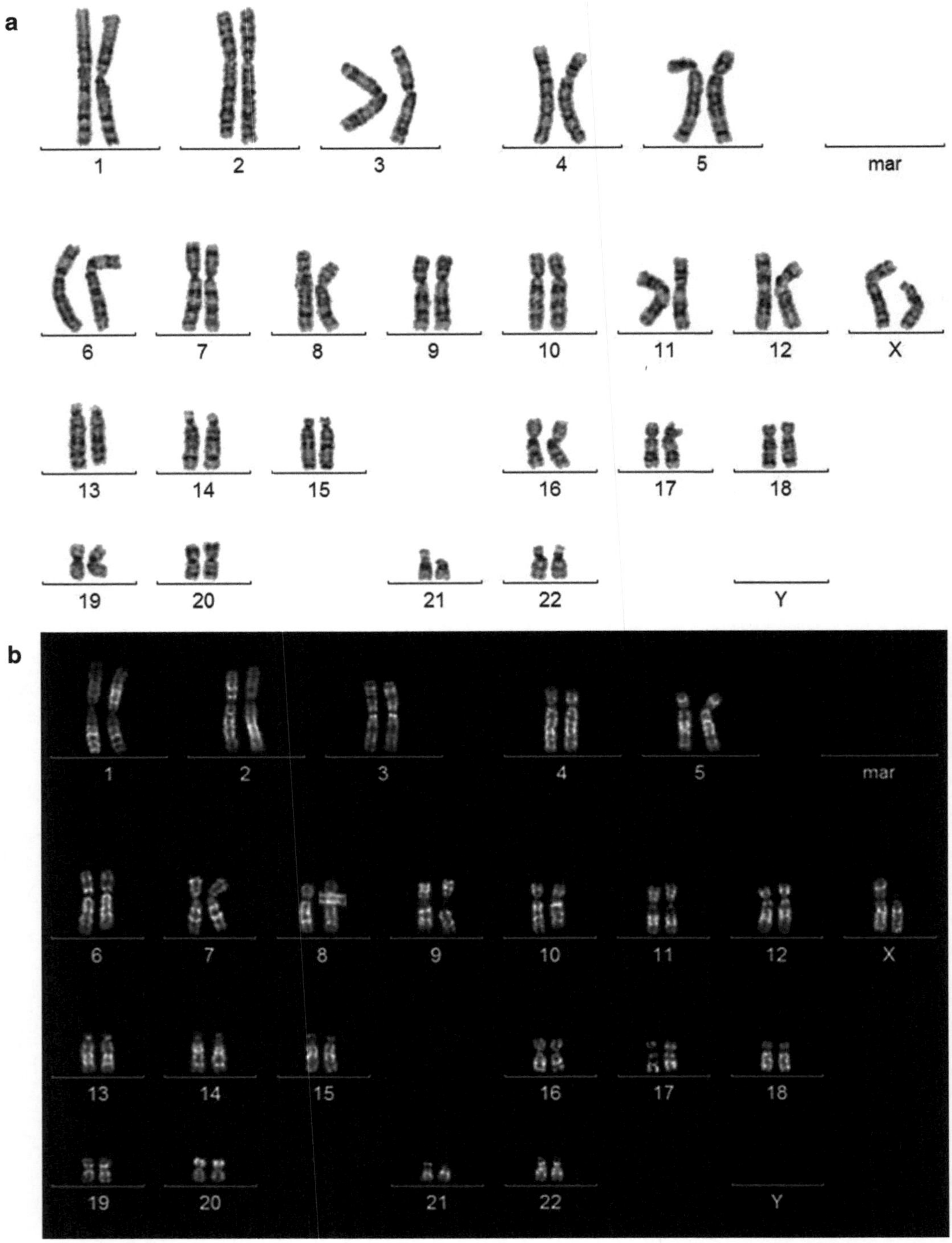

Fig. 7.15 Karyotype 46,X,del(X)(p11.23). **a** GTG-, **b** QFQ-banding, 500 and 400 bands per genome, **c** GTG-banding, chromosomes from different mitoses, *left* normal, *right* deleted X-chromosome, **d** ideograms of the X-chromosome, large deletion in the short arm

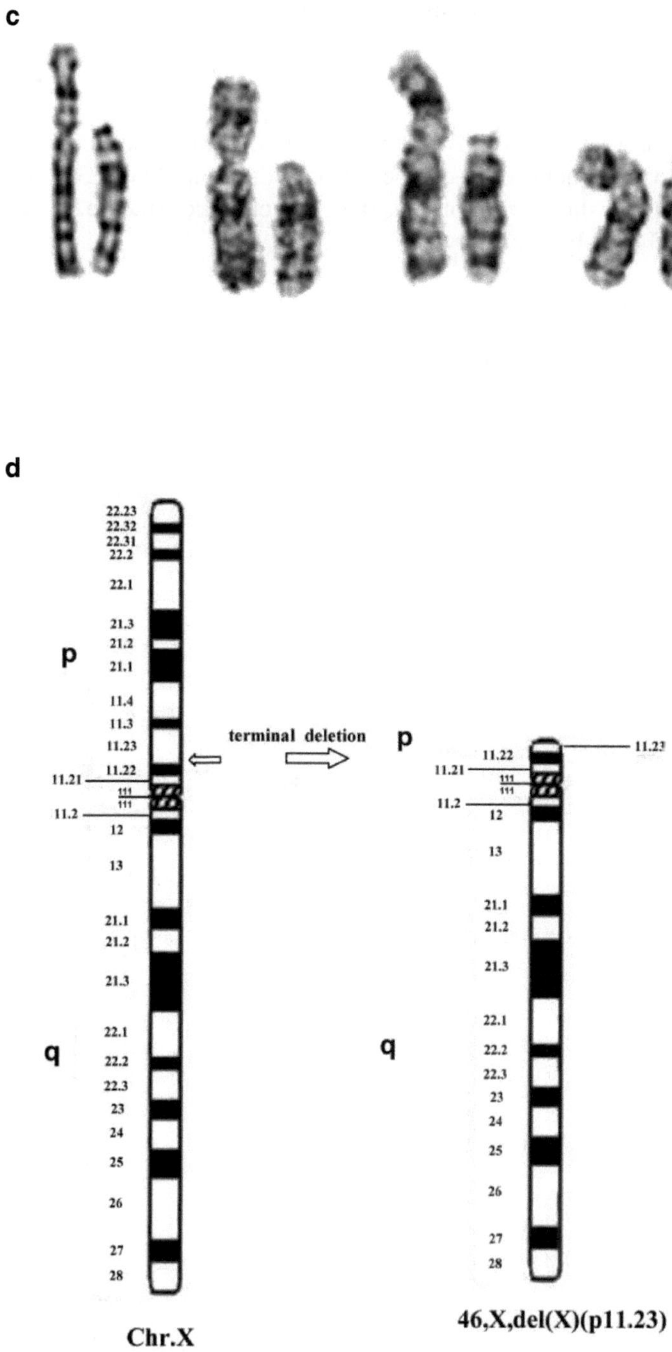

Fig. 7.15 (continued)

■■ Case 56: Karyotype: 46,X,del(X)(p22.1)

Indication for chromosome analysis: Sterility.

■ Morphology

The deleted X-chromosome has the breakpoint in p22.1 a well-known frequent locus of aberration induction. About 30% of the distal short arm got lost, the characteristic bands p21.1 to p21.3 have been preserved.

■ Cave

Chromosome investigation in the mother before genetic counselling is indicated (◘ Fig. 7.16).

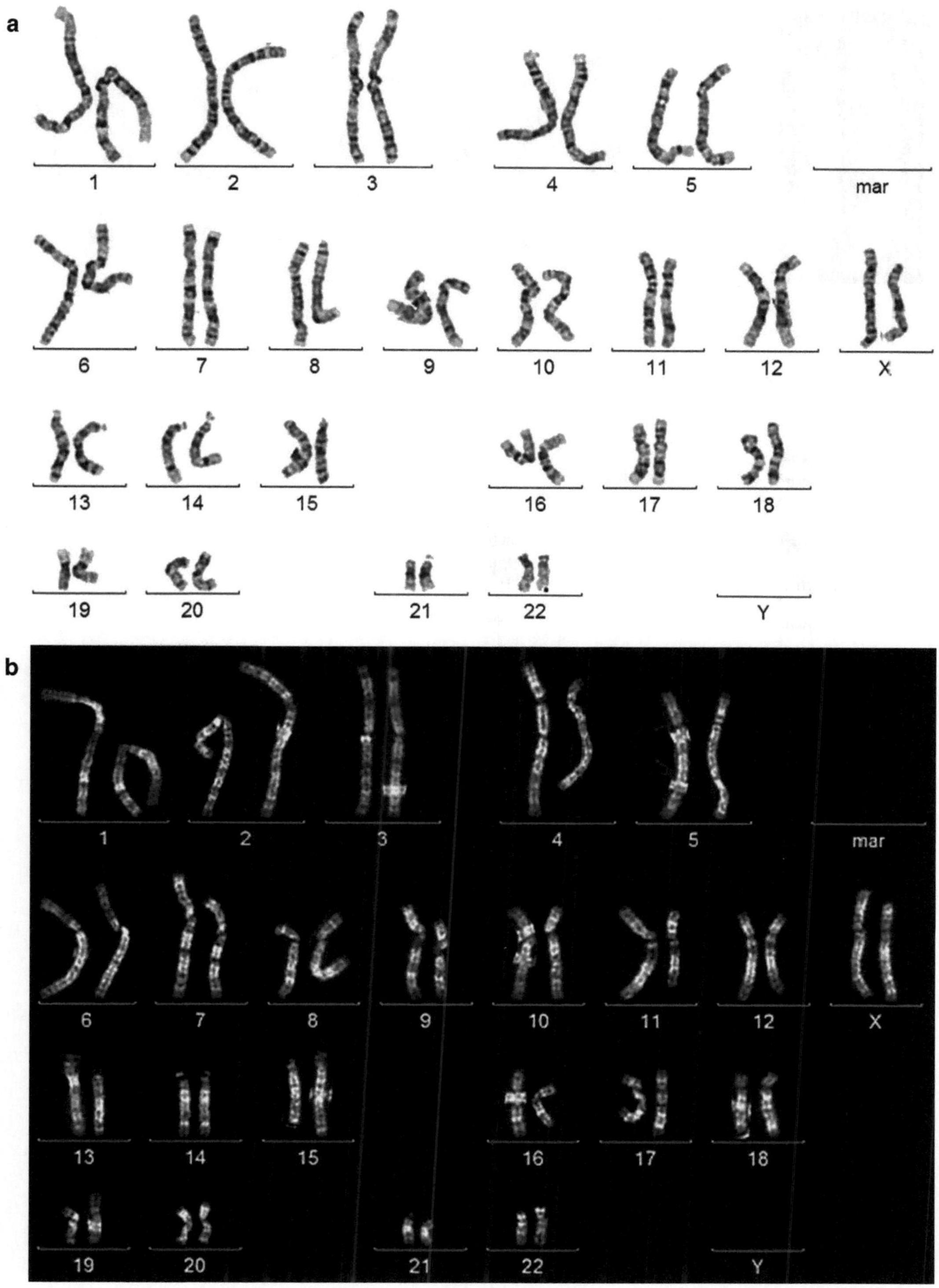

Fig. 7.16 Karyotype 46,X,del(X)(p22.1). **a** GTG-, **b** QFQ-banding, 700 and 500 bands per genome, **c** QFQ-, **d** GTG-banding, chromosomes from different mitoses, *left* normal, *right* deleted X-chromosome, **e** ideograms of the X-chromosome, terminal short arm deletion

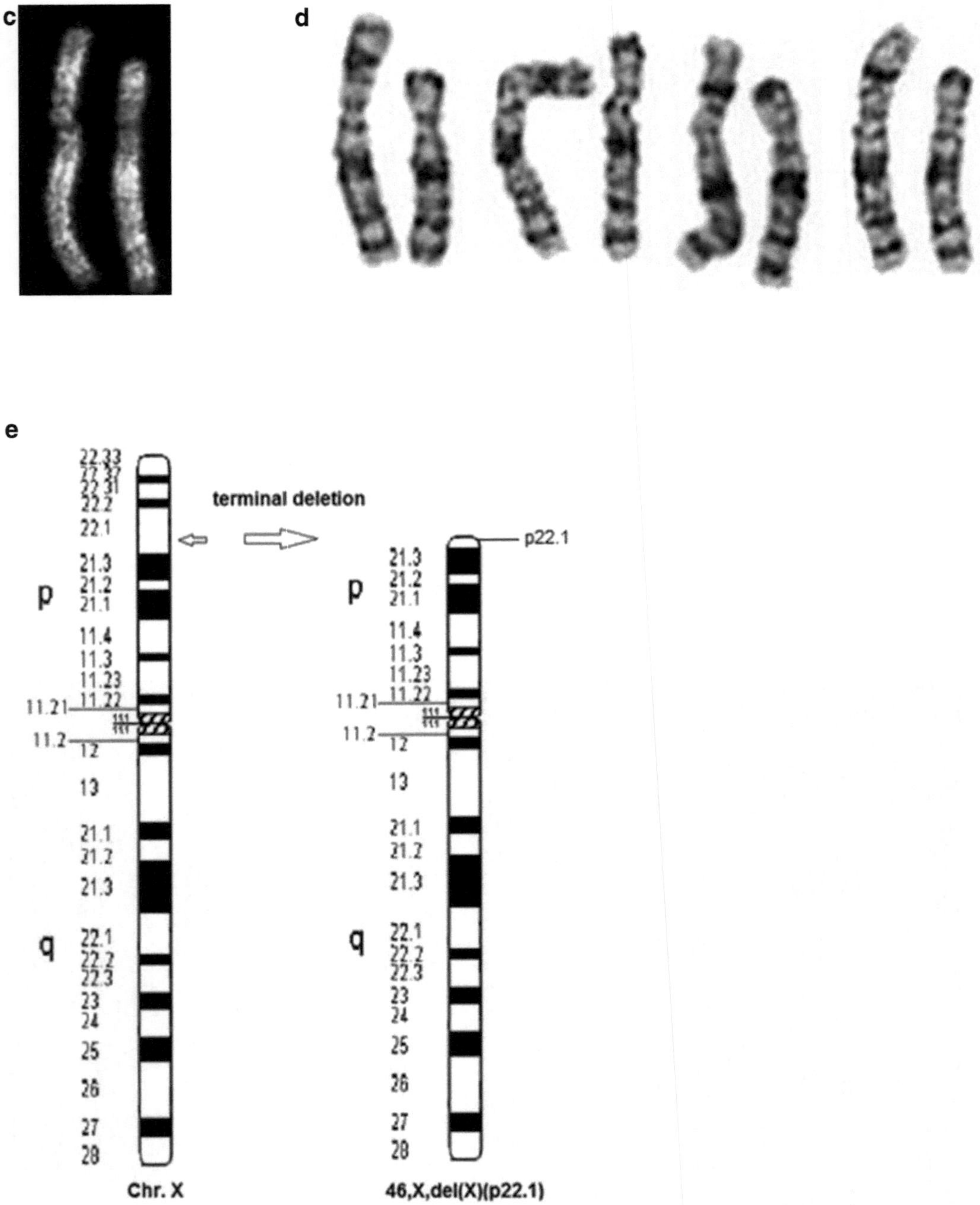

■ Fig. 7.16 (continued)

Case 57: Karyotype: 46,X,del(X)(p22.11p22.2),inv(4)(p12q11.2)

Indication for chromosome analysis: Increased number of abortions, normal phenotype, uterus bicornis.

Morphology

1. del(X)(p22.11p22.2)
   Interstitial deletion in the distal region of the short arm of the X-chromosome. About 20% of its total length got lost. The deleted segment is not easy to localise; therefore, a resolution of at least 550 bands is required.
2. inv(4)(p12q11.2)
   (see: ► Sect. 15.6)

Cave

The deleted region in Xp comprises 17 relevant genes. As the phenotype of the patient is normal, a screwed inactivation of the deleted chromosome can be expected. Therefore, genetic counselling has to take into account that 50% of male offspring will be carrier of a 13.2 Mb deletion which can be expected to be lethal. 50% of the female offspring will be carrier of the deletion as their mother, but if not a screwed inactivation of the aberrant chromosome takes place in the same way, it can present with a pathologic phenotype (◘ Fig. 7.17).

7

a

b

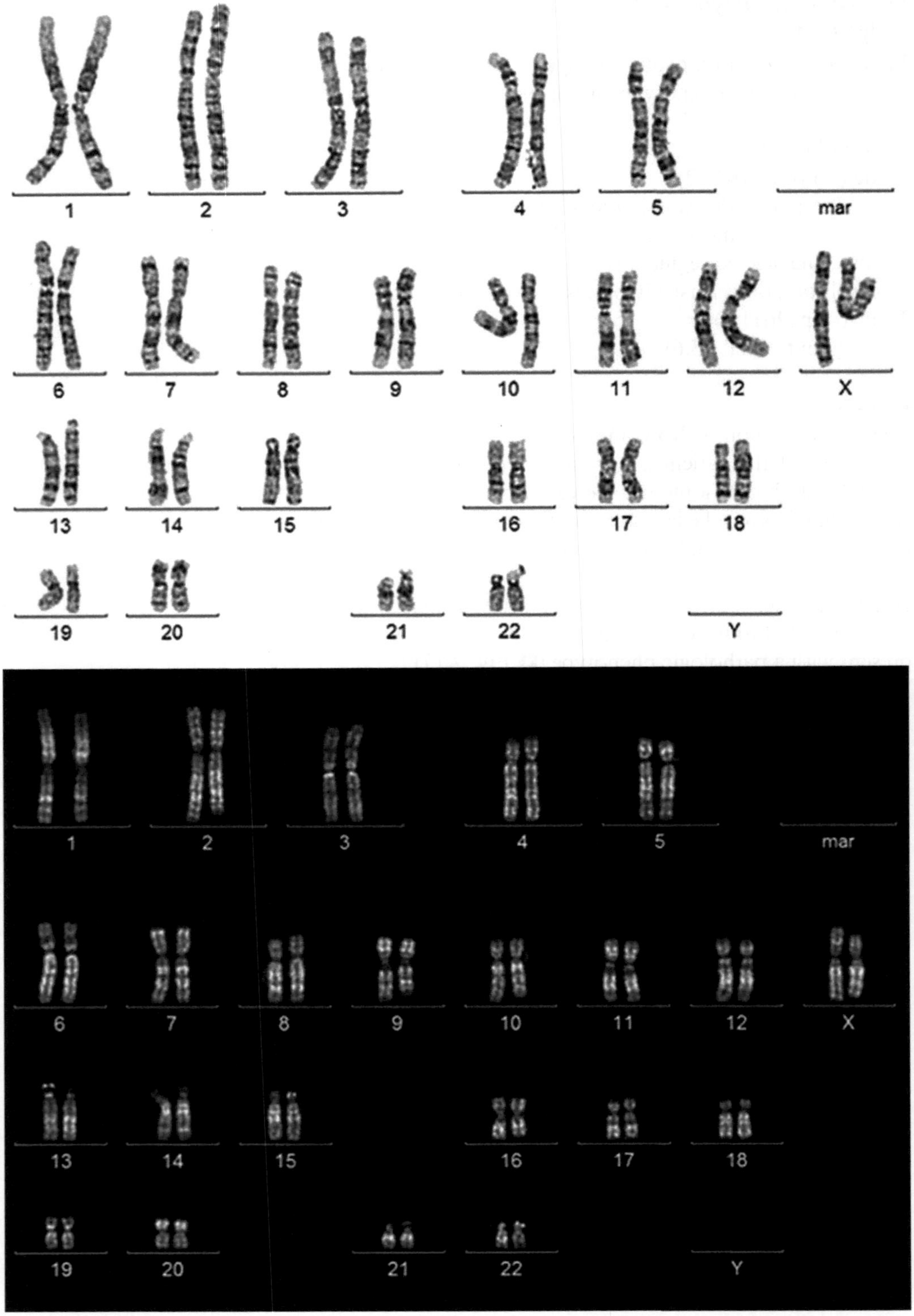

Fig. 7.17 Karyotype 46,X,del(X)(p22.1p22.2),inv(4)(p12q11.2). **a** GTG-, **b** QFQ-banding, 600 and 500 bands per genome, **c** QFQ-, **d** GTG-banding, chromosomes from different mitoses, *left* normal, *right* deleted X-chromosome, inv(4) (see Sect. 15.6), **e** ideograms from the X-chromosome, small interstitial deletion in the short arm

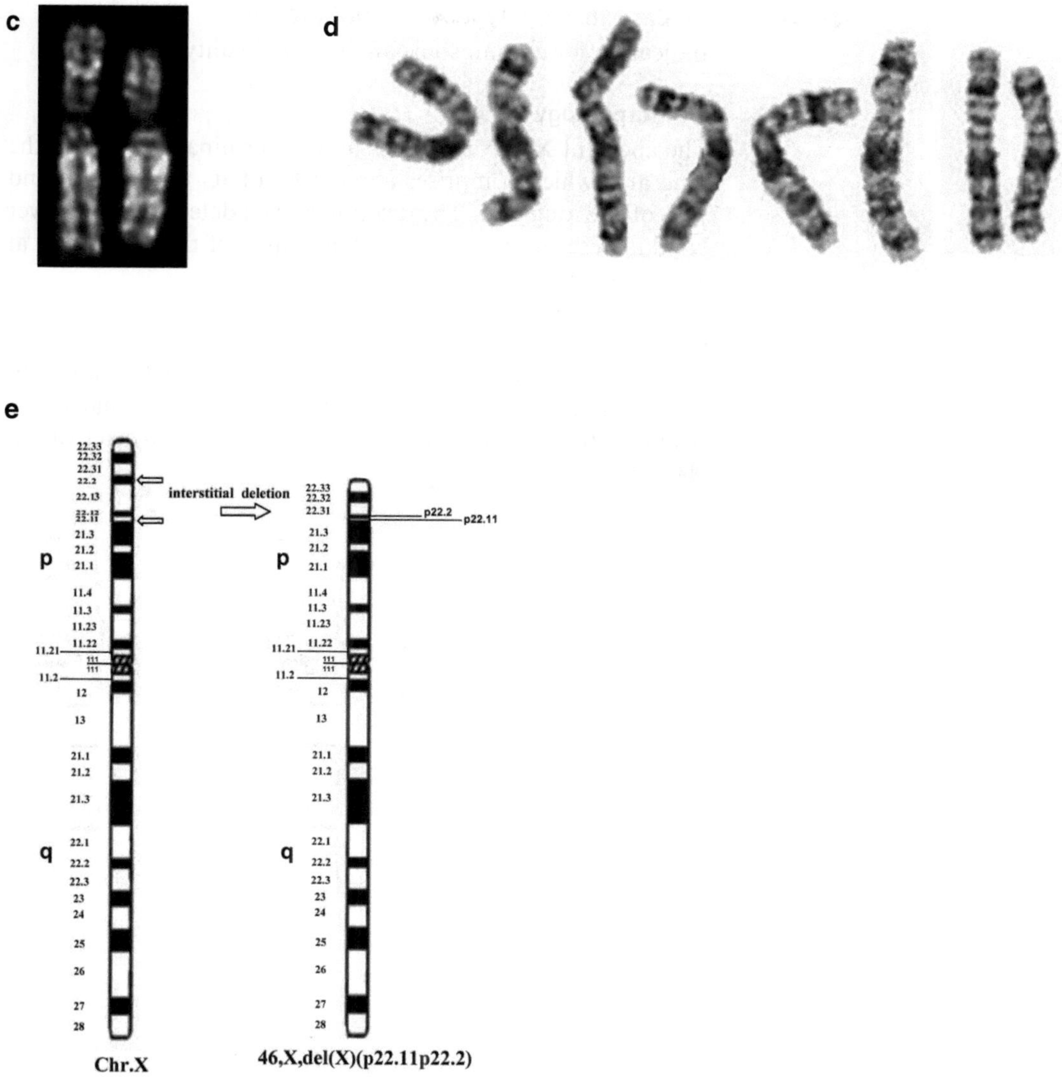

Fig. 7.17 (continued)

▪▪ Case 58: Karyotype: 46,X,del(X)(q21.2)

Indication for chromosome analysis: Infertility.

▪ Morphology

The aberrant X-chromosome shows a terminal deletion in the long arm which comprises about 50% of its total length and 70% of the long arm. The deletion is even detectable at a lower banding resolution but for exact definition of the breakpoint at least 550 bands per genome are required.

▪ Cave

Genetic counselling should be offered to discuss the possibilities of infertility treatment, including after additional investigations the possible occurrence of screwed X-inactivation (◘ Fig. 7.18).

a

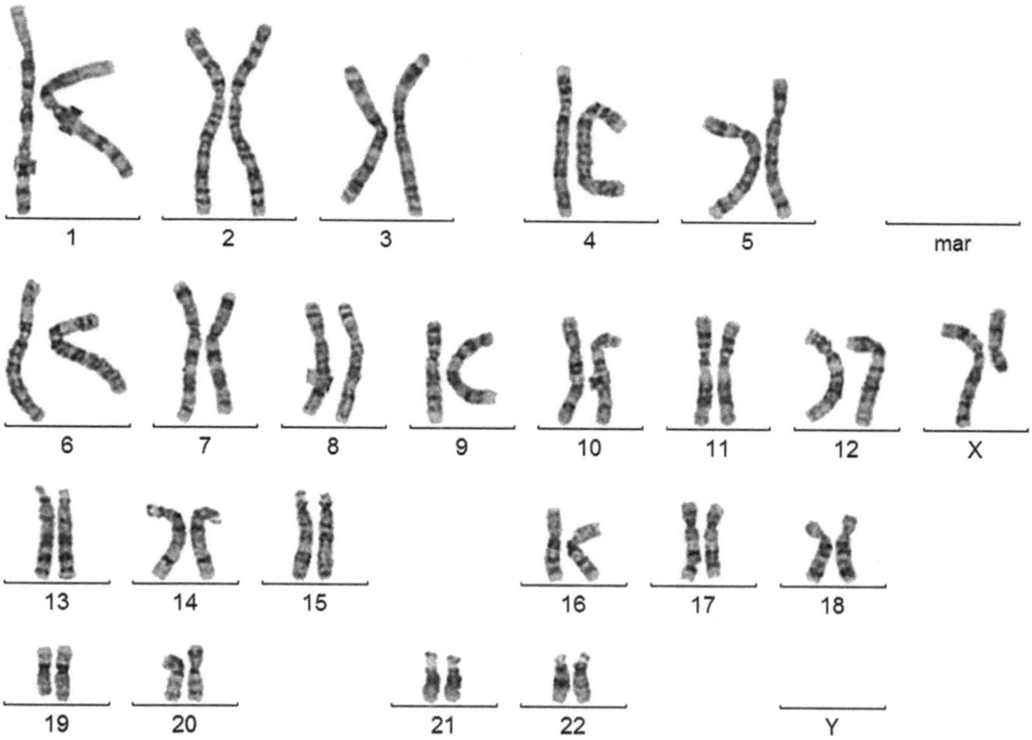

b

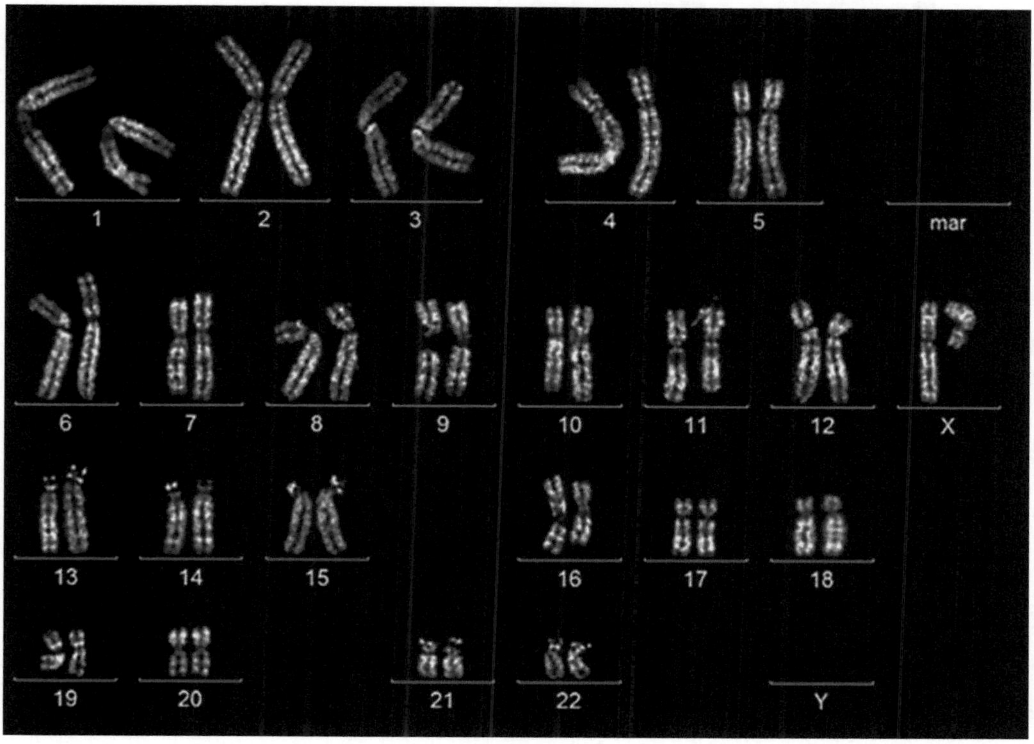

Fig. 7.18 Karyotype 46,X,del(X)(q21.2). **a** GTG-, **b** QFQ-banding, 700 and 500 bands per genome, **c** QFQ-, **d** GTG-banding chromosomes from different mitoses, *left* normal, *right* deleted X-chromosome, **e** ideograms of the X-chromosome, large distal deletion in the long arm

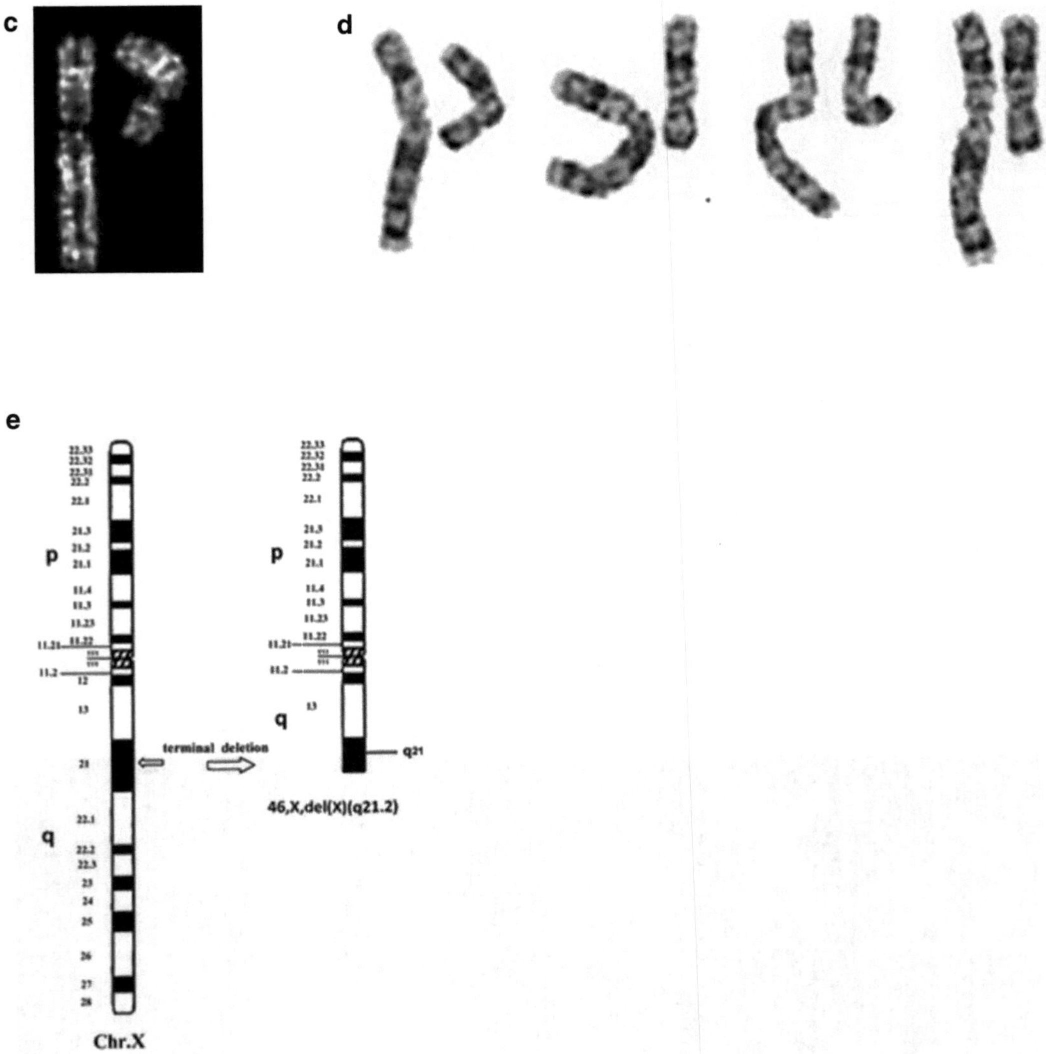

Fig. 7.18 (continued)

■■ Case 59: Karyotype: 46,X,del(X)(q27.2)
Indication for chromosome analysis: Infertility.

■ Morphology

The terminal part of the long arm of the X-chromosome is deleted. It comprises about 10% of the total length of the long arm and about 6% of the total length of the X-chromosome. In this case, a resolution of about 850 bands per genome was necessary to define the breakpoint in the subband q27.2.

■ Cave

If pregnancy can be induced the patient should be informed about possible risk factors in male offspring (◘ Fig. 7.19).

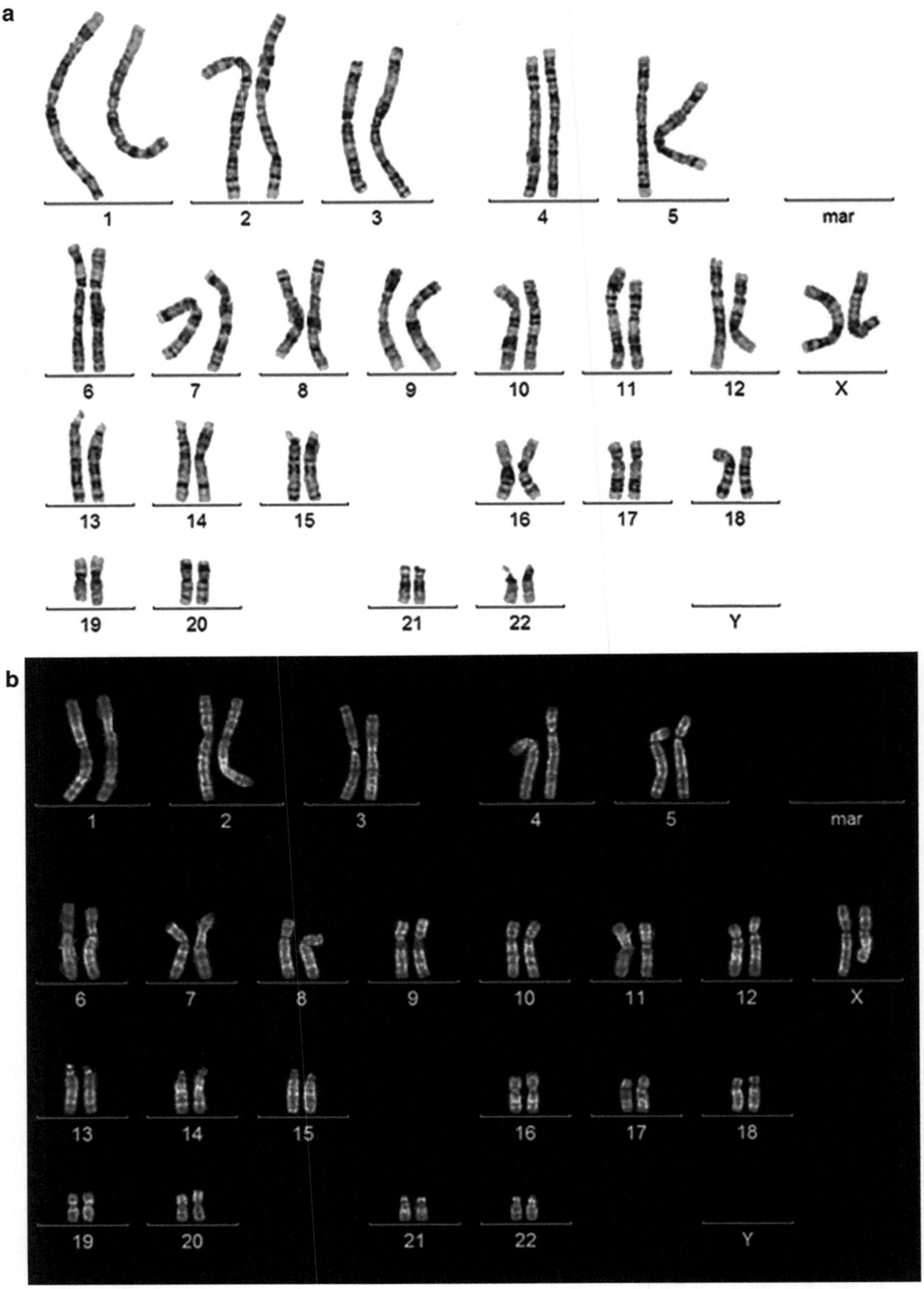

Fig. 7.19 Karyotype 46,X,del(X)(q27.2). **a** GTG-, **b** QFQ-banding, 700 and 400 bands per genome, **c** QFQ-, **d** GTG-banding, chromosomes from different mitoses, *left* normal, *right* deleted X-chromosome, **e** ideograms of the X-chromosome, small terminal long arm deletion

c d

e

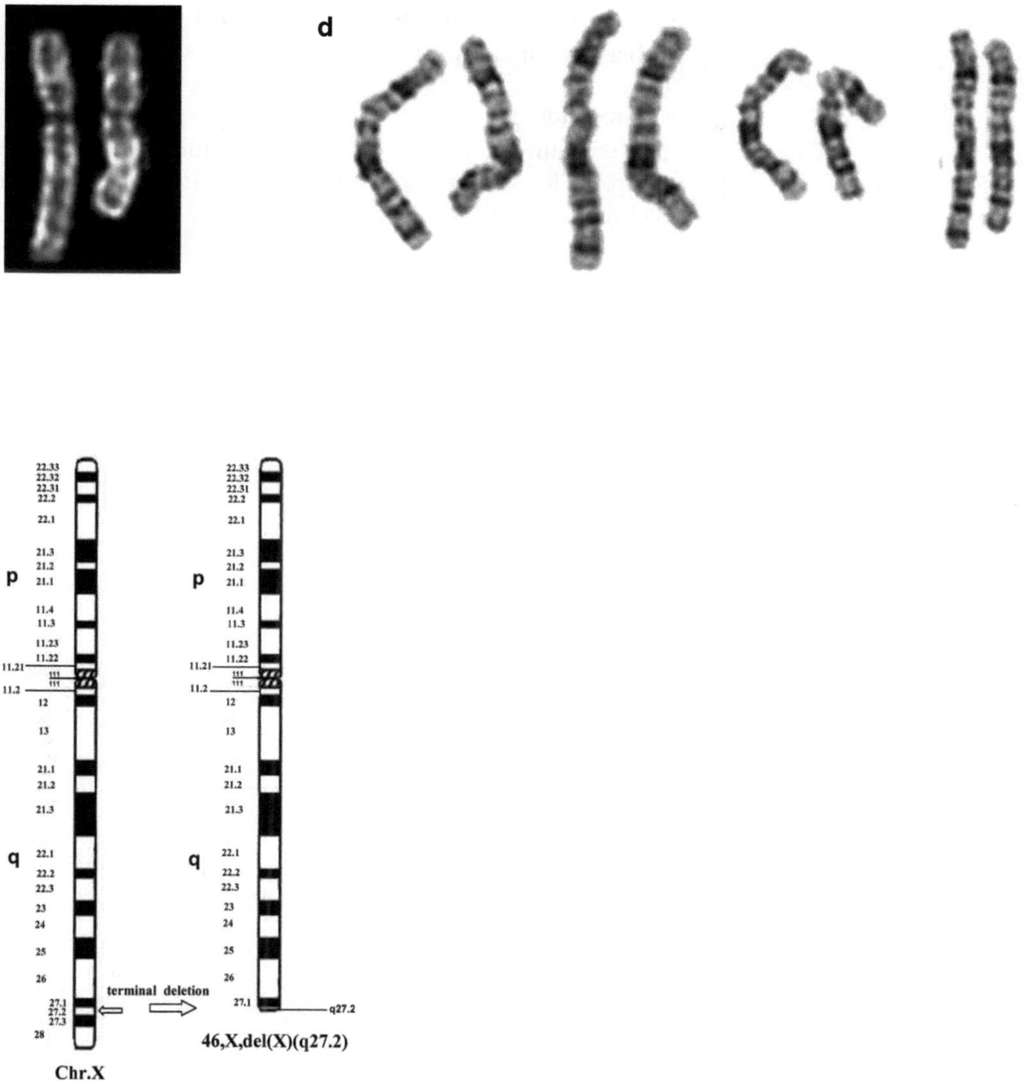

Fig. 7.19 (continued)

▪▪ Case 60: Karyotype: 46,X,del(Y)(q11.22)

Indication for chromosome analysis: Azoospermia.

▪ Morphology

The aberrant Y-chromosome shows a terminal deletion of the complete heterochromatic block in q12 and of the adjacent euchromatin with breakpoint in q11.22.

▪ Cave

The relevance of the deletion of the AZF-region has to be explained to the patient (◘ Fig. 7.20).

a
b
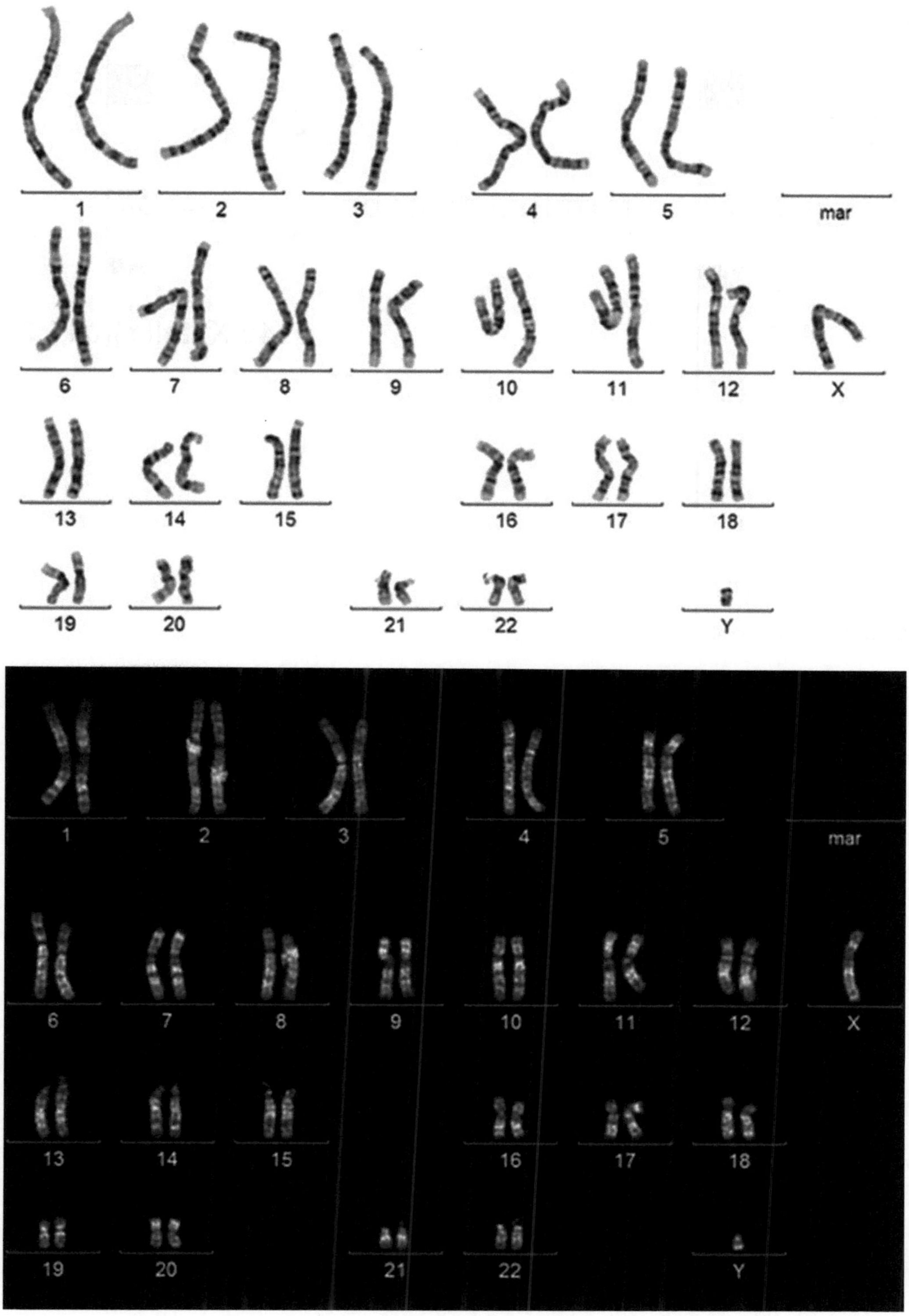

Fig. 7.20 Karyotype 46,X,del(Y)(q11.22). **a** GTG-, **b** QFQ-banding, 500 and 400 bands per genome, **c** ideograms of the Y-chromosome, deletion in the distal long arm

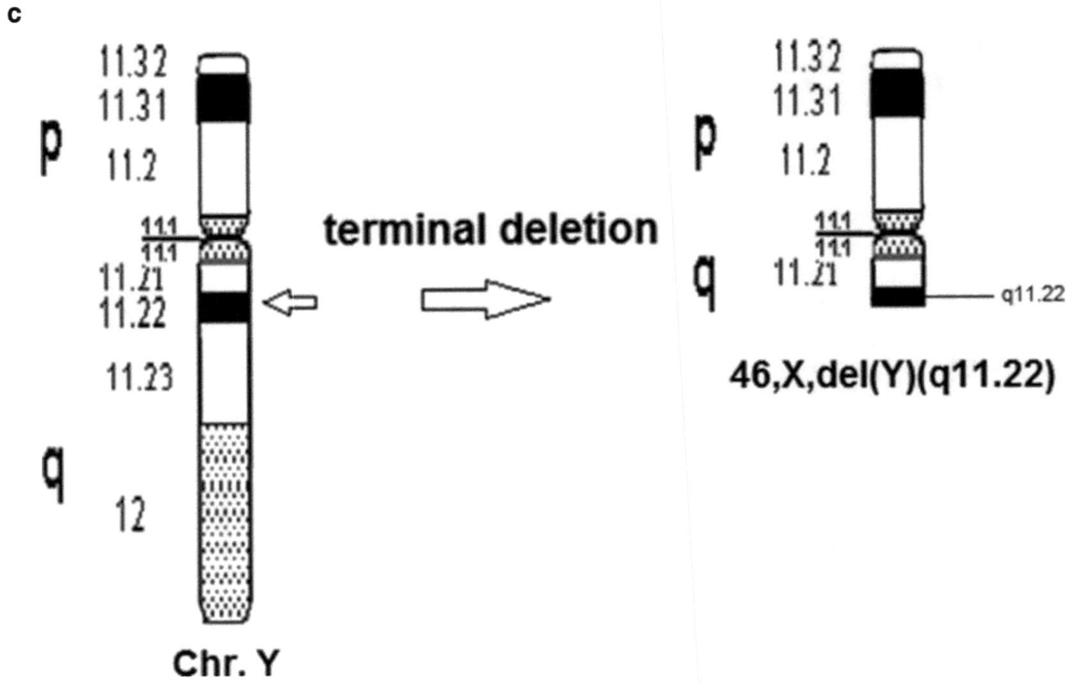

Fig. 7.20 (continued)

7.2 Summary

Out of the own investigation group 20 case reports were presented in detail. The interchromosomal distribution of the aberrant chromosomes is relatively equal as to the autosomes though the small chromosomes 19–22 were not involved. The gonosomes X and Y were overrepresented compared to the autosomes, which is caused by the main investigation group of sterility patients (70:30%).

It was an unexpected observation that in two cases (6 and 13) the aberration was not an isolated deletion. Therefore, these cases belong to the group of complex chromosome rearrangements (CCR), and this group requires according to recent findings additional molecular genetic investigations.

Case 6 shows beside mitoses with a deletion of chromosome 8 one metaphase with a normal karyotype. This leads to the conclusion that the deletion originated in the postzygotic phase, and this has to be regarded as a very rare process. Case 13 additional molecular genetic investigations show that the patient has not only a deletion: his chromosome abnormality is the unbalanced form of a reciprocal translocation 7/18 of parental origin. This case makes it clear how relevant are the extended investigations even in simple appearing cases of structural changes of the chromosome.

If the localisation of the breakpoints in the chromosomes of the investigation group is analysed, 15 of the deletions were terminal and 5 interstitial aberrations. Even if the deletions are separated into autosomal and gonosomal abnormalities, this distribution is the same. But the small number of cases has to be taken into account.

The relevance of parental analyses before genetic counselling could be demonstrated.

Duplications

Contents

8.1 Duplications – 230

8.2 Summary – 240

C. Behrend et al., *Human Chromosome Atlas*, https://doi.org/10.1007/978-3-031-10588-3_8

8.1 Duplications

61. 46,XX,dup(9)(p11.2)
62. 46,XY,dic(9)(pter→q11::p12 → qter)
63. ♀ 46,X,dup(X)(p11.4p21.2)X.
    ♂ 46,dup(X)(p11.4p21.2)y

Case 61: Karyotype: 46,XX,dup(9)(p11.2)

Indication for chromosome analysis: Increased rate of abortions.

Morphology

The additional band in p11.2 is dark staining (AT-rich) after GTG banding and negative after CBG staining. In the aberrant chromosome 9, the arm index has changed the chromosome into a more metacentric structure. The FISH analysis defined the additional band as a euchromatic variant without genetic relevance (see: ► Sect. 15.1).

Cave

In the pathologic course of a future pregnancy, the karyotype of the conceptus should be analysed to exclude meiotic abnormalities by abnormal crossing over in Pachytene (◘ Fig. 8.1)

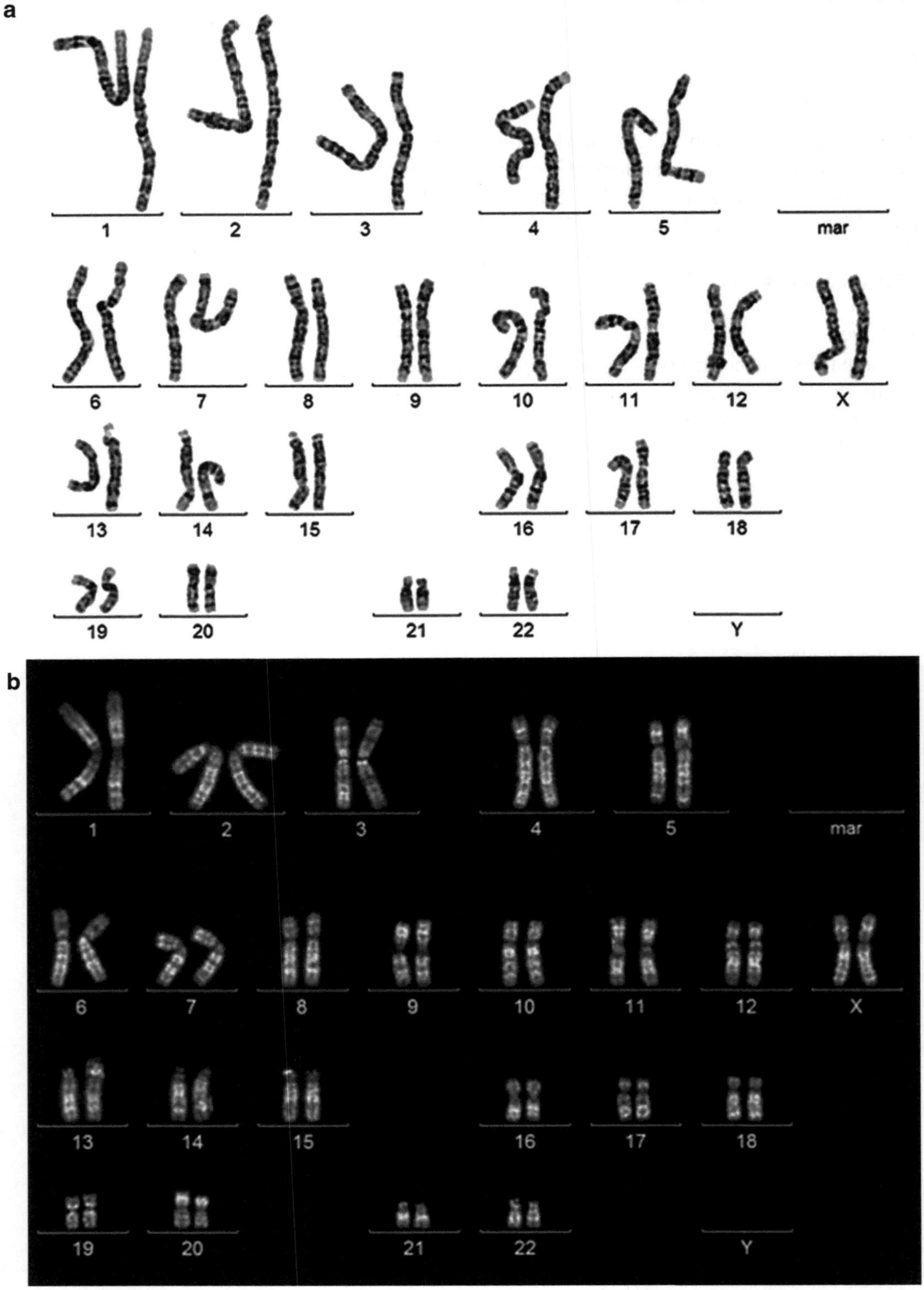

Fig 8.1 Karyotype 46,XX,dup(9)(p11.2). **a** GTG-, **b** QFQ-banding, 700 and 400 bands per genome, **c** CBG-, **d** QFQ-, **e** GTG-banding, chromosomes from different mitoses, *left* normal chromosomes 9, *right* derivate 9, **f** ideograms of chromosome 9; left normal, right duplicated chromosome, small interstitial duplication

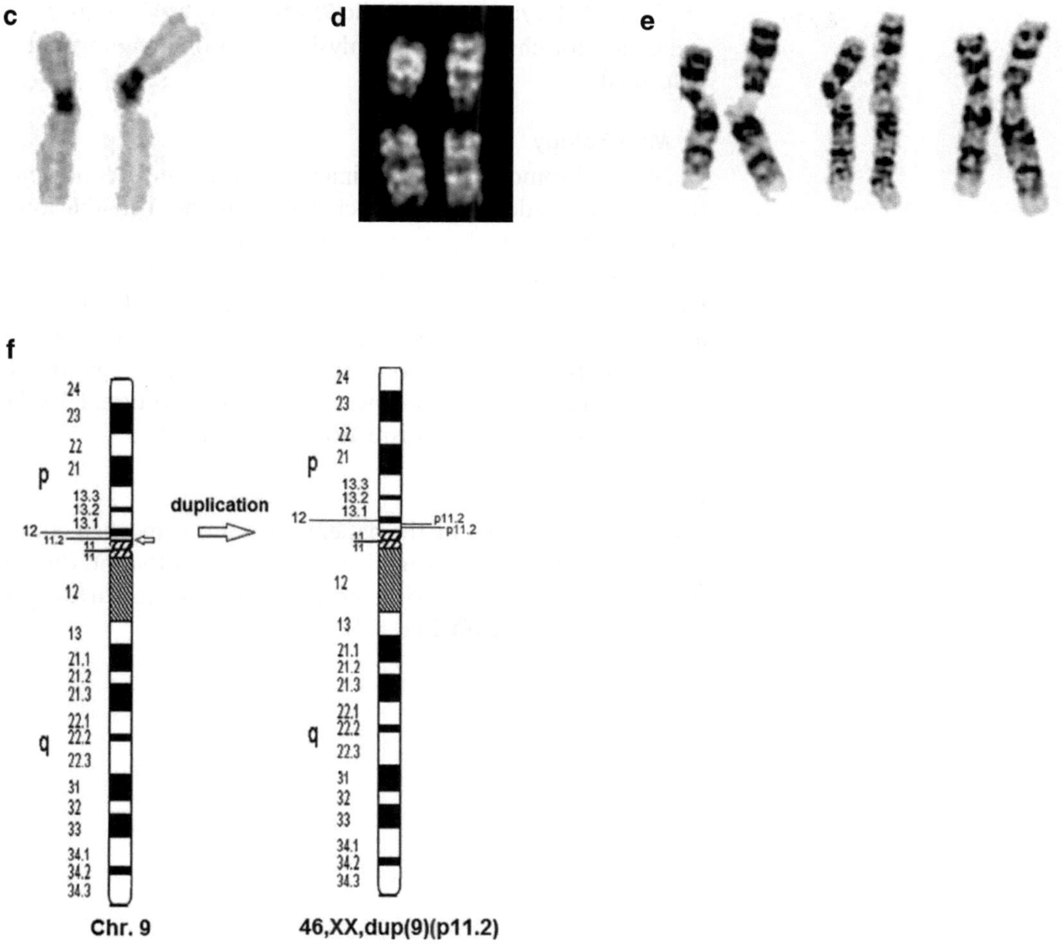

Fig. 8.1 (continued)

■■ Case 62: Karyotype: 46,XY,dic(9)(pter→q11::p12 → qter)

Indication for chromosome analysis: Gonosomal abnormality suspected.

■ Morphology

Dicentric chromosome 9 with inactivation of the second centromere. The additional AT-rich band proximal in the long arm comprises furthermore a pericentromeric region with the second centromere inactivated. The additional region is located in the middle of the heterochromatic block q12. The additional segment is CBG negative. The inactivation of the second centromere seems to be stable, as no malsegregation of the aberrant chromosome 9 was observed. Additional FISH investigations after microdissection are indicated.

■ Cave

The minor anomalies of the patient are a problem when counselling the family as the genetic risk is less than 1% and due to the almost normal phenotype indication for prenatal investigation is questionable (◘ Fig. 8.2).

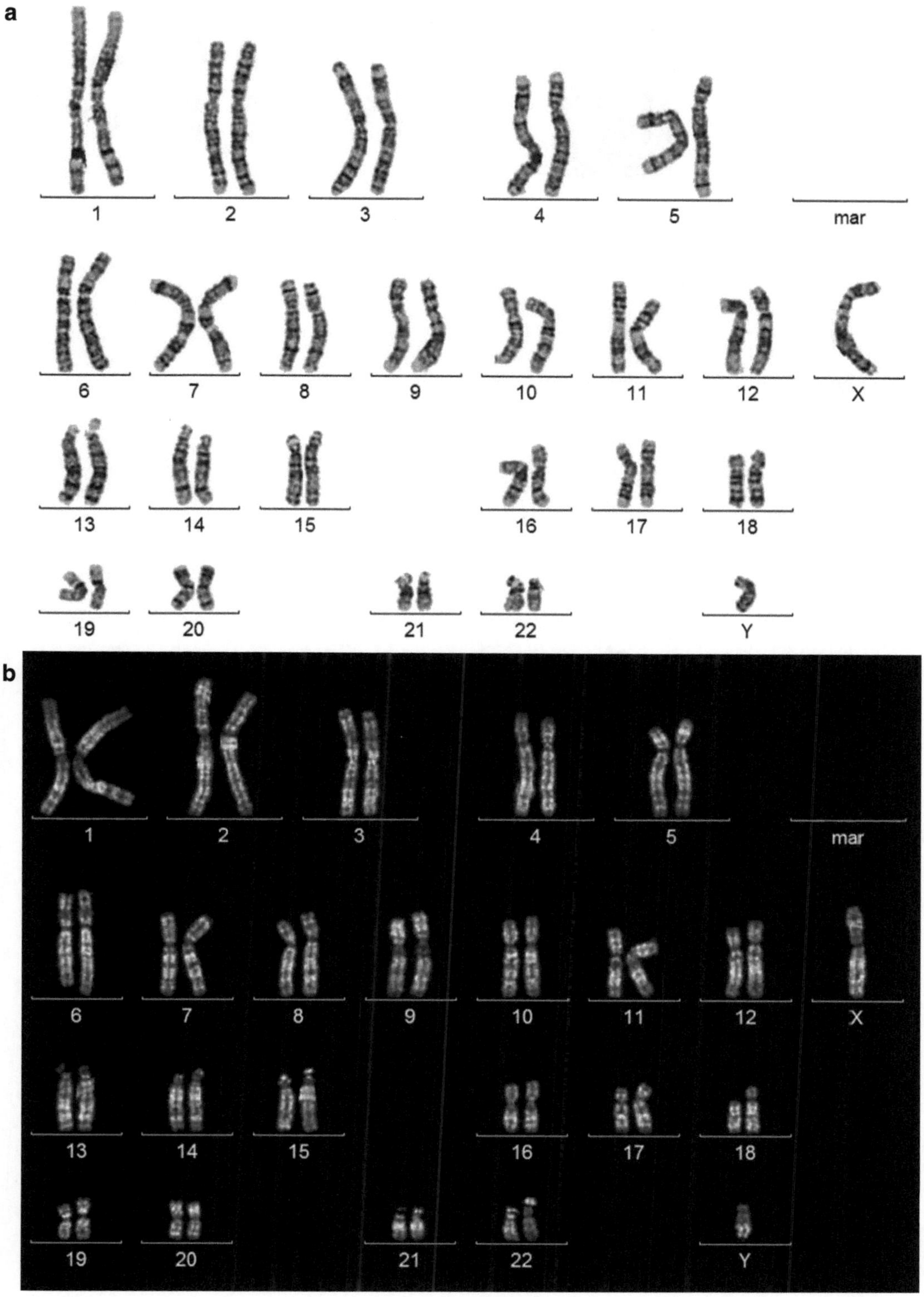

Fig 8.2 Karyotype 46,XY,dup(9),dic(9)(pter→q11::p12 → qter). Single pathologic cell with deletion of the short arm of chromosome 18. **a** GTG-, **b** QFQ-banding, 700 and 500 bands per genome, **c** CBG-, **d** QFQ-, **e** GTG-banding, chromosomes from different mitoses, *left* normal, *right* aberrant chromosome 9, **f** ideograms of chromosome 9; *left* normal, *right* duplicated chromosome, small interstitial duplication

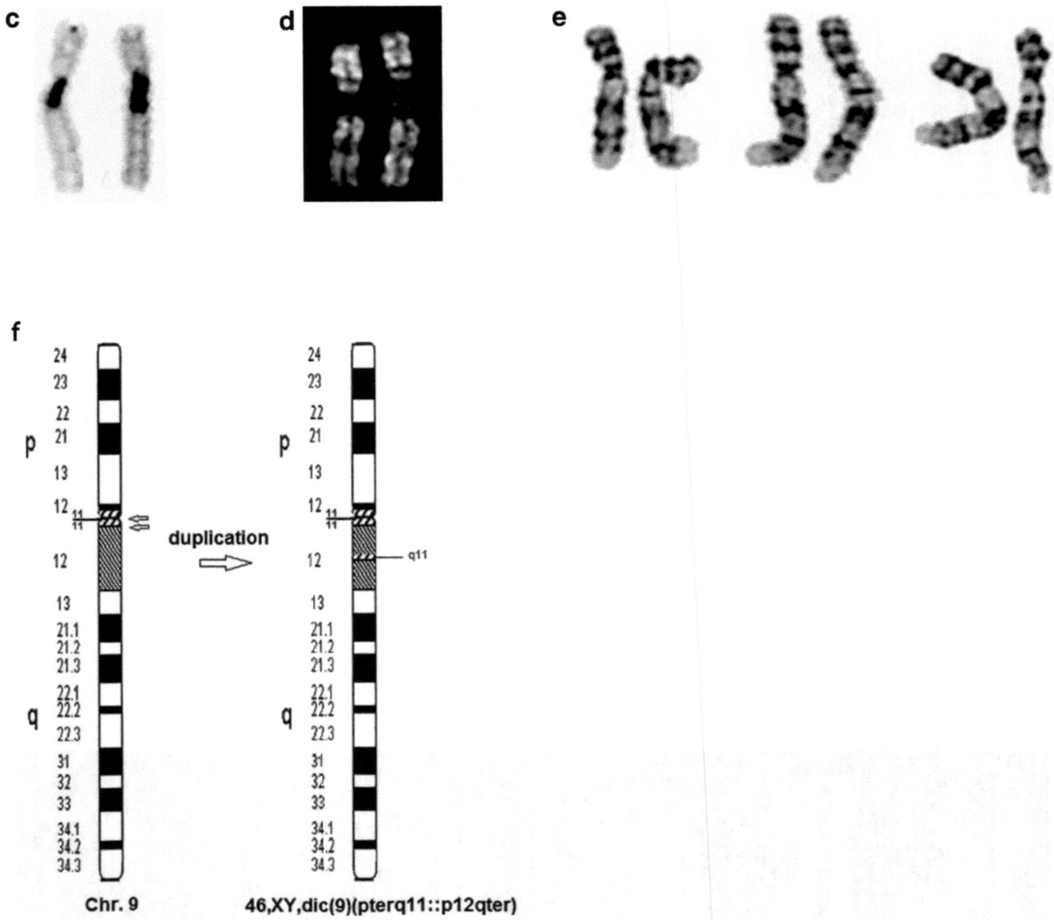

Fig. 8.2 (continued)

■■ Case 63: Familial Duplication

- *Mother: Karyotype: 46,X,dup(X)(p11.4p21.2).*
- *Son: Karyotype: 46,dup(X)(p11.4p21.2)y.*

Indication for chromosome analysis: congenital abnormalities in the son especially ambiguous genitalia.

■ Morphology

The additional band in Xp is AT-rich and identical in both mother and son. It involves about 5% of the total length of the X-chromosome. The proximal banding pattern in Xp shows characteristic differences to the normal one.

■ Cave

Prenatal diagnosis has to be offered to the consultant to exclude partial additional duplications or deletions in the conceptus, and in female foetuses the inactivation pattern of normal and aberrant X-chromosome should be analysed (■ Fig. 8.3).

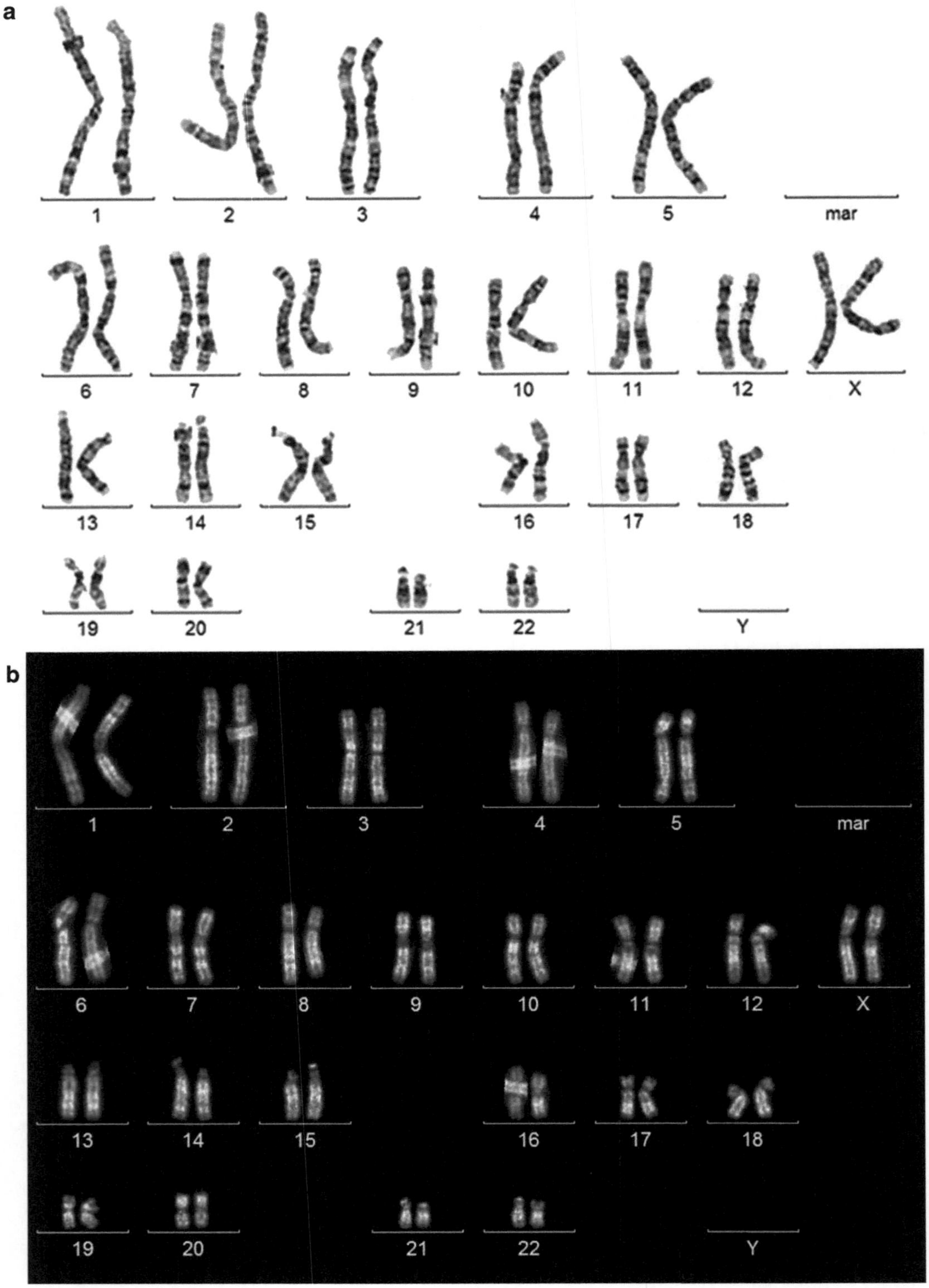

Fig. 8.3 Karyotype 46,X,dup(X)(p11.4p21.2). **a** GTG-, **b** QFQ-banding, 500 and 400 bands per genome, **c** QFQ-, **d** GTG-banding, chromosomes from different mitoses, *left* normal, *right* aberrant chromosome X, **e** ideograms of the X-chromosome, proximal duplication in the short arm

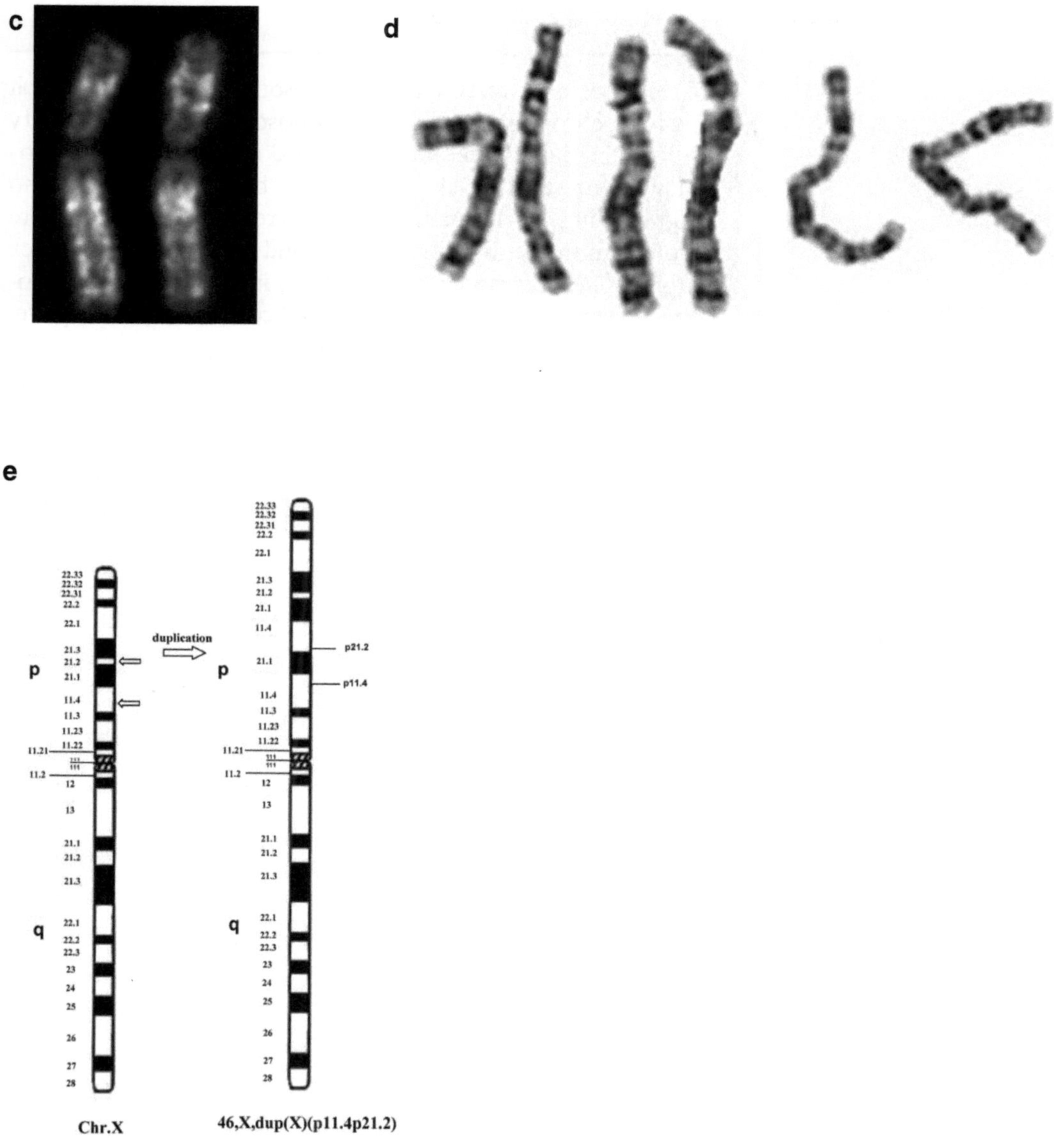

Fig. 8.3 (continued)

8.2 Summary

Three cases are presented, two autosomal duplications, each involving chromosome 9 and one gonosomal with abnormality of Xp. The duplication 9p12 is defined as a euchromatic variant without genetic risk. But it must be guaranteed that no neighbouring euchromatic, genetic relevant segments are involved in the duplication. The second case shows a complex chromosome rearrangement resulting in a dicentric chromosome 9. The inactivation of the second centromere is complete; a second cell line is not present. The third case, a partial duplication of the short arm of the X-chromosome leads to a normal phenotype and fertility in the female carrier but to a malformation syndrome in the hemizygote son. The relevance of a screwed X inactivation in female carriers of an X abnormality must always be taken into account when counselling carriers and their relatives.

Ring Chromosomes

Contents

9.1 Ring Chromosomes – 242

9.2 Summary – 251

C. Behrend et al., *Human Chromosome Atlas*, https://doi.org/10.1007/978-3-031-10588-3_9

9.1 Ring Chromosomes

64. 46,XX,r(18)(p11.23q21.33)
65. mos45,X[25]/46,X,r(X)(p22.1q21)[29],
66. 46,X,r(Y)(p11.32q11.21).

■■ Case 64: Karyotype: 46,XX,r(18)(p11.23q21.33)

Indication for chromosome analysis: Developmental retardation, morphologic peculiarities according to de Grouchy syndrome II.

■ Morphology

The size of the ring chromosome is not identical in all metaphases analysed, part of the mitoses shows a duplication of the ring. This abnormality is caused by a pathologic course of replication of the ring (S-phase).

■ Cave

A mosaic karyotype of the patient with malsegregation of the ring has to be excluded. The chromosomes of the parents should be analysed before genetic counselling to exclude a balanced rearrangement (inversion) or a low-grade mosaic (◘ Fig. 9.1).

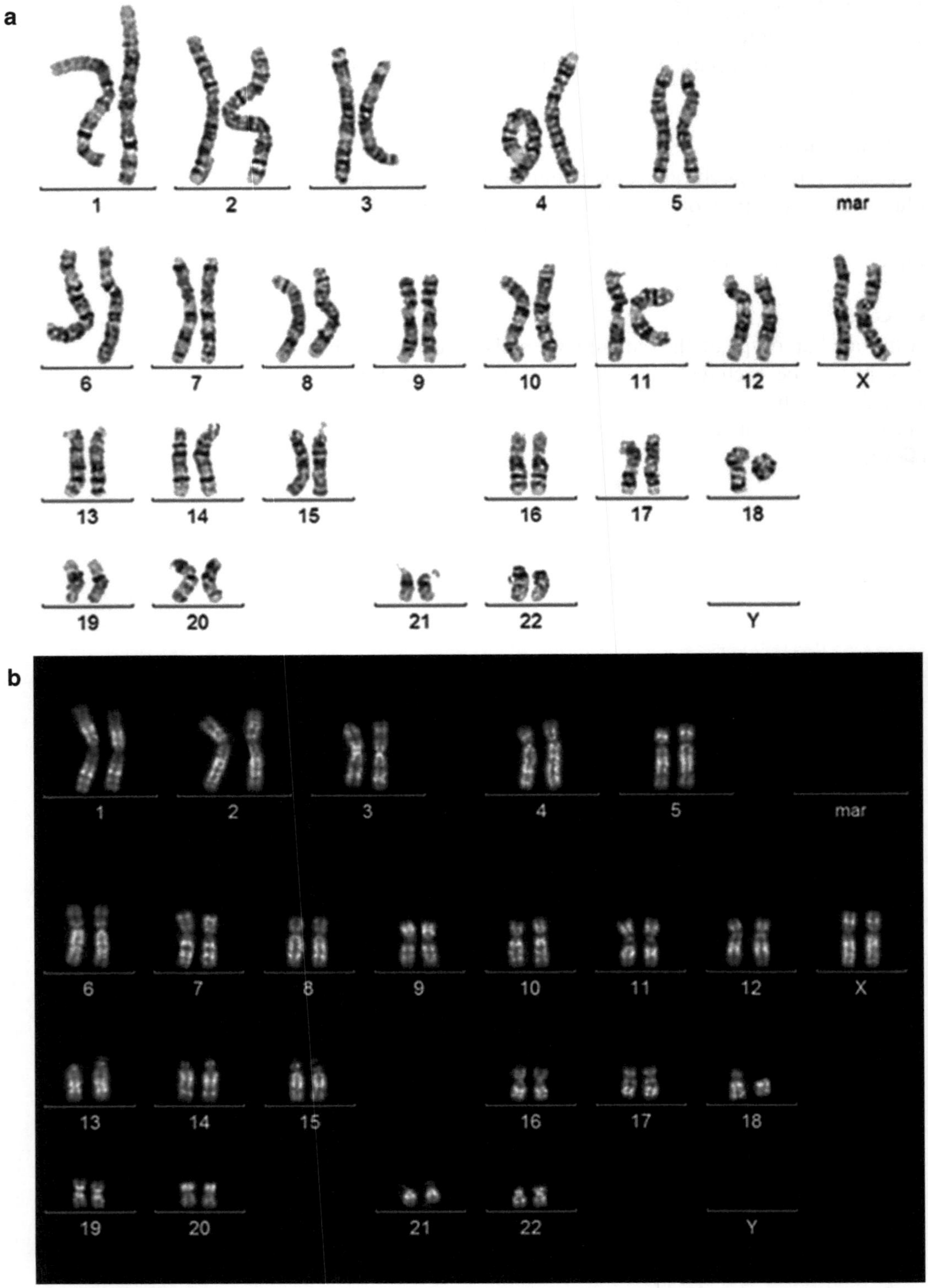

◘ Fig. 9.1 Karyotype 46,XX,r(18)(p11.23q21.33). **a** GTG-, **b** QFQ-banding, 500 and 400 bands per genome, **c** CBG-, **d** QFQ-, **e** GTG-banding, showing the ring in different positions and a duplication in 2 of them; chromosomes from different mitoses, **d**, **e** *left* normal, *right* ring chromosome 18, **f** ideograms and schematic illustration of the ring formation of chromosome 18. The presentation is completed by the findings after microarray analysis

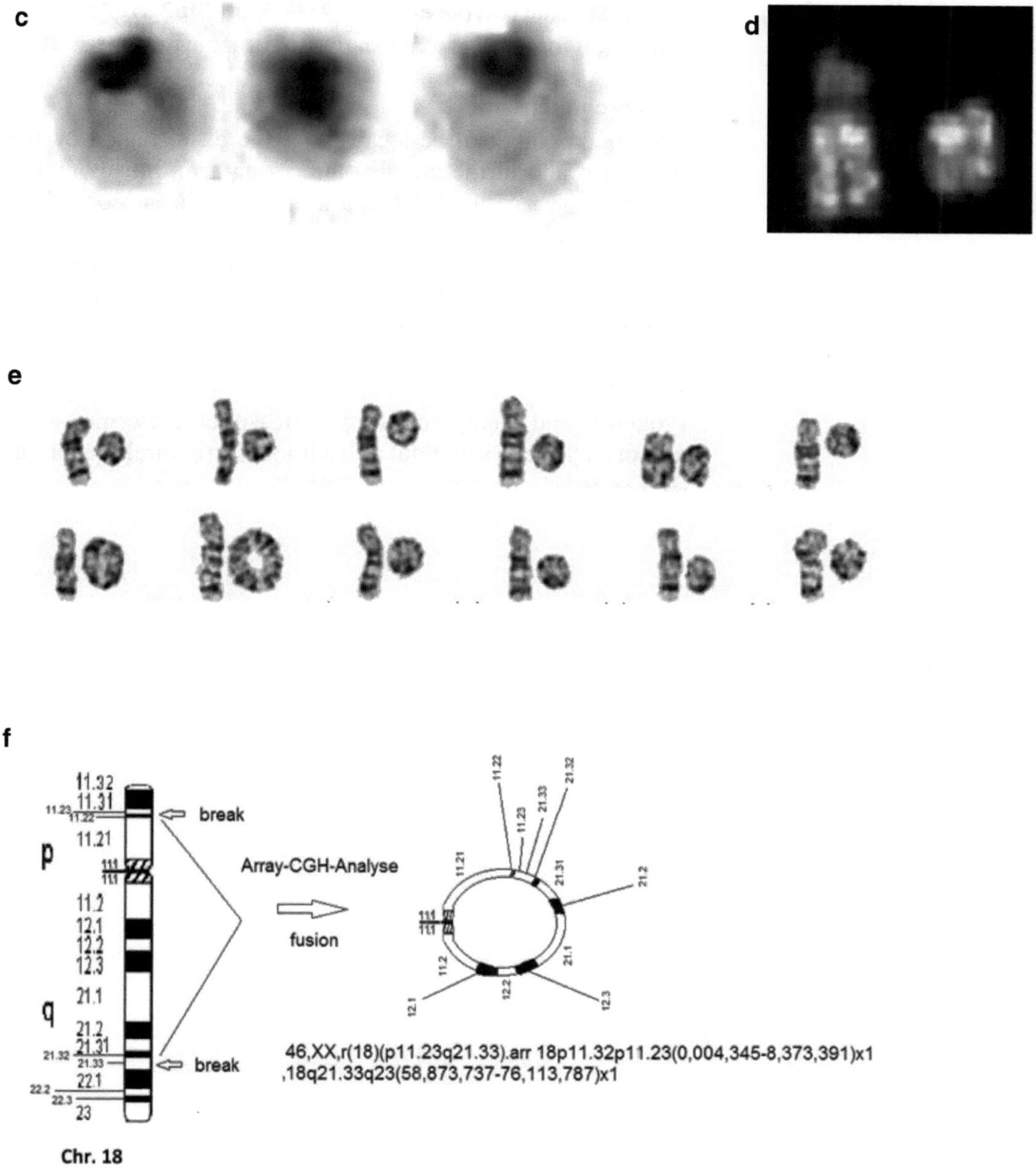

■ **Fig. 9.1** (continued)

▪▪ Case 65: Karyotype: mos45,X[25]/46,X,r(X)(p22.1q21)[29]

Indication for chromosome analysis: Turner syndrome.

▪ Morphology

The breakpoints can be localised in the distal region of the short arm and in the middle of the long arm, thus comprising about 38% of the total length of the X-chromosome. The ring is unstable in the course of cell divisions and got lost in about 50% of the mitoses analysed. Besides, partial and complete duplications of the ring by abnormalities of chromosome replication (S-phase of the cell cycle) were documented.

▪ Cave

Prognosis and therapy of the patient will be performed as in monosomy X cases without an additional structural aberration (◘ Fig. 9.2).

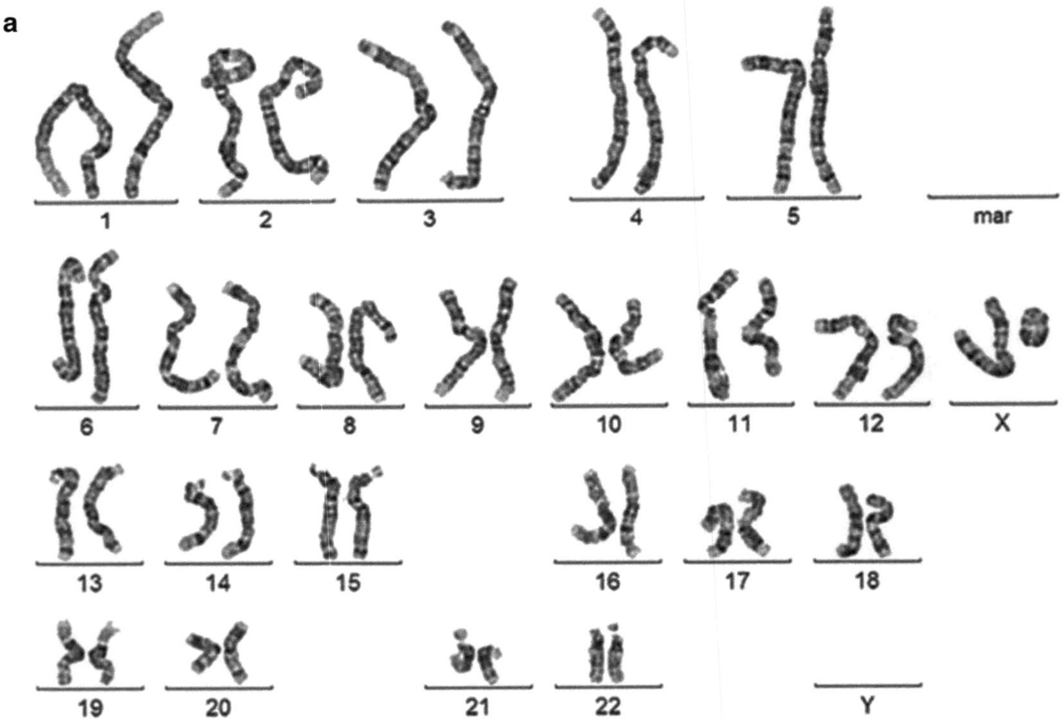

◘ **Fig. 9.2** Karyotype mos 45,X[25]/46,X,r(X)(p22.1q21)[29]. **a** GTG-banding, 850 bands per genome, **b** CBG-, **c** GTG-banding, **b** right figure shows a ring duplication, **c** *left* normal, *right* ring chromosome X, **d** ideograms of the X-chromosome with illustration of the ring formation

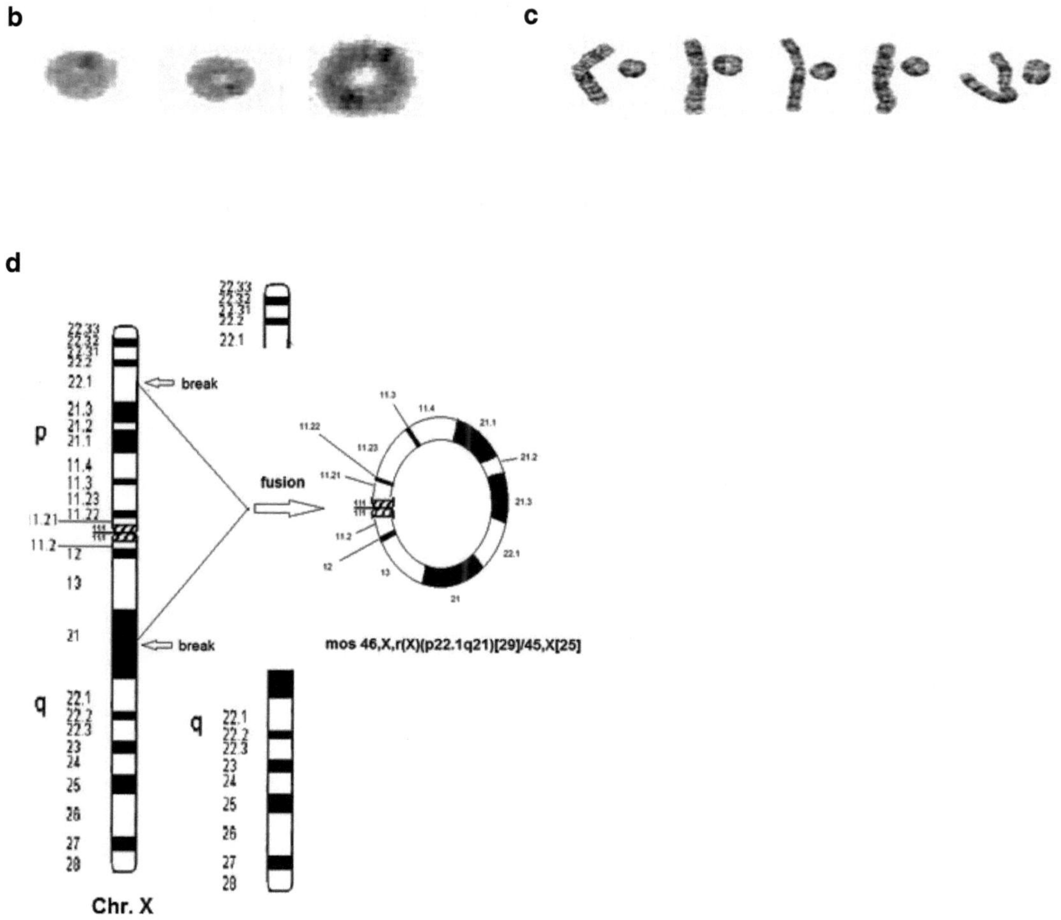

Fig. 9.2 (continued)

▪▪ Case 66: Karyotype: 46,X,r(Y)(p11.32q11.21)

Indication for chromosome analysis: Male infertility.

▪ Morphology

The ring chromosome is characterised by the missing heterochromatic region (Yq12). The ring is stable in the course of cell divisions; therefore, mosaic formation could be excluded.

▪ Cave

none (◘ Fig. 9.3).

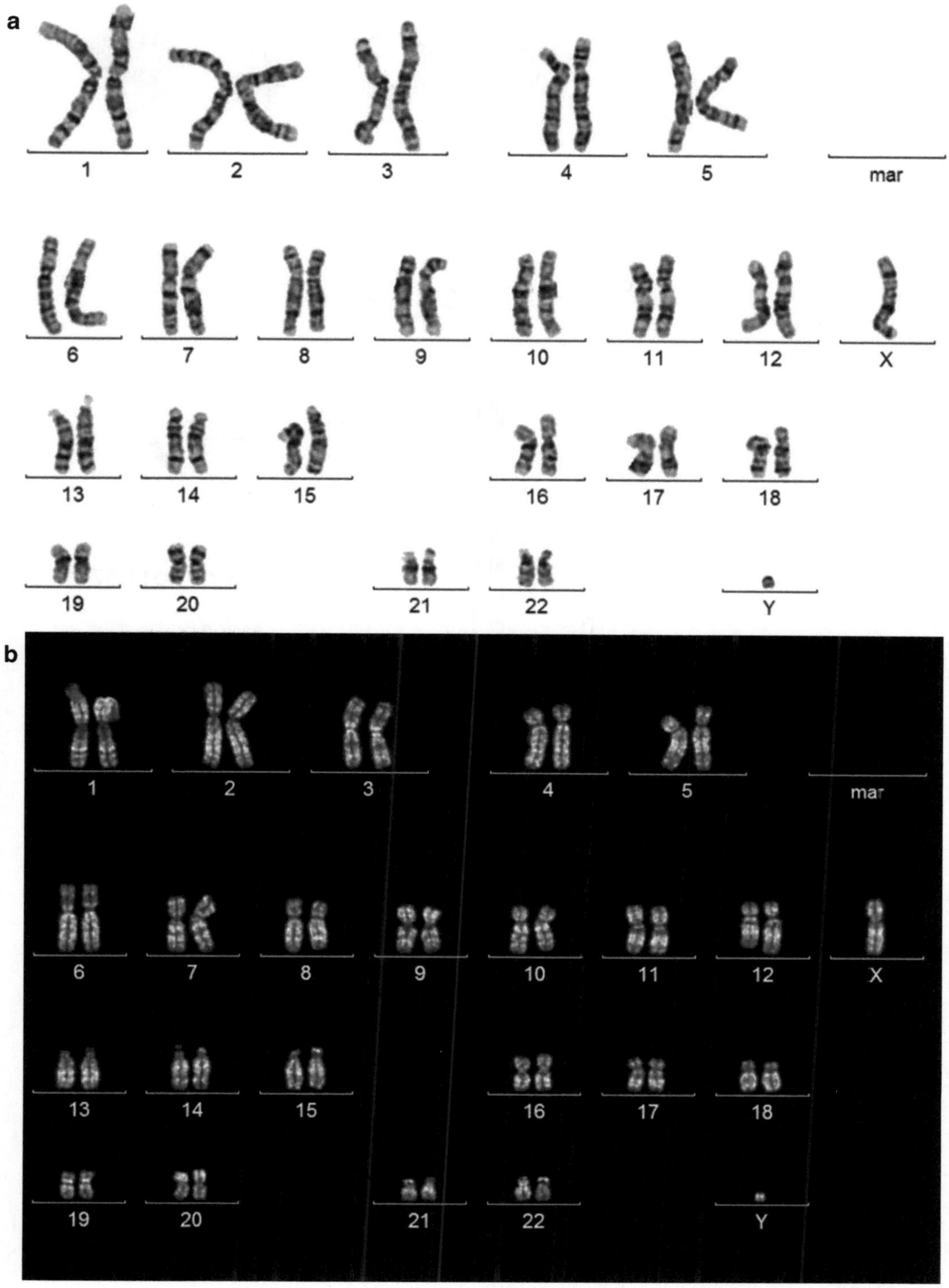

Fig. 9.3 Karyotype 46,X,r(Y)(p11.32q11.21). **a** GTG-, **b** QFQ-banding, 550 and 400 bands per genome, **c** GTG-banding, chromosomes from different mitoses, duplicated ring, **d** ideograms of ring Y-chromosome with scheme of the ring formation

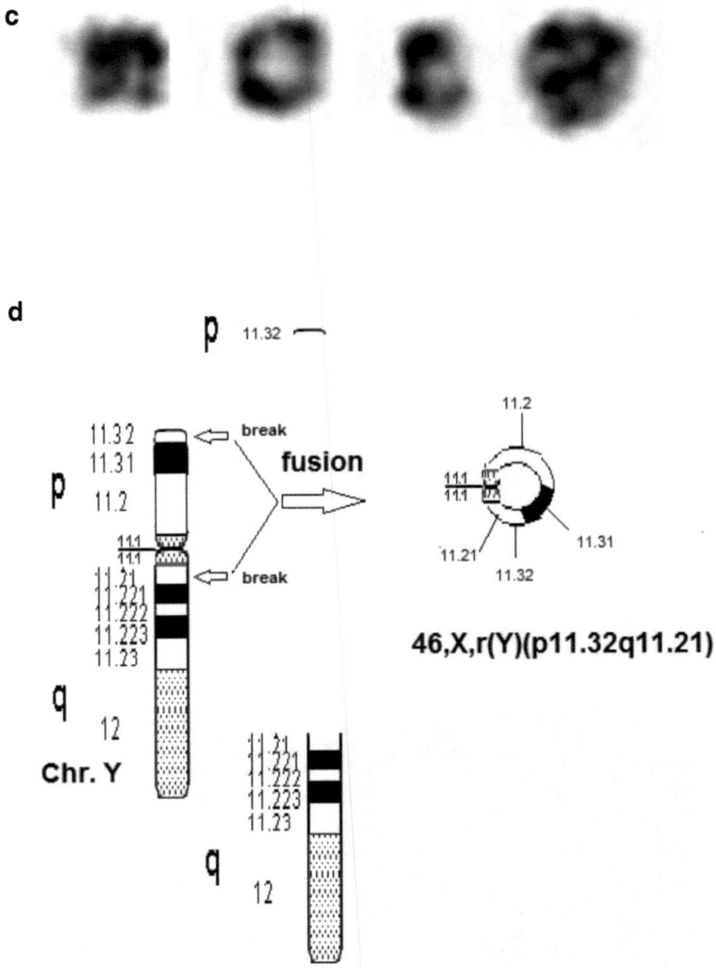

Fig. 9.3 (continued)

9.2 Summary

Three cases are described in detail, which make the problems clear when ring chromosomes must be analysed. Chromosomes 18, X and Y are involved in ring formation. In all three cases, additional investigations by FISH and microarray were necessary to define exactly the breakpoints. As in the present case of r(18), this aberration is, according to the literature, characterised by a frequently pathologic course of the cell cycle, especially the S-phase, were partial and complete duplications of the ring chromosome arise. In the second case of a ring chromosome X, the karyotype is characterised by a second cell line with monosomy X (45,X), which developed by loss of the ring chromosome in part of the cells. These secondary X0-mosaics are frequent observations in different types of structural aberrations of the X-chromosome. As in the first case of r(18) also in the patient with r(X) duplications of the ring chromosomes caused by replication errors were analysed. In the third case of a ring chromosome Y, the ring was stable and inherited by the somatic cell divisions. Duplications of the ring were not observed. But from the literature it is well known that a cell line with monosomy X as well as partial and complete duplications of r(Y) can occur.

The three cases presented here, make it obvious, that in all cases of ring chromosomes at least 50 mitoses have to be analysed, to find out if the carrier presents with secondary mosaic formation by aneuploidisation and/or secondary structural aberrations. In cases of extremely instable ring chromosomes, the indication for additional investigations of the complex karyotype might be given. These can be performed either by the analysis of a second somatic cell system or by the investigation of the same cell system after an interval of 1 or 2 years.

Isochromosomes

Contents

10.1 Isochromosomes – 254

10.2 Summary – 269

C. Behrend et al., *Human Chromosome Atlas*, https://doi.org/10.1007/978-3-031-10588-3_10

10.1 Isochromosomes

67. 47,XY,+i(18)(p10)
68. 46,XX,i(18)(q10)
69. 46,X,i(X)(q10)
70. mos 45,X[26]/46,X,idic(X)(p11.23)[24],
71. 46,X,idic(Y)(q11.23).

■■ **Case 67: Karyotype: 47,XY,+i(18)(p10)**

Indication for chromosome analysis: Developmental retardation, minor facial dysmorphisms.

■ **Morphology**

The additional isochromosome shows a characteristic structure. Only in the presence of a second centromere the existence of a second cell line without the isochromosome that is with normal karyotype has to be taken into account. The frequency of this aberration is about 1:45.000.

■ **Cave**

The majority of additional isochromosomes 18 are de novo and paternal in origin. A few familial cases have been published. Therefore, chromosome investigations in the parents should be performed before genetic counselling (■ Fig. 10.1).

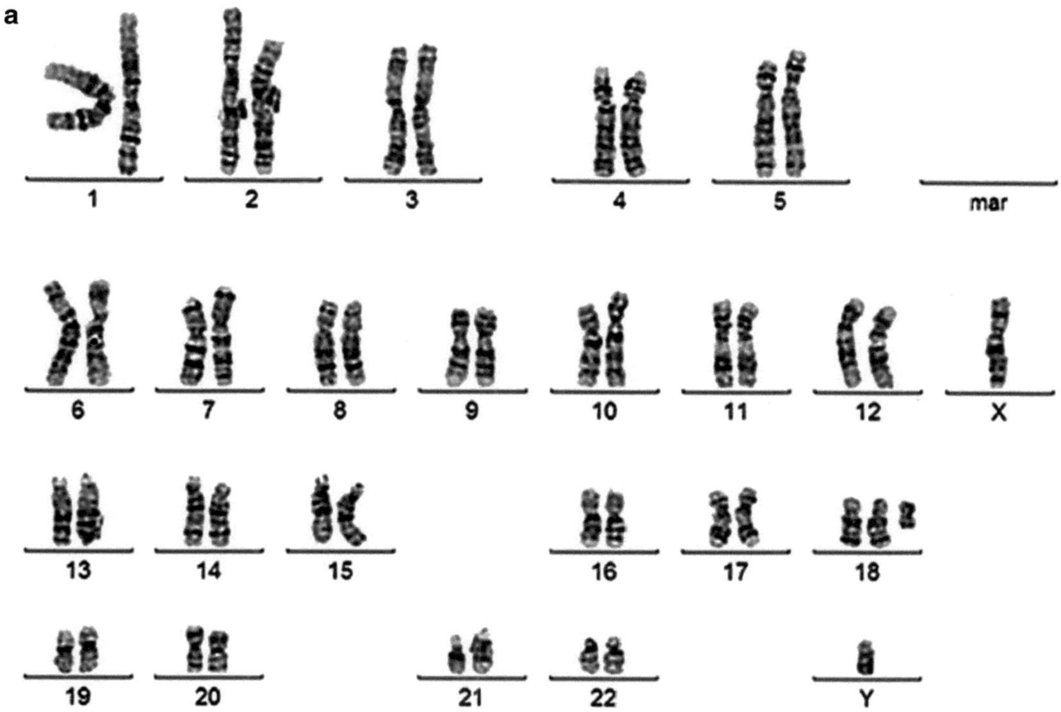

■ **Fig. 10.1** **a** Karyotype 47,XY,+i(18)(p10), GTG-banding, 500 bands per genome, **b** ideograms of normal and aberrant chromosomes 18; aneuploidy by additional isochromosome 18p or tetrasomy 18p, *left* normal, *right* isochromosome p

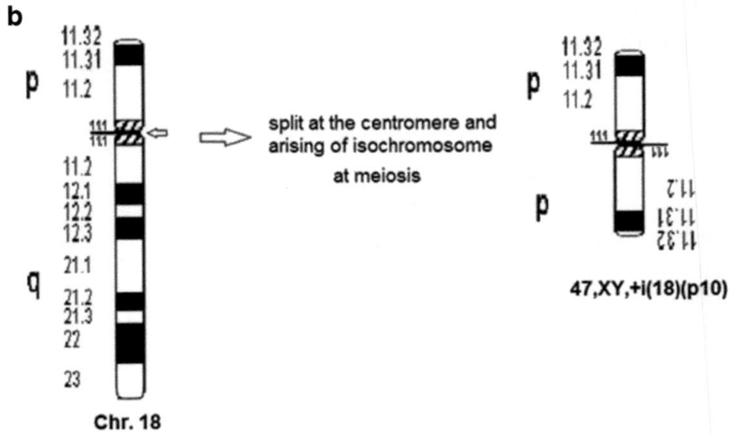

Fig. 10.1 (continued)

▪▪ Case 68: Karyotype: 46,XX,i(18)(q10)

Indication for chromosome analysis: Prenatal growth retardation, severe malformations, mild trisomy 18 features.

▪ Morphology

The aberrant structure i(18q) is so easy so recognize, that even a resolution of less than 550 bands per genome permits the analysis. This is relevant, as prenatal investigations after CVS are performed and the ultrasound shows early abnormalities of the conceptus. The frequency of the aberration is less than 1:90.000.

▪ Cave

Single familial cases have been observed, one with a parental inversion giving rise to the rearrangement. Before genetic counselling the parents should be investigated (◘ Fig. 10.2).

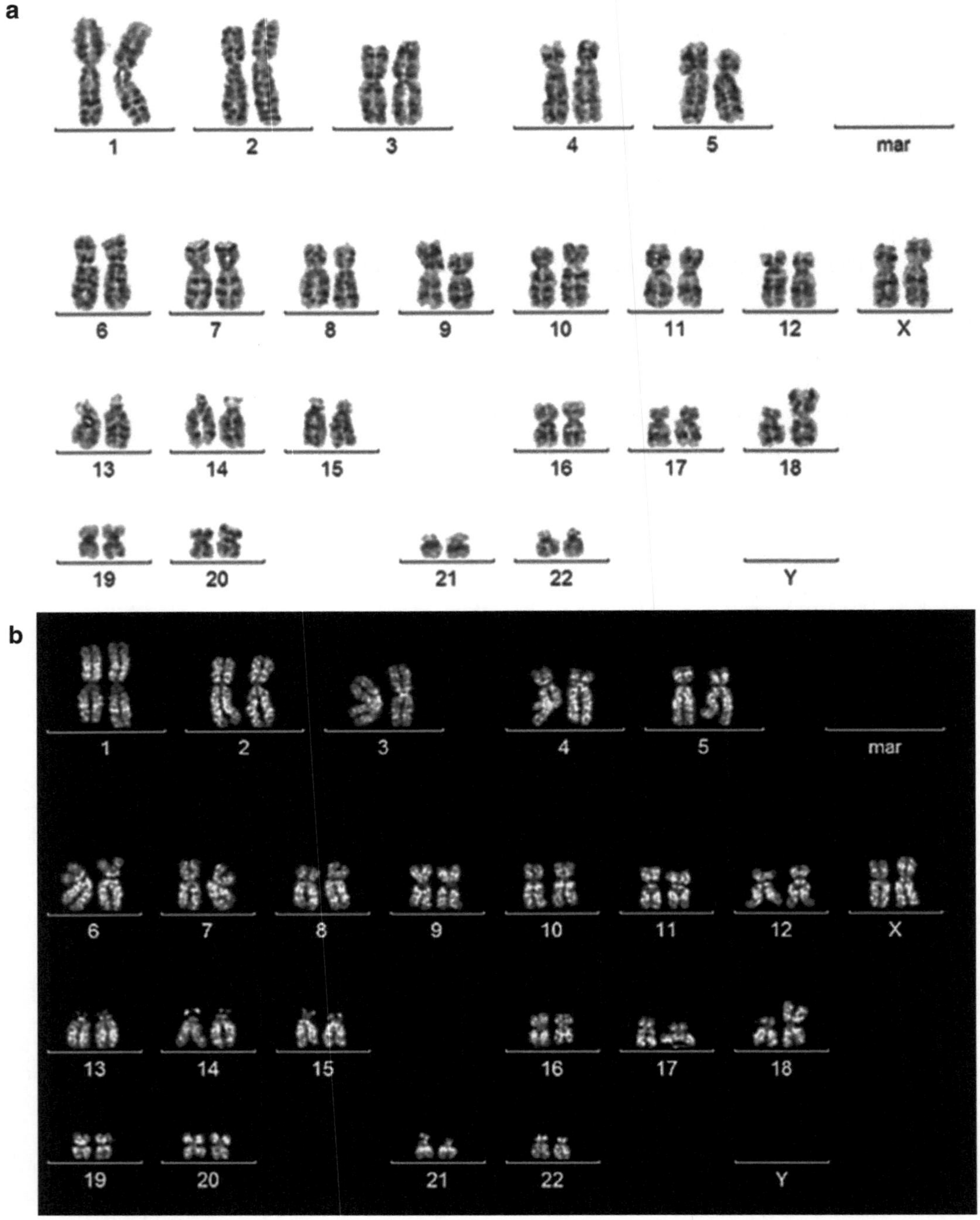

■ **Fig. 10.2** Karyotype 46,XX,i(18)(q10) **a** GTG-, **b** QFQ- banding, each 400 bands per genome, **c** QFQ-, **d** GTG-banding, chromosomes from different mitoses, *left* normal, *right* aberrant chromosome 18, **e** ideograms of normal (*left*), aberrant chromosomes 18 (*right*)

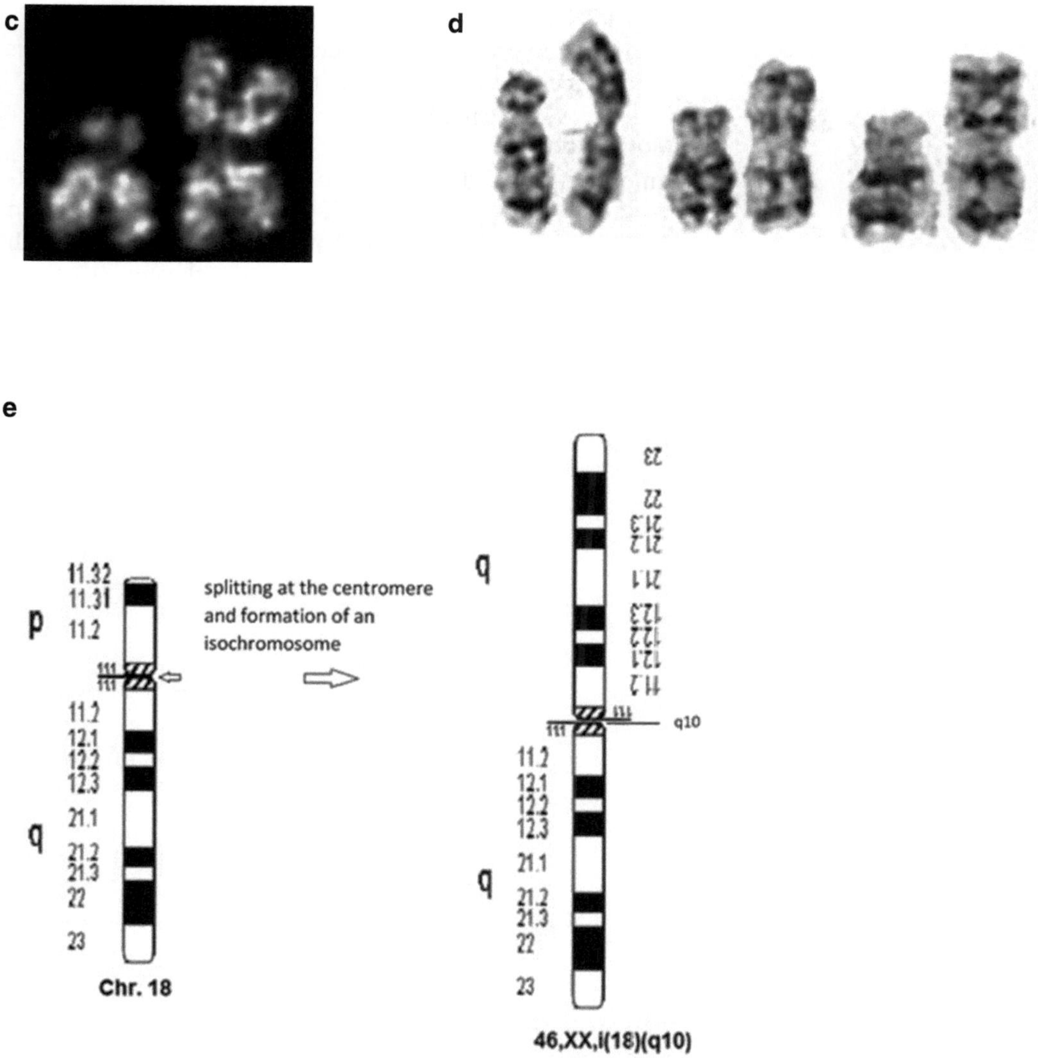

Fig. 10.2 (continued)

▪▪ Case 69: Karyotype: 46,X,i(X)(q10)

Indication for chromosome analysis: Prenatal diagnosis because of increased maternal age (40 years).

▪ Morphology

The monocentric and metacentric isochromosome is consisting of the two long arms of the X-chromosome. This leads to a trisomy Xq and a monosomy Xp. The isochromosome with one centromere was inherited unchanged in the course of cell divisions.

▪ Cave

The child will show the phenotype of Turner syndrome (◘ Fig. 10.3).

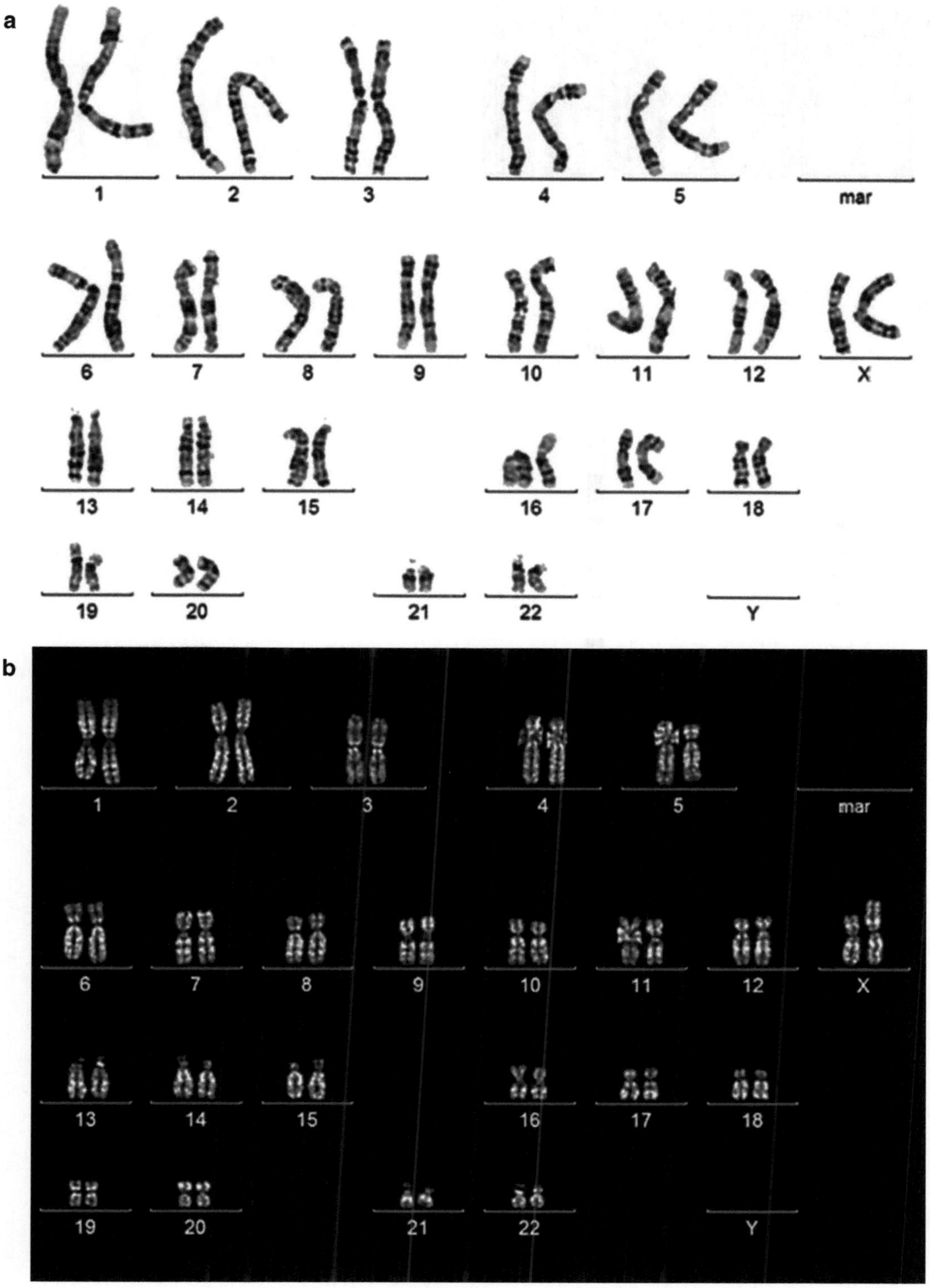

Fig. 10.3 Karyotype 46,X,i(X)(q10). **a** GTG-, **b** QFQ-banding, 550 and 400 bands per genome, **c** QFQ-, **d** GTG-banding, chromosomes from different mitoses, **c** *left* normal, *right* aberrant X-chromosome, **e** ideograms of normal and aberrant X-chromosomes

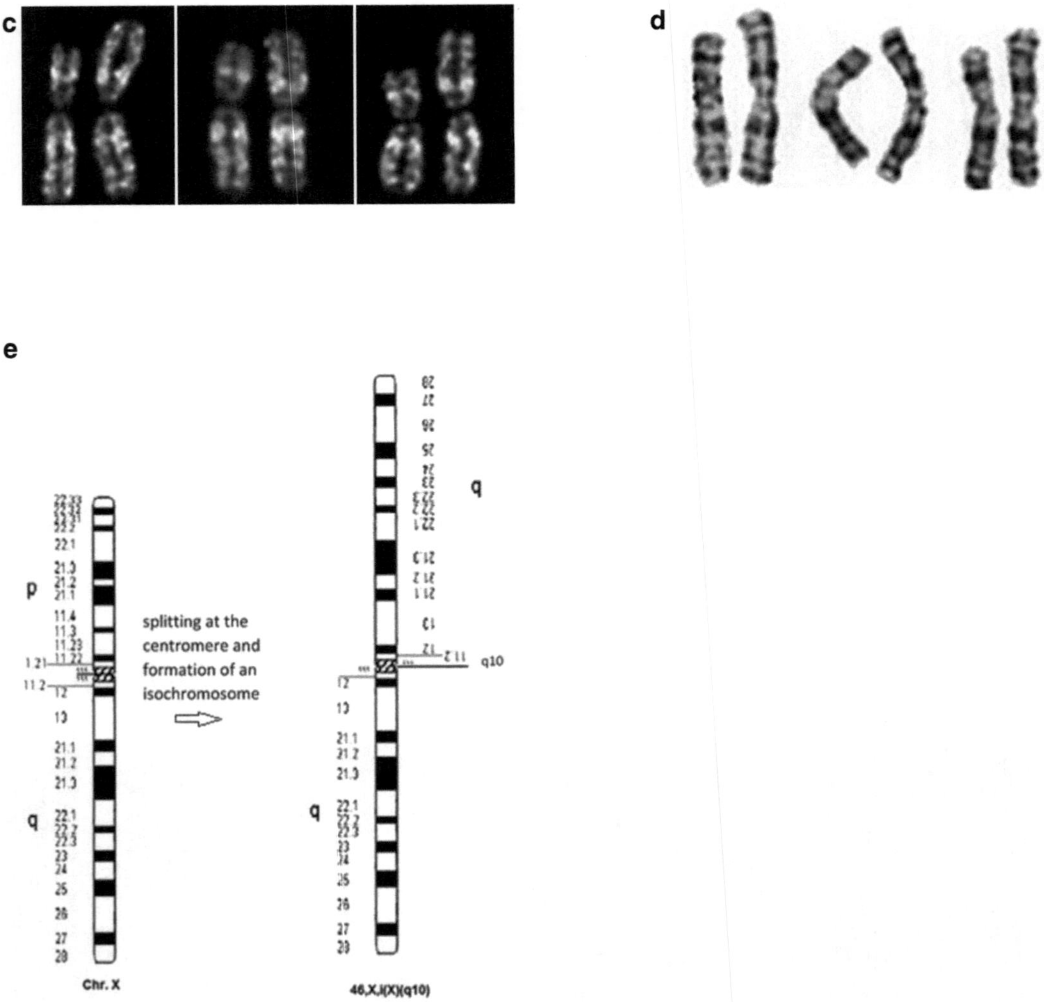

■ **Fig. 10.3** (continued)

■■ Case 70: Karyotype: mos 45,X[26]/46,X,idic(X)(p11.23)[24]

Indication for chromosome analysis: Developmental retardation of a 7-year-old girl.

■ Morphology

The isochromosome X is dicentric with breakpoint in the proximal part of the short arm. The inactivation of the second centromere (see also C-banding) is incomplete leading to a second cell line with monosomy X.

■ Cave

The characteristic phenotype of Turner syndrome is often not recognisable in young girls. But to the parents the prognosis has to be explained and especially the given possibilities of therapy have to be discussed (◘ Fig. 10.4).

a

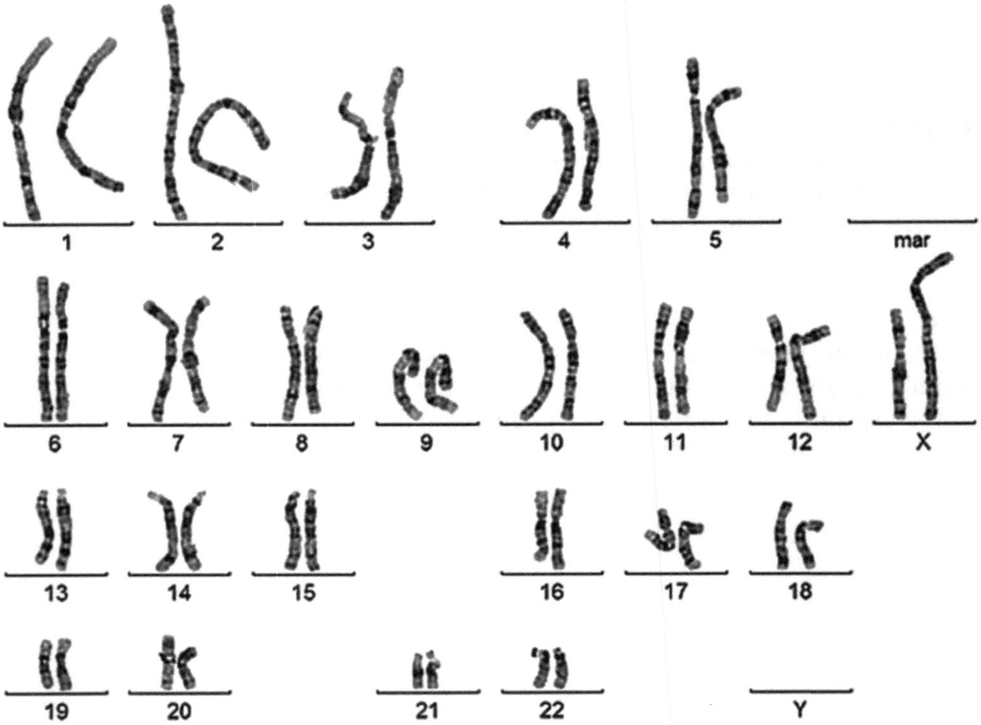

b

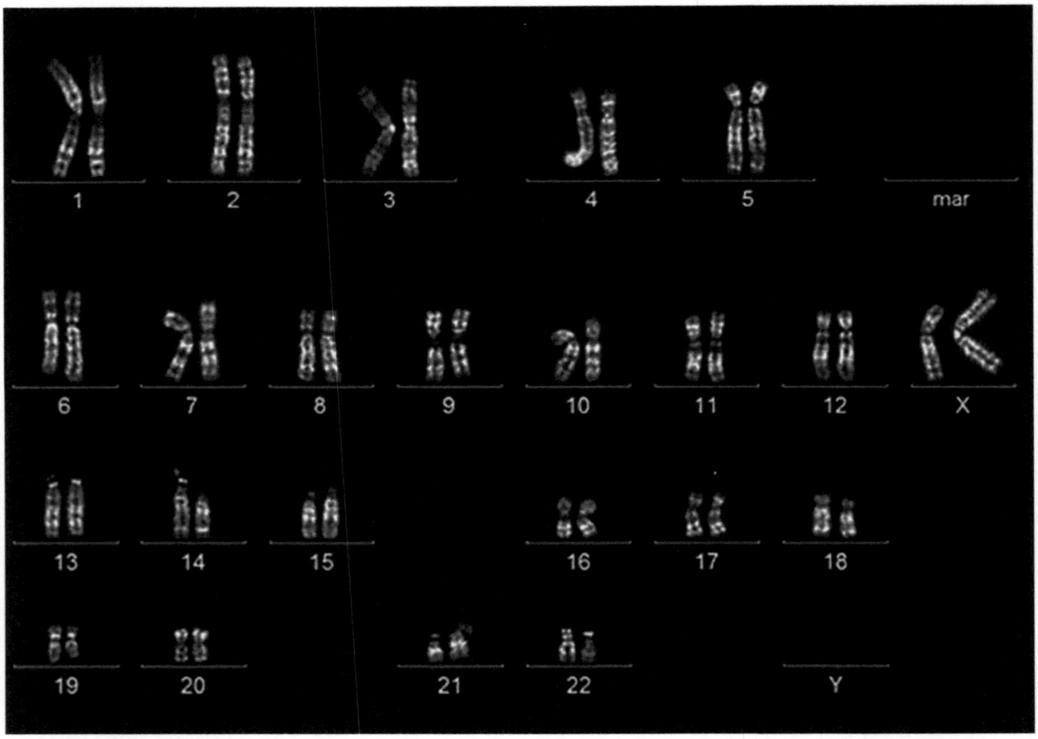

■ **Fig. 10.4** Karyotype mos 45,X[26]/46,X,idic(X)(p11.23)[24]. **a** GTG-, **b** QFQ-banding, 700 and 400 bands per genome, **c** CBG-, **d** QFQ-, **e** GTG-banding, chromosomes from different mitoses, left normal, right aberrant X-chromosome, **f** ideograms of the X-chromosome deletion of the short and duplication of the long arm. Breakpoint in the proximal short arm, resulting in a dicentric derivate

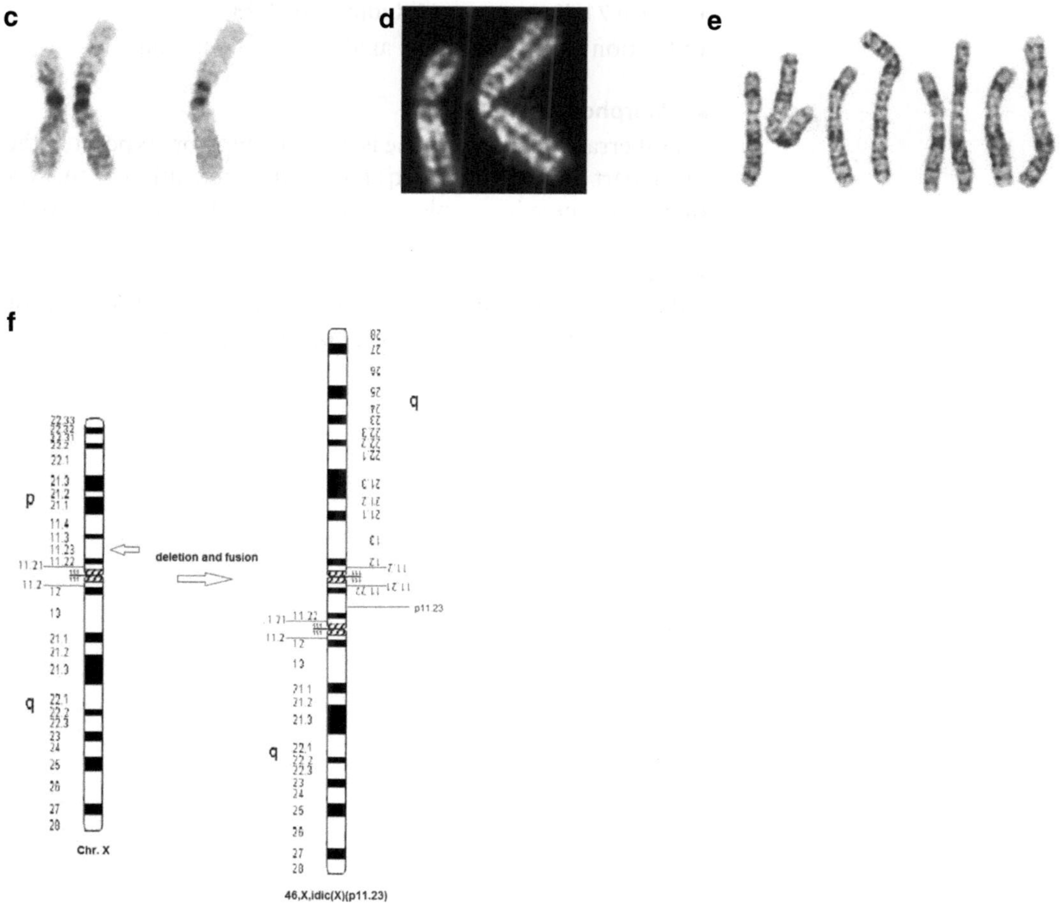

Fig. 10.4 (continued)

Case 71: Karyotype: 46,X,idic(Y)(q11.23)

Indication for chromosome analysis: Azoospermia.

Morphology

The aberrant Y-chromosome is dicentric with breakpoint in the distal part of the long arm (q11.23). The inactivation of the second centromere is complete, no further cell line was observed.

Cave

Additional molecular investigations revealed deletions of sY254, sY255 and the AZFc-region (◼ Fig. 10.5).

a

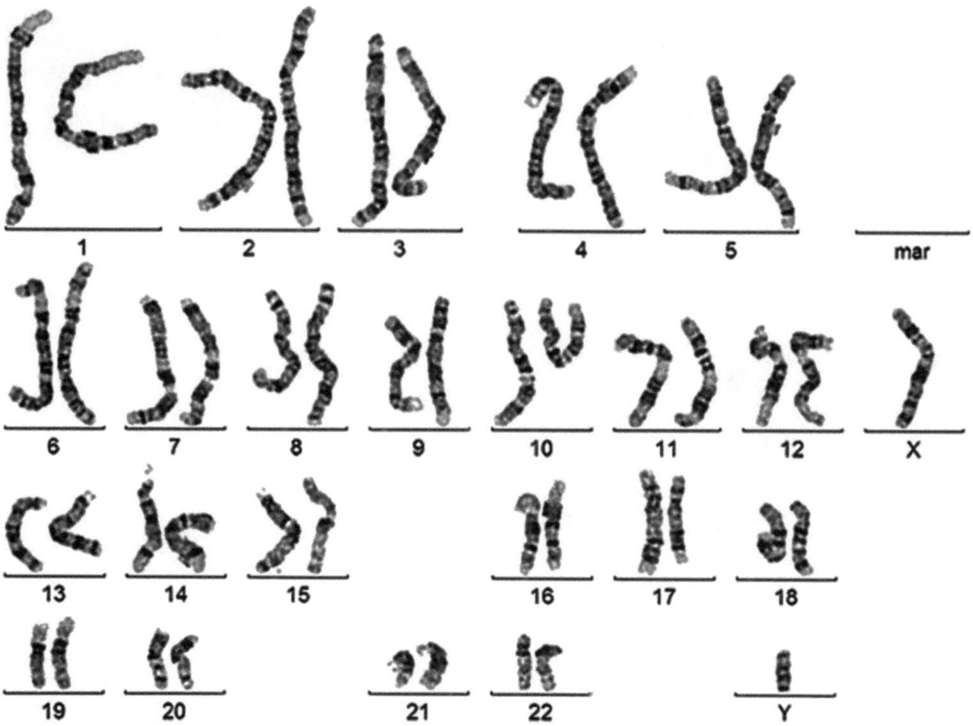

b

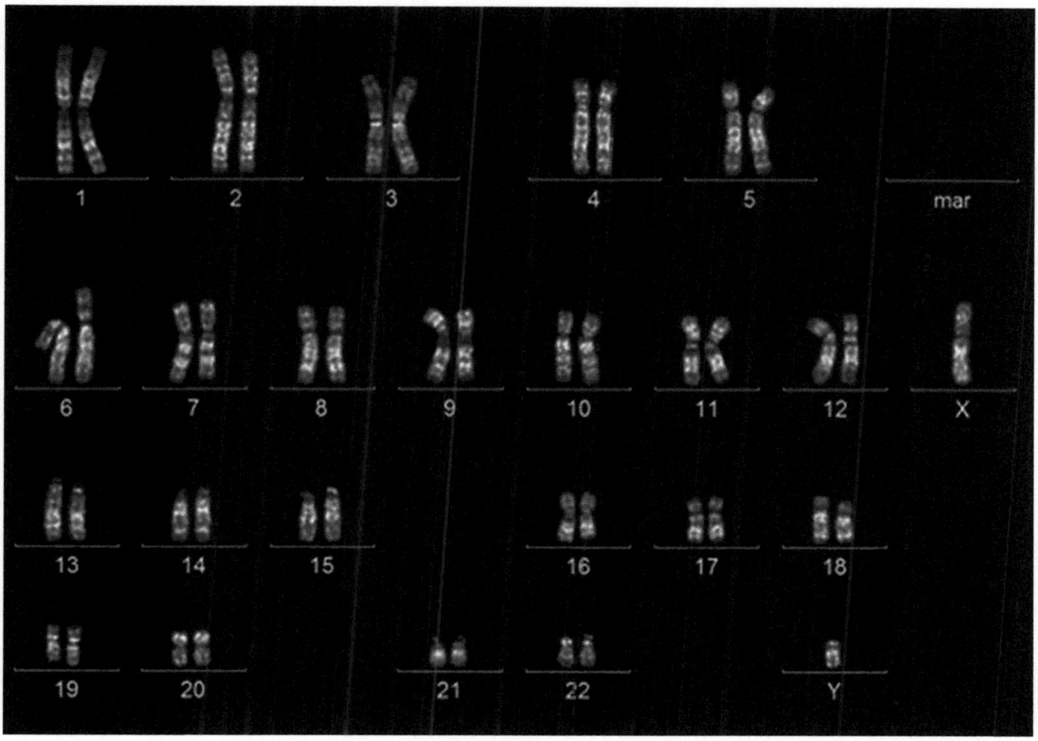

◘ Fig. 10.5 Karyotype 46,X,idic(Y)(q11.23). **a** GTG-, **b** QFQ-banding, 700 and 400 bands per genome, **c** CBG-, **d** QFQ-, **e** GTG-banding, chromosomes from different mitoses, **d/1**: normal Y-chromosome from a control person, **d/2**: aberrant Y-chromosome with 1 centromere inactivated, **f** ideograms of the Y-chromosome, deletion of the terminal long arm and duplication of the short and proximal long arm, one centromere inactivated

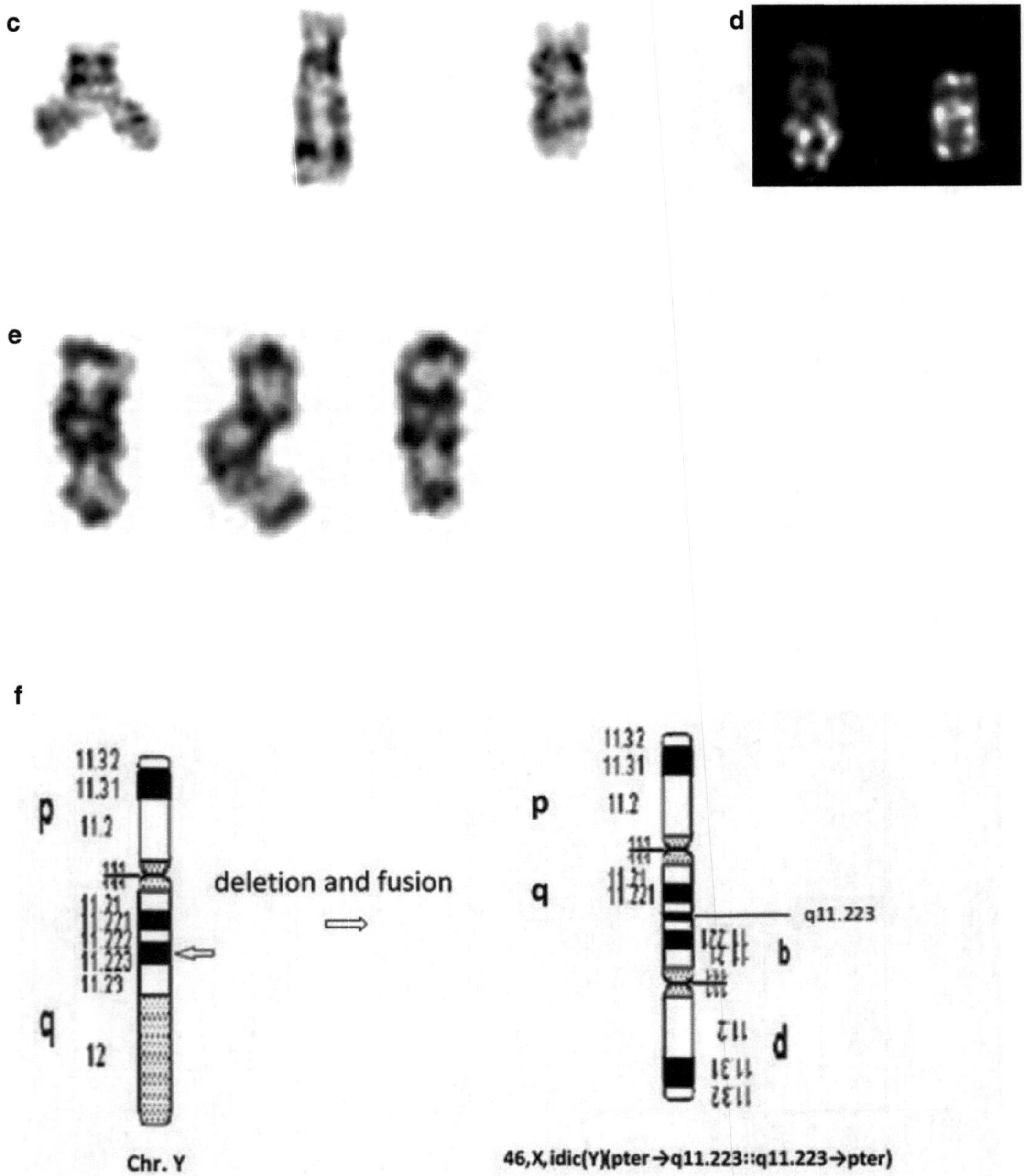

Fig. 10.5 (continued)

10.2 Summary

Exemplary five cases of isochromosomes—two of the autosomes and three of the gonosomes are presented in detail. These are two aberrations of chromosome 18, two of the X- and one of the Y-chromosome. The first case is an additional isochromosome of the short arms of chromosome 18. This type of supernumerary derivate 18 is a frequent abnormality, of which about 100 cases have been reported. This isochromosome is usually monocentric and thus stable inherited by mitoses. The second case is an isochromosome of the long arms of chromosome 18, again monocentric and stable when inherited. The chromosome number is 46, thus it combines a monosomy 18p with trisomy 18q. As it leads to a severe malformation complex in the carrier, life expectancy is significantly reduced. The third case concerns a monocentric isochromosomes X of the long arms (q). This is a frequent structural aberration of the X-chromosome. Monocentric isochromosomes X are in the majority of cases stable in mitosis, in contrast to dicentric rearrangements. The development of mosaicism by a second cell line with the karyotype 45,X has not to be expected. But this feature was observed in the second case of chromosome X aberration presented here, where the dicentric isochromosome X with breakpoint in Xp11.23 leads to the formation of a second cell line with karyotype 45,X. The reason for this instability is the incomplete inactivation of the second centromere. In the last case, a dicentric Y-chromosome is presented with breakpoint in the distal euchromatic region of the long arm (Yq11.23). By CBG-banding it was clearly analysable that the second centromere is always completely inactivated. Therefore, this isochromosome is stable inherited during mitoses. A second cell line does not exist.

The cases presented here, make it obvious, that the inactivation of centromeres in dicentric chromosomes can lead to a defect in the process of the cell cycle and therefore always must be analysed by CBG-banding. If abnormalities of inactivation are stated, at least 50 mitoses have to be analysed, to find out if quantitative and qualitative changes of the karyotype exist, caused by mosaic formation.

Single Case Presentations of Interchromosomal Rearrangements

Contents

Chapter 11 Translocations – 273

Chapter 12 Insertions –309

Chapter 13 Complex Chromosome Rearrangements (CCR) –315

Chapter 14 Marker Chromosomes –327

Chapter 15 Heteromorphismes: Mutations in Non-Coding-DNA Regions –339

Translocations

Contents

11.1 Reciprocal Translocations – 274

11.2 Summary – 303

11.3 Robertsonian Translocations – 304

11.4 Summary – 308

C. Behrend et al., *Human Chromosome Atlas*, https://doi.org/10.1007/978-3-031-10588-3_11

11.1 Reciprocal Translocations

72. 46,XY,t(2;18)(p25.3;p11.2)
73. 46,XX,t(4;16)(q31.3;q13)
74. 46,XX,t(9;11)(q11;p11.2), ish t(9;11)(cep9+;cep11+)
75. 46,XY,der(9),t(9;12)(p24.3;q24.13),arr 12q24.13q24.33 (114,023,119-133,773,499)x3 de novo,
76. 46,XX,t(11;22)(q23.3;q11.2)
77. 46,XX,t(12;18)(q10;p11.2)
78. 45,XY,der(18),t(18;22);(p11.32;q11.2),−22
79. 46,X,t(X;6)(q26.1;p25.1)
80. 46,X,t(Y;5)(q11.21;p15.33).

■■ Case 72: Karyotype: 46,XY,t(2;18)(p25.3;p11.2)

Indication for chromosome analysis: five handicapped children in the family (see pedigree).

■ Morphology

The two translocation chromosomes are not easy to identify as the exchanged regions are small. This requires a high banding resolution for identification (at least 700 bands per genome) and in prenatal diagnosis an additional microarray investigation is required.

■ Cave

The phenotype of the carriers of the same type of unbalanced translocation is different in the severity of clinical symptoms. Thus, a prognosis during the prenatal period must include this restriction (■ Fig. 11.1).

a

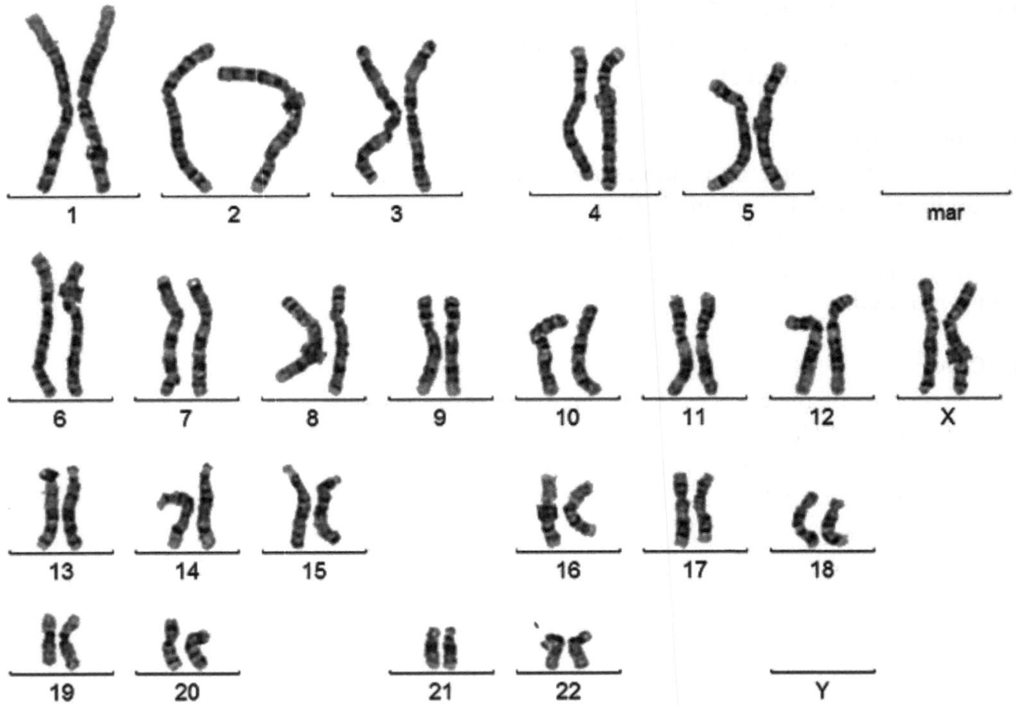

b

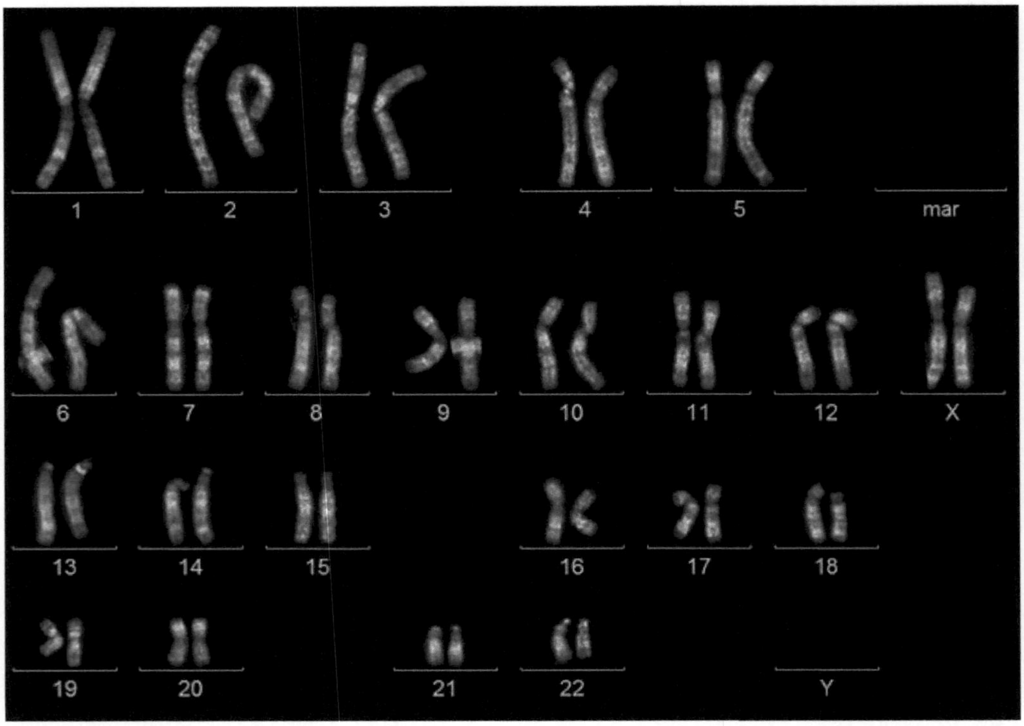

Fig. 11.1 Karyotype 46,XY,t(2;18)(p25.3;p11.2). **a** GTG-, **b** QFQ-banding, 600 and 500 bands per genome, **c** QFQ-, **d** GTG-banding, chromosomes from different mitoses, *left* normal, *right* aberrant chromosomes 2 and 18, **e** ideograms of normal and aberrant chromosomes 2 and 18, balanced translocation, **f** Pedigree showing the inheritance of 2/18 translocation in four generations of the family

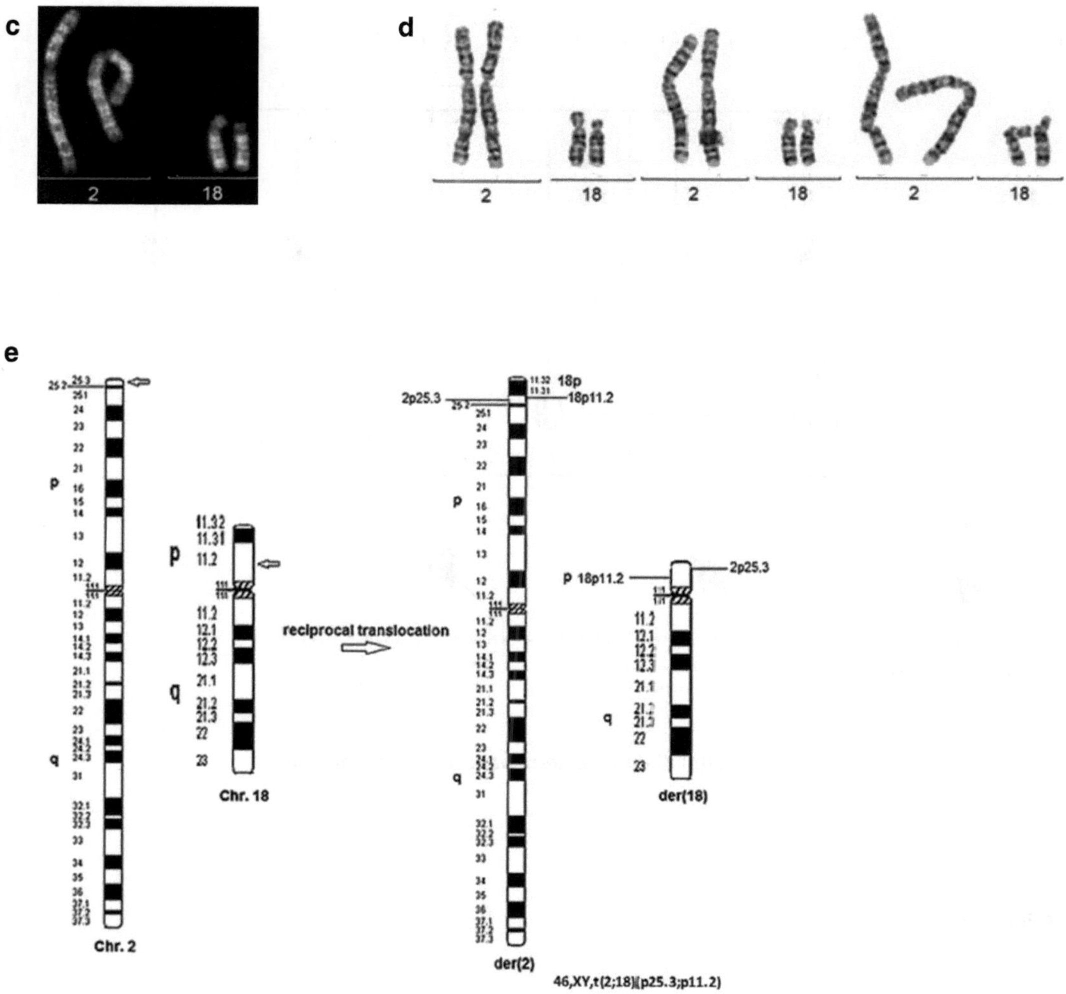

Fig. 11.1 (continued)

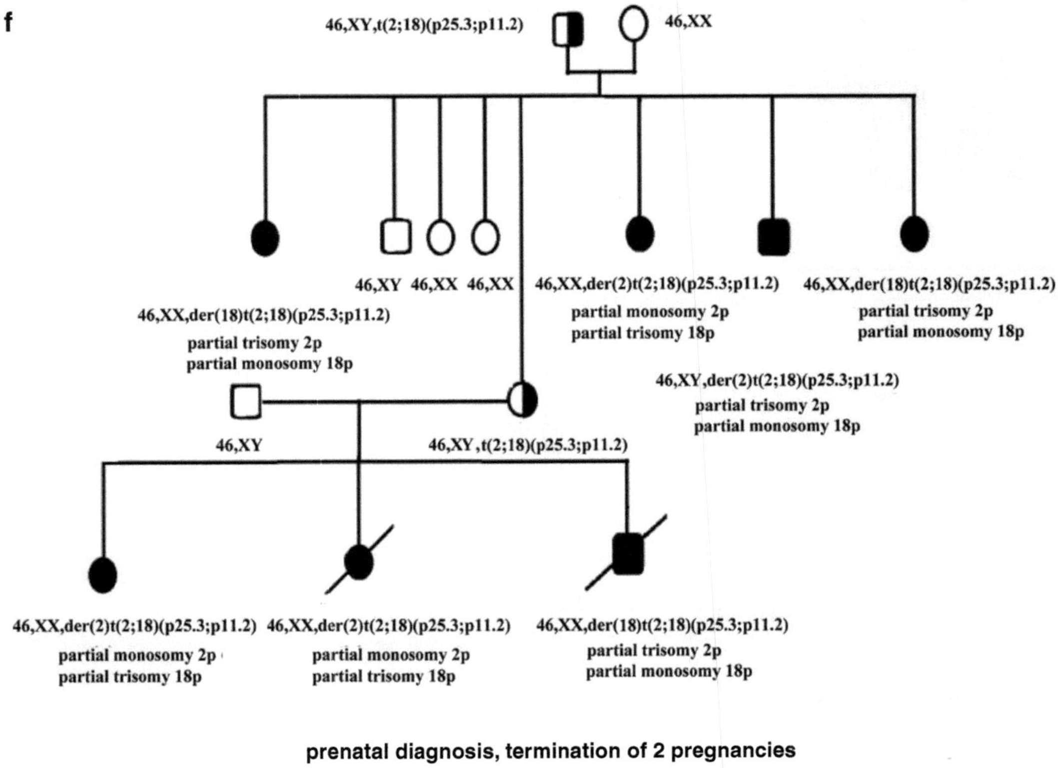

Fig. 11.1 (continued)

▪▪ Case 73: Karyotype: 46,XX,t(4;16)(q31.3;q13)

Indication for chromosome analysis: Sterility.

▪ Morphology

The regions exchanged comprise about 20% (chromosome 4) and 30% (chromosome 16) of the two translocation chromosomes. Thus, they can be easily recognised by their changed size and altered centromere index. For exact localisation of the breakpoints, a banding resolution of at least 700 bands per genome is recommended.

▪ Cave

Family investigations should be offered before genetic counselling as the translocation might be familiar in origin. The different types of unbalanced translocation and their developmental possibilities must be discussed (◘ Fig. 11.2).

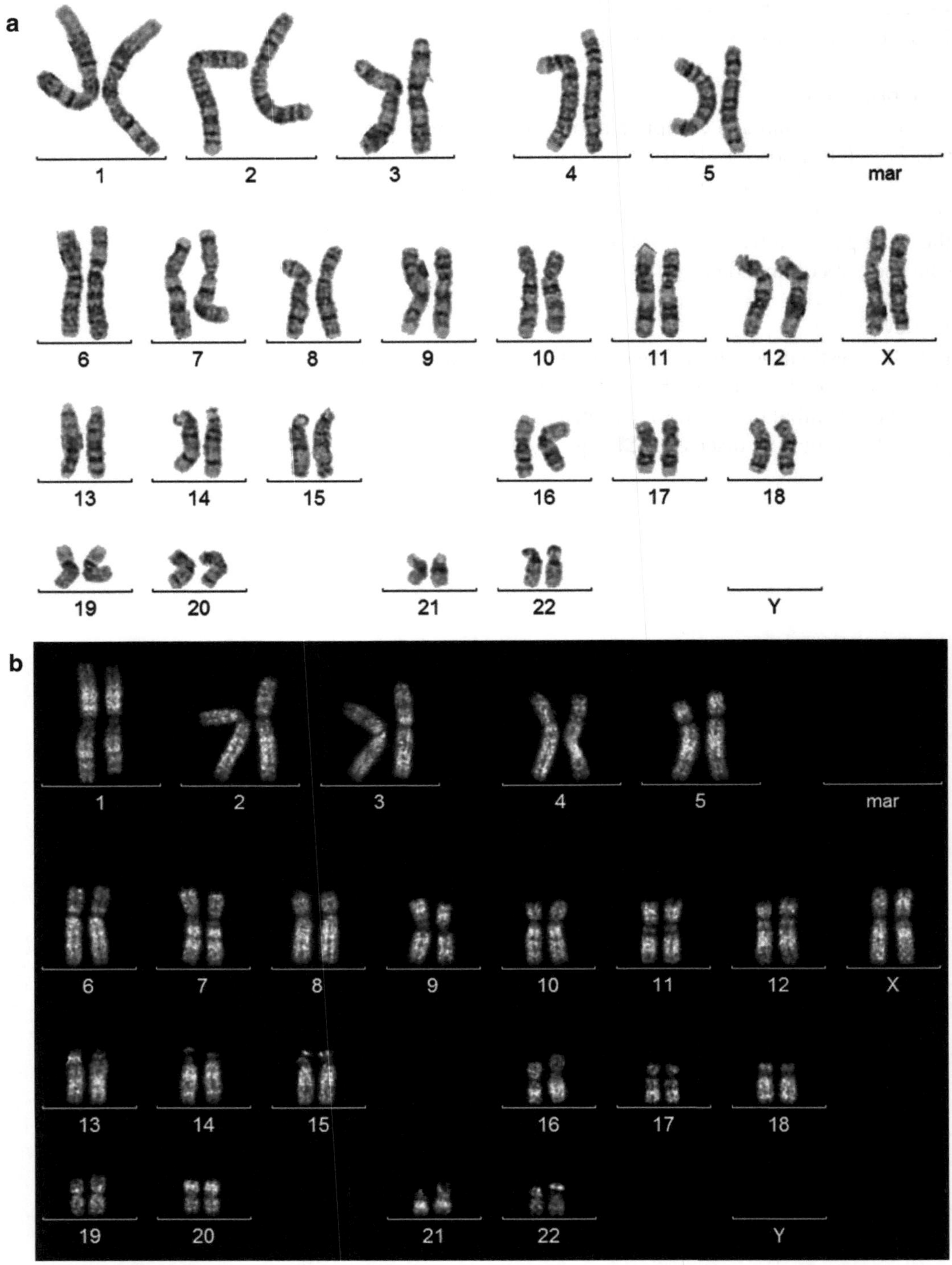

Fig. 11.2 Karyotype 46,XX,t(4;16)(q31.3;q13). **a** GTG-, **b** QFQ-banding, 550 and 400 bands per genome, **c** GTG-banding, chromosomes from different mitoses, *left* normal, *right* aberrant chromosomes 4 and 16, **d** ideograms of normal and aberrant chromosomes 4 and 16, balanced translocation

c
d
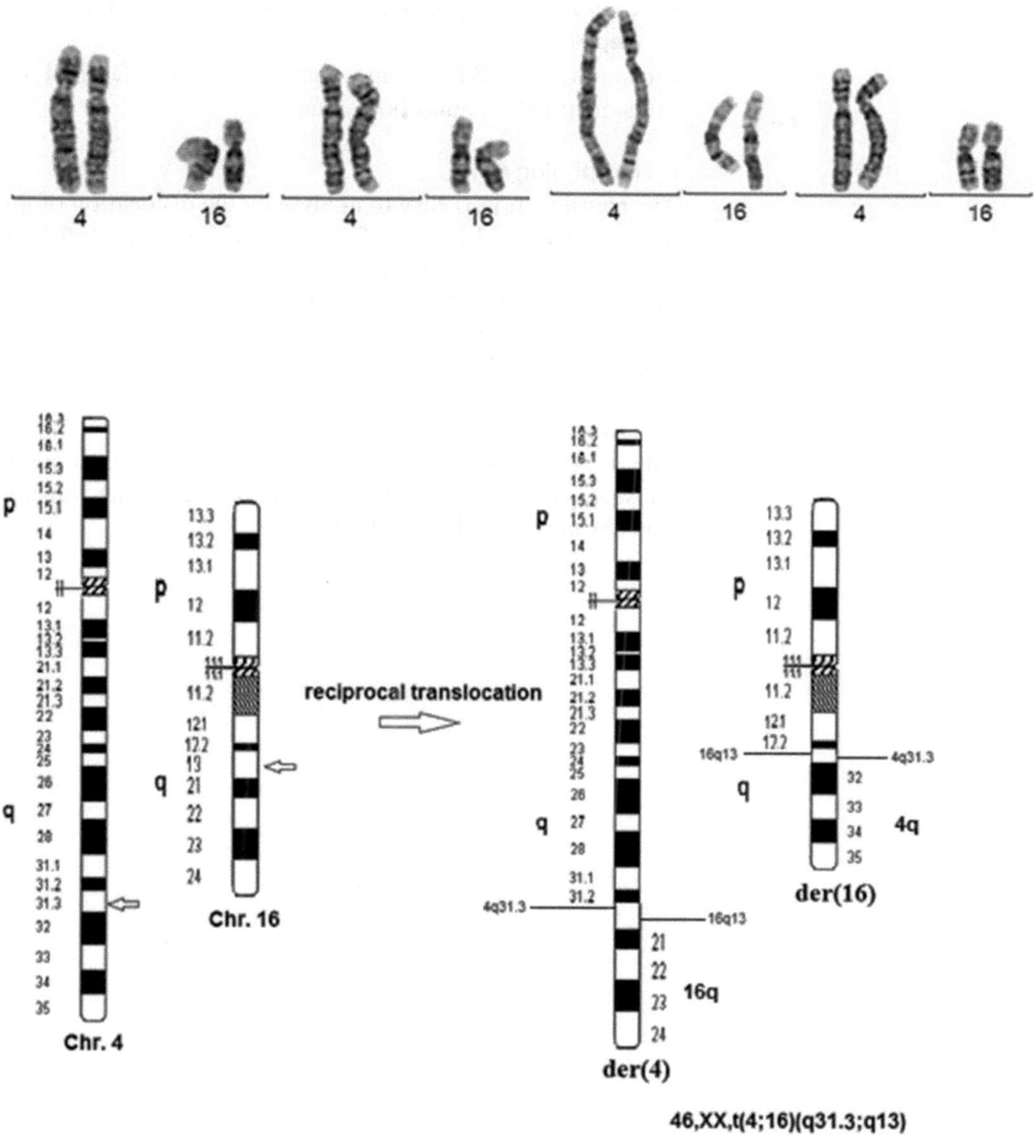

Fig. 11.2 (continued)

▪▪ Case 74: Karyotype: 46,XX,t(9;11)(q11;p11.2) ish t(9;11)(cep9+;cep11+)

Indication for chromosome analysis: Known familial rearrangement before polar body analysis.

▪ Morphology

The translocation is easy to analyse by the dislocation of 9q12 into the proximal region of 11p (see C-banding) and by the changed size and centromere index of the translocation chromosomes, which both appear almost metacentric. The size of the regions exchanged is about 66% in chromosome 9 and 35% in chromosome 11.

▪ Cave

The risk of an unbalanced karyotype is increased by a possible malsegregation, i.e. aneuploidisation, of the smaller translocation chromosome (◘ Fig. 11.3).

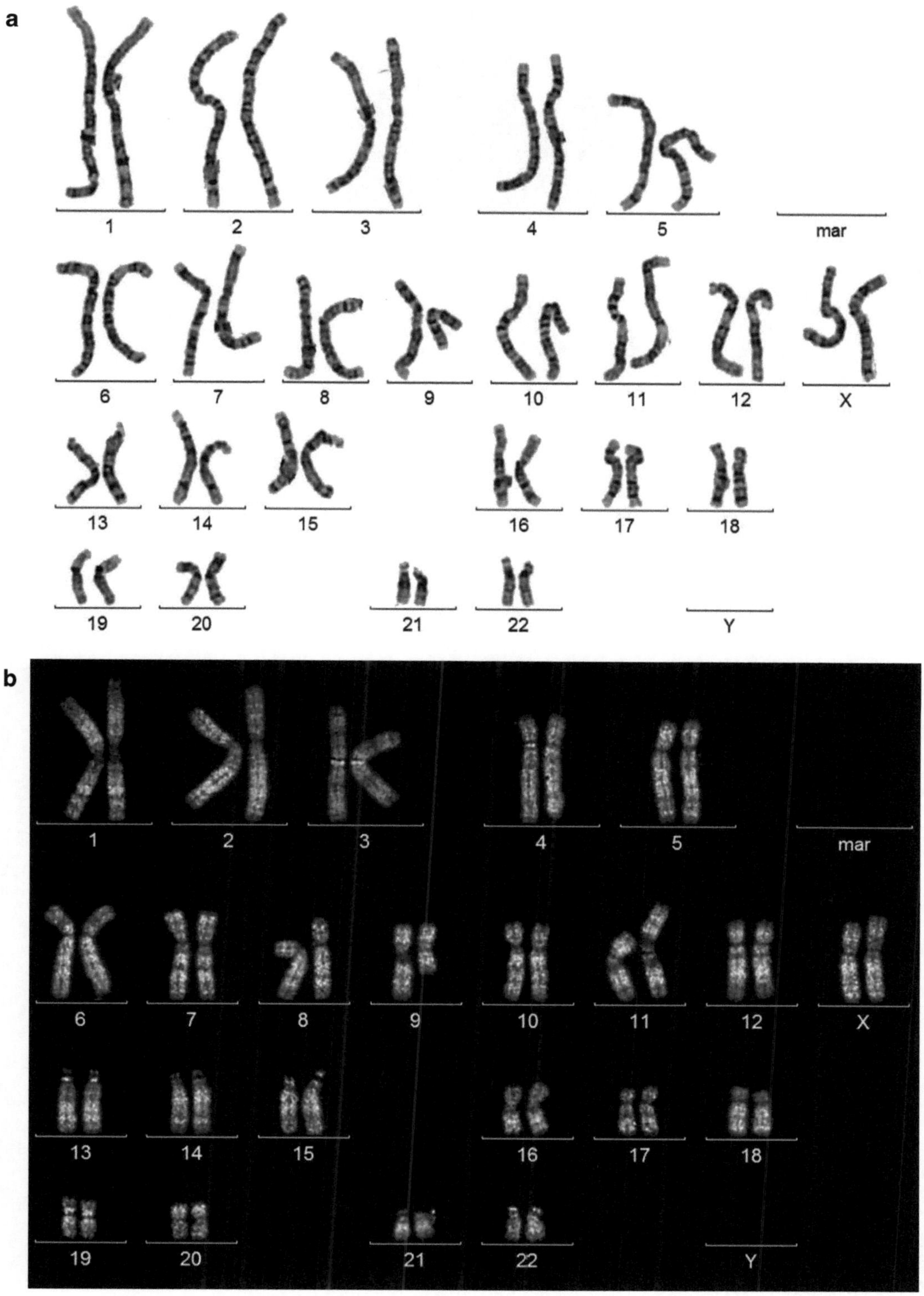

Fig. 11.3 Karyotype 46,XX,t(9;11)(q11;p11.2). **a** GTG-, **b** QFQ-banding, 700 and 400 bands per genome, **c** CBG-banding, **d** QFQ-banding, **e** GTG-banding chromosomes from different mitoses; **d, e** *left* normal, *right* aberrant chromosomes 9 and 11, **f** ideograms of normal and aberrant chromosomes 9 and 11, balanced translocation

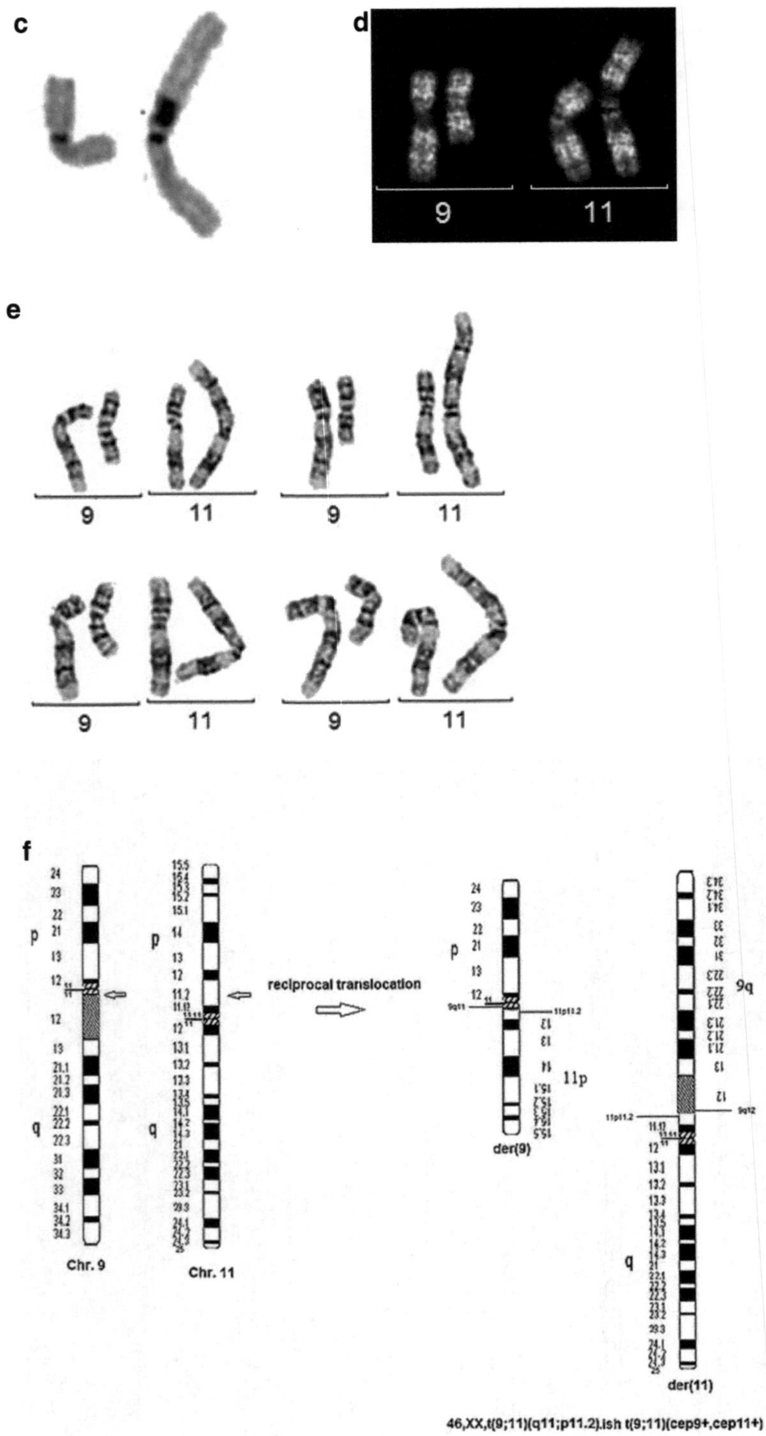

Fig. 11.3 (continued)

▪▪ Case 75: Karyotype: 46,XY,der(9),t(9;12)(p24.3;q24.13),arr 12q24.13q24.33(114,023,119-133,773,499)x3 de novo

Indication for chromosome analysis: Developmental retardation, dysmorphic features.

▪ Morphology

One chromosome 9 is elongated in the distal region of the short arm by about 10% of its total length. It can be recognised by the altered banding pattern and the more metacentric structure.

▪ Cave

After analyses of normal parental karyotypes, this unbalanced translocation could only be identified exactly by microarray investigation (◘ Fig. 11.4).

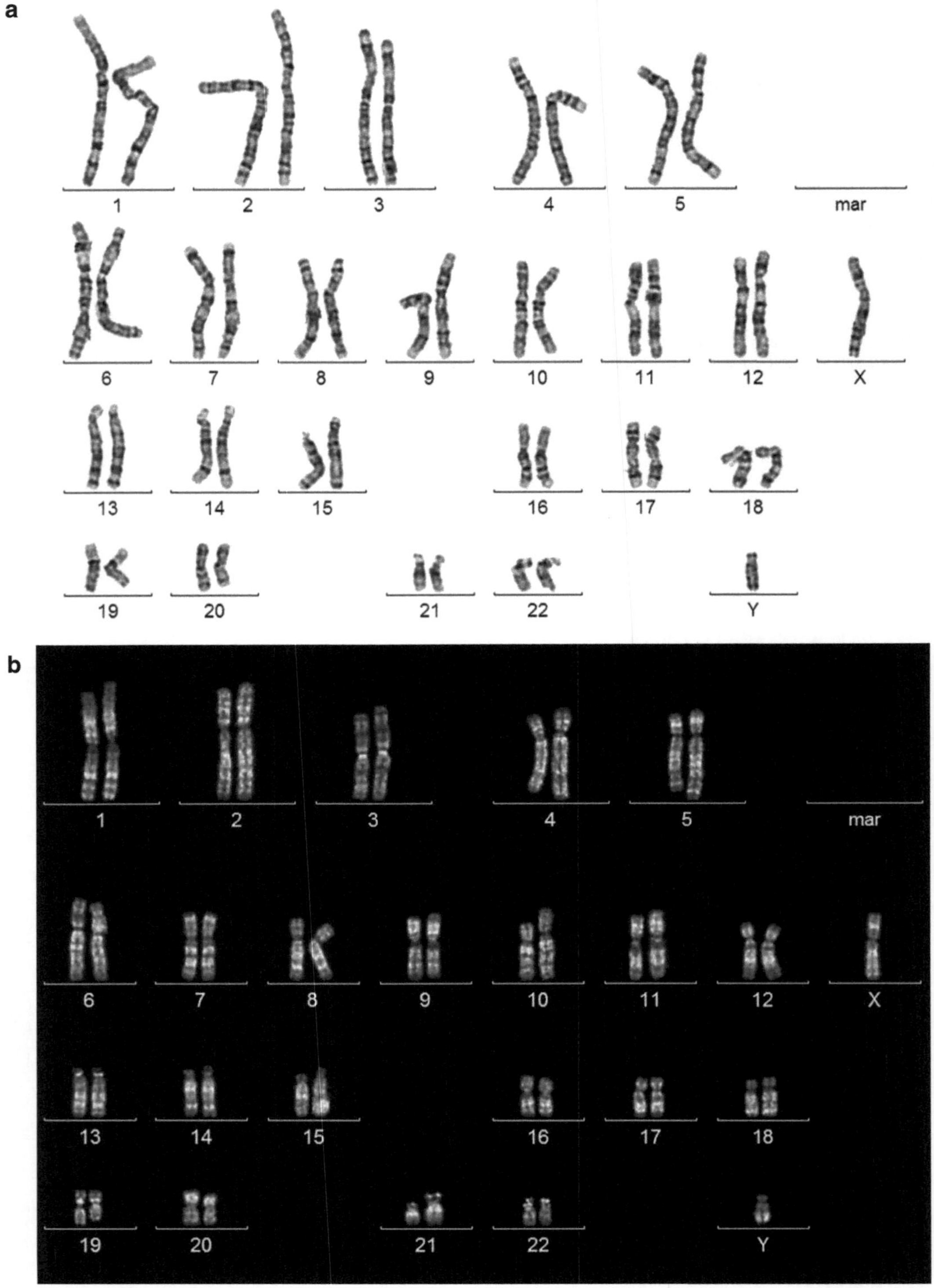

Fig. 11.4 Karyotype 46,XY,der(9),t(9;12)(p24.3;q24.13). **a** GTG-, **b** QFQ-banding, 700 and 400 bands per genome, **c** CBG-, **d** QFQ-, **e** GTG-banding chromosomes from different mitoses, *left* normal, *right* aberrant chromosome 9, **f** ideograms from normal 9 and 12 and aberrant chromosome der(9), unbalanced translocation

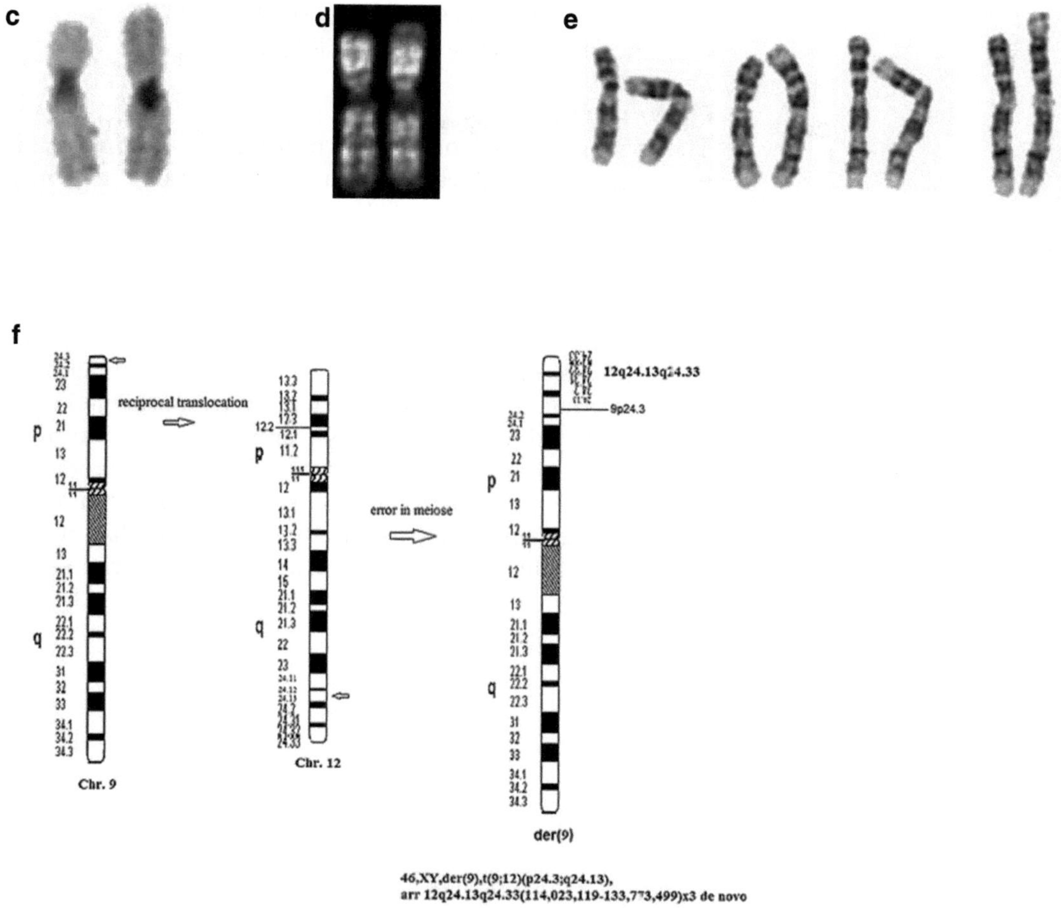

Fig. 11.4 (continued)

■■ Case 76: Karyotype: 46,XX,t(11;22)(q23.3;q11.2)

Indication for chromosome analysis: Abortions.

■ Morphology

The terminal regions of the two translocation chromosomes have been exchanged leading to an altered banding pattern in these regions, and the size of them has changed.

■ Cave

This is the most frequent reciprocal translocation in humans. The unbalanced translocation usually leads to abortion. But the derivative chromosome 22 can be malsegregated in meiosis thus occurring as supernumerary marker chromosome in the descendant. This aberration is comparable with life and presents with characteristic clinical picture (Emanuel syndrome).

Besides, in single carriers of the balanced translocation the breakpoint in chromosome 11 disrupts a gene leading to the monogenic defect of acute intermittent porphyria (AIP) (◘ Fig. 11.5).

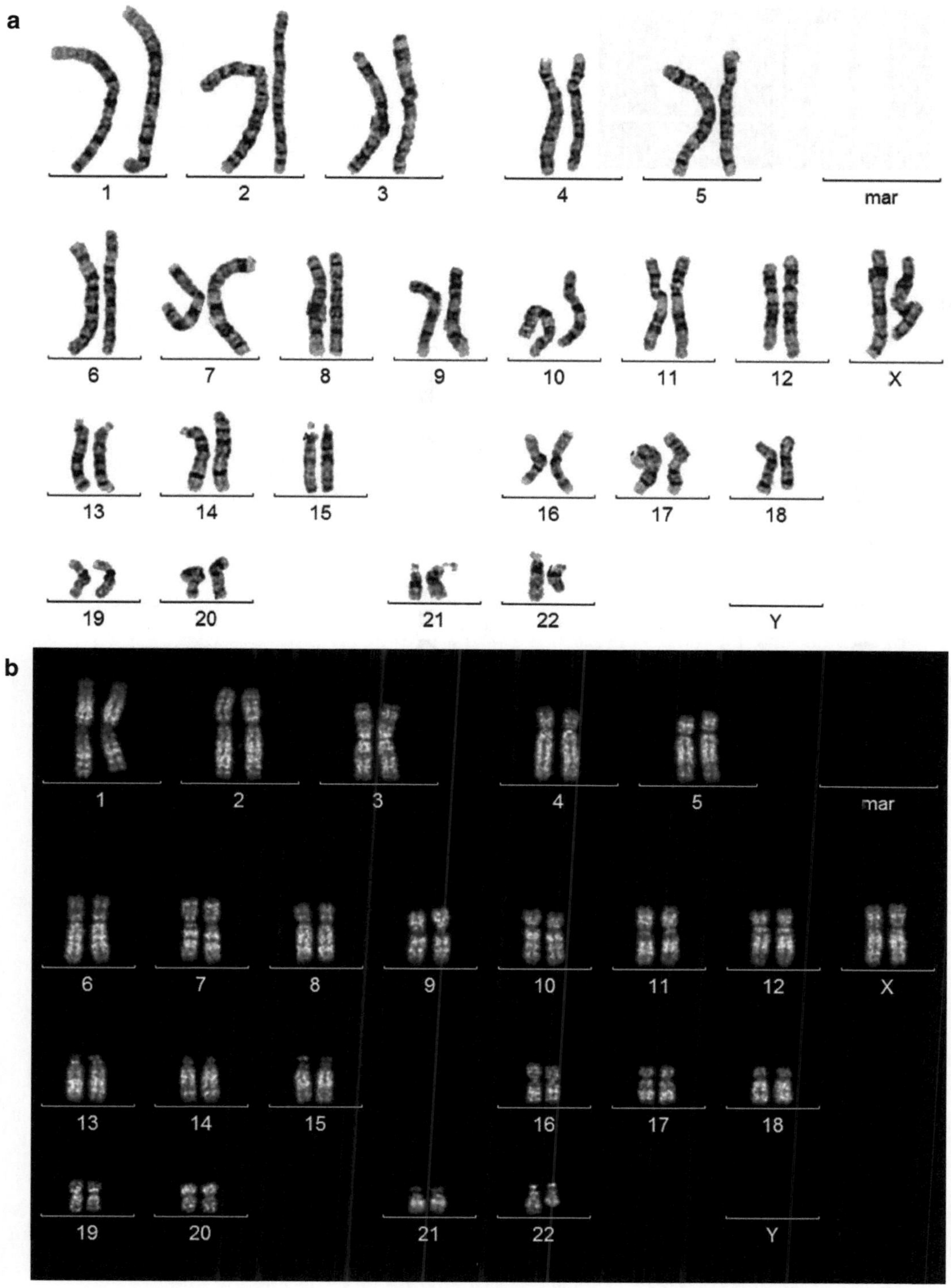

Fig. 11.5 Karyotype 46,XX,t(11;22)(q23.3;q11.2). **a** GTG-, **b** QFQ-banding, 700 and 400 bands per genome, **c** QFQ-banding, **d** GTG-banding chromosomes from different mitoses; *left* normal, *right* aberrant chromosomes 11 and 22, **e** ideograms of normal and aberrant chromosomes 11 and 22, balanced translocation

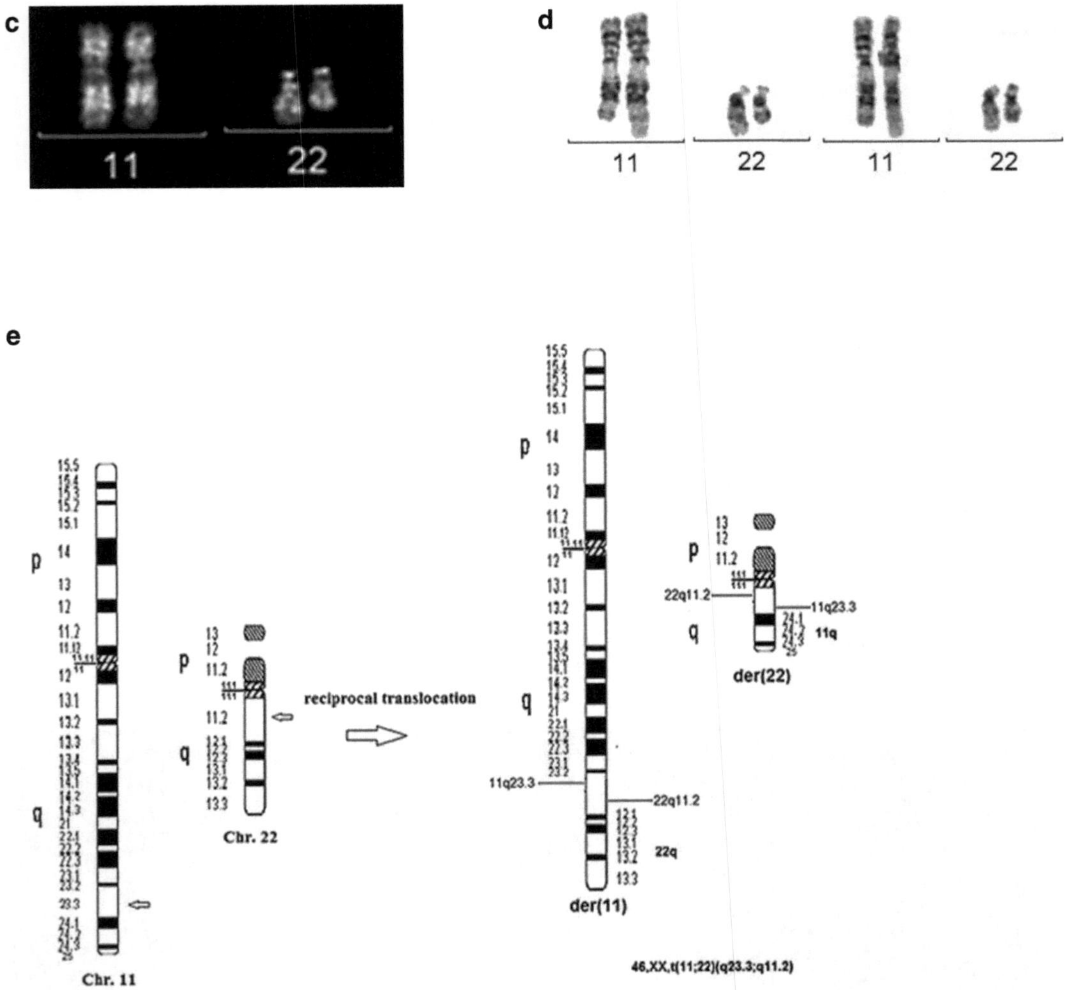

Fig. 11.5 (continued)

▪▪ Case 77: Karyotype: 46,XX,t(12;18)(q10;p11.2)

Indication for chromosome analysis: Investigation of polar body is planned, because of reduced fertility.

▪ Morphology

In chromosome 12, the centromere region is split, thus the larger translocation chromosome has become dicentric and its remains are almost undetectable (see C-banding). Both translocation chromosomes changed size and centromeric index compared to their normal homologues.

▪ Cave

In spite of the centromere splitting in chromosome 12 and a dicentric derivate the translocation chromosomes proved to be stable in transmission. Prenatal chromosome investigation will be performed, because meiotic errors cannot be excluded. This was discussed by genetic counselling as well the given indication for family investigations (◘ Fig. 11.6).

a

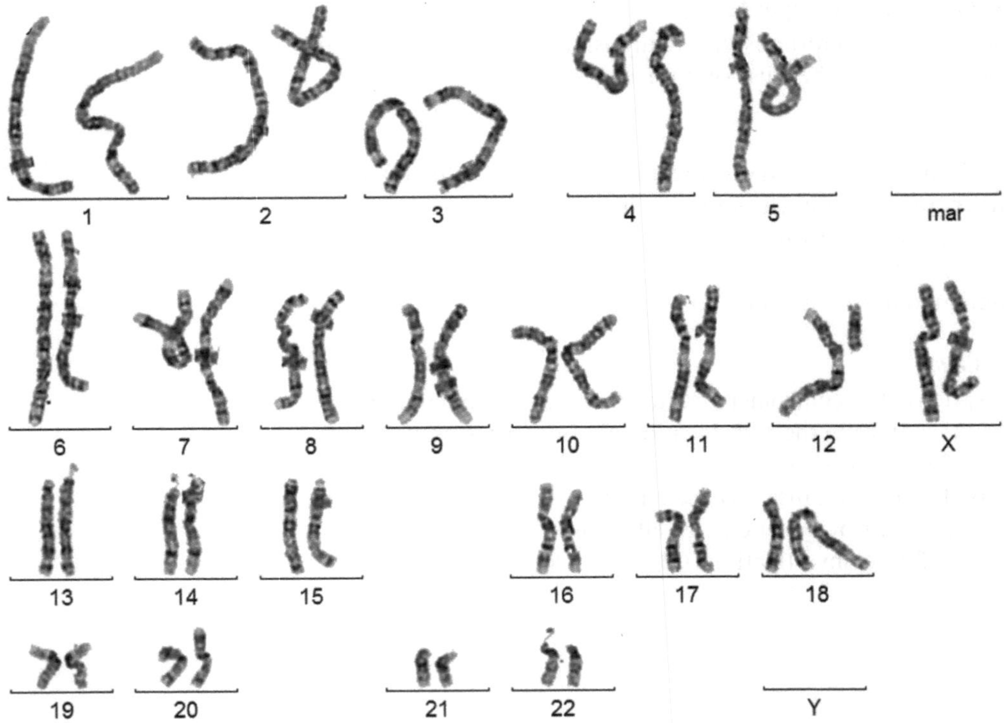

b

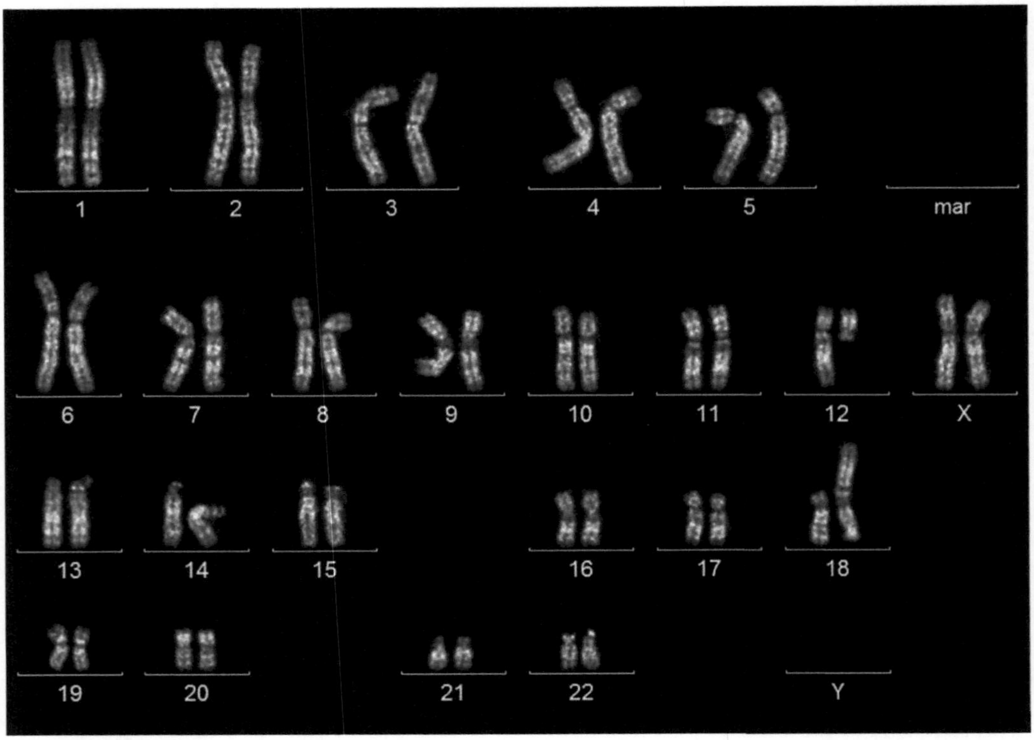

Fig. 11.6 Karyotype 46,XX,t(12;18)(q10;p11.2). **a** GTG-, **b** QFQ-banding, 850 and 400 bands per genome, **c** CBG-banding, **d** QFQ-banding, **e** GTG-banding chromosomes from different mitoses; **d**, **e** *left* normal, *right* aberrant chromosomes 12 and 18, **f** ideogram of the translocation

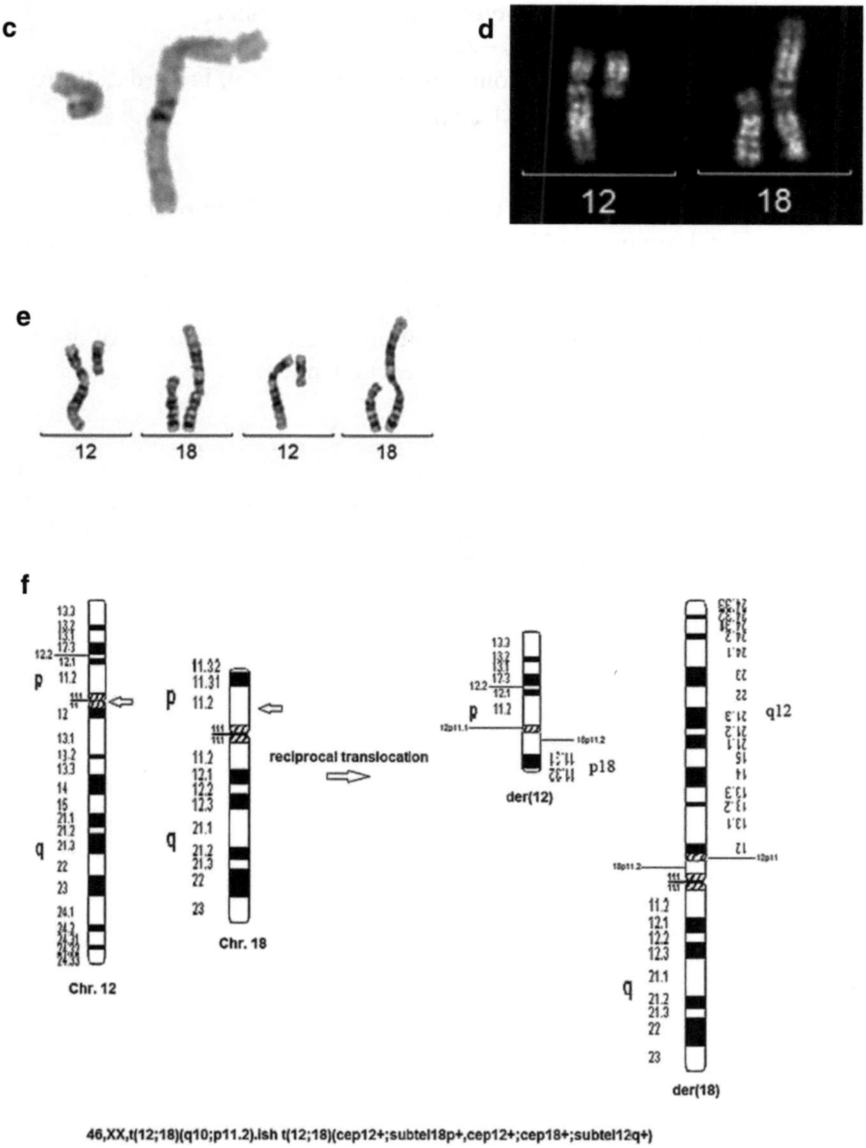

Fig. 11.6 (continued)

▪▪ Case 78: Karyotype: 45,XY,der(18),t(18;22)(p11.32;q11.2), −22

Indication for chromosome analysis: Mental retardation, congenital heart defect.

▪ Morphology

The karyotype of the patient shows 45 chromosomes with one chromosome 22 missing. The long arms of it have been translocated on the proximal short arm region of chromosome 18. Thus, the abnormal chromosome 18 is easy to identify but for exact breakpoint identification a banding resolution of at least 700 bands per genome is recommended.

▪ Cave

Before genetic counselling the chromosomes of the parents should be investigated. The clinical picture is caused by deletion 18p (de Grouchy syndrome I) and deletion 22q11.2 (di George syndrome) (◘ Fig. 11.7).

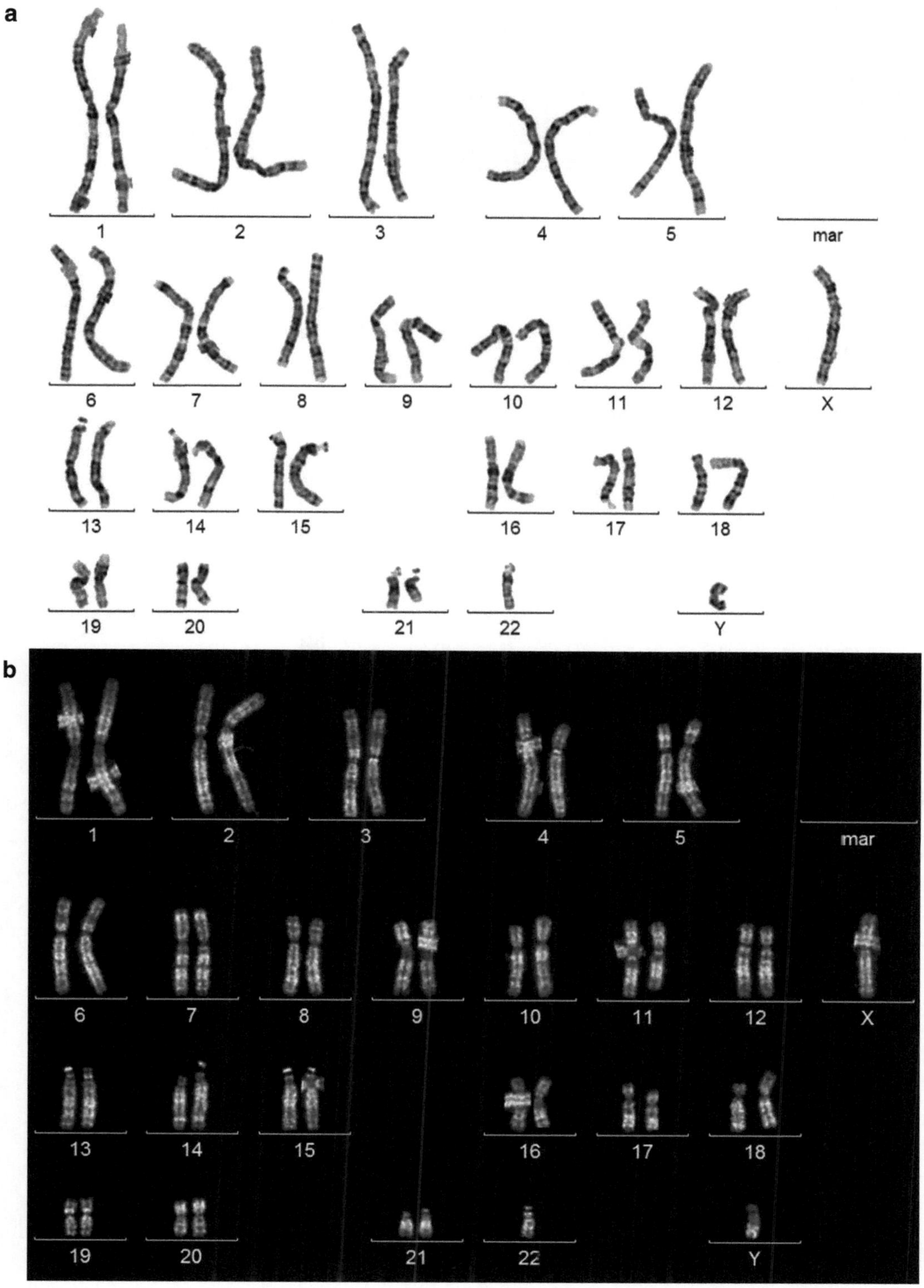

Fig. 11.7 Karyotype 45,XY,der(18),t(18;22)(p11.32;q11.2), −22. **a** GTG-, **b** QFQ-banding, 700 and 400 bands per genome, **c** normal chromosomes 22 in QFQ- and GTG-banding, **d** QFQ-, **e** GTG-banding, chromosomes from different mitoses, *left* normal, *right* aberrant chromosome 18, **f** ideograms from normal chromosomes 18 and 22 and the translocation product, unbalanced translocation 18/22

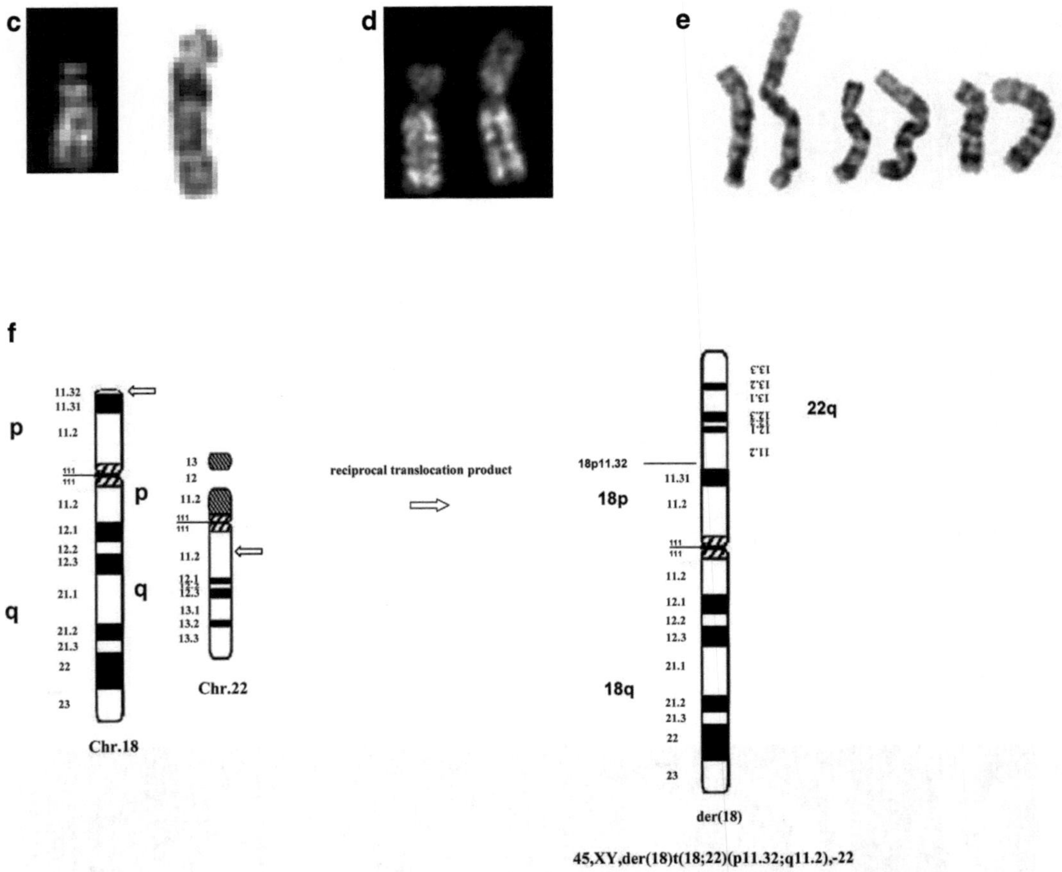

■ **Fig. 11.7** (continued)

■ ■ Case 79: Karyotype: 46,X,t(X;6)(q26.1;p25.1)

Indication for chromosome analysis: Infertility.

■ Morphology

The regions exchanged are very small (about 4 and 3% of the total chromosome length. Thus, their identification requires a high banding resolution of more than 700 bands per genome.

■ Cave

An unbalanced karyotype has to be especially discussed in a male carrier because of his X-chromosome hemizygosity (◘ Fig. 11.8).

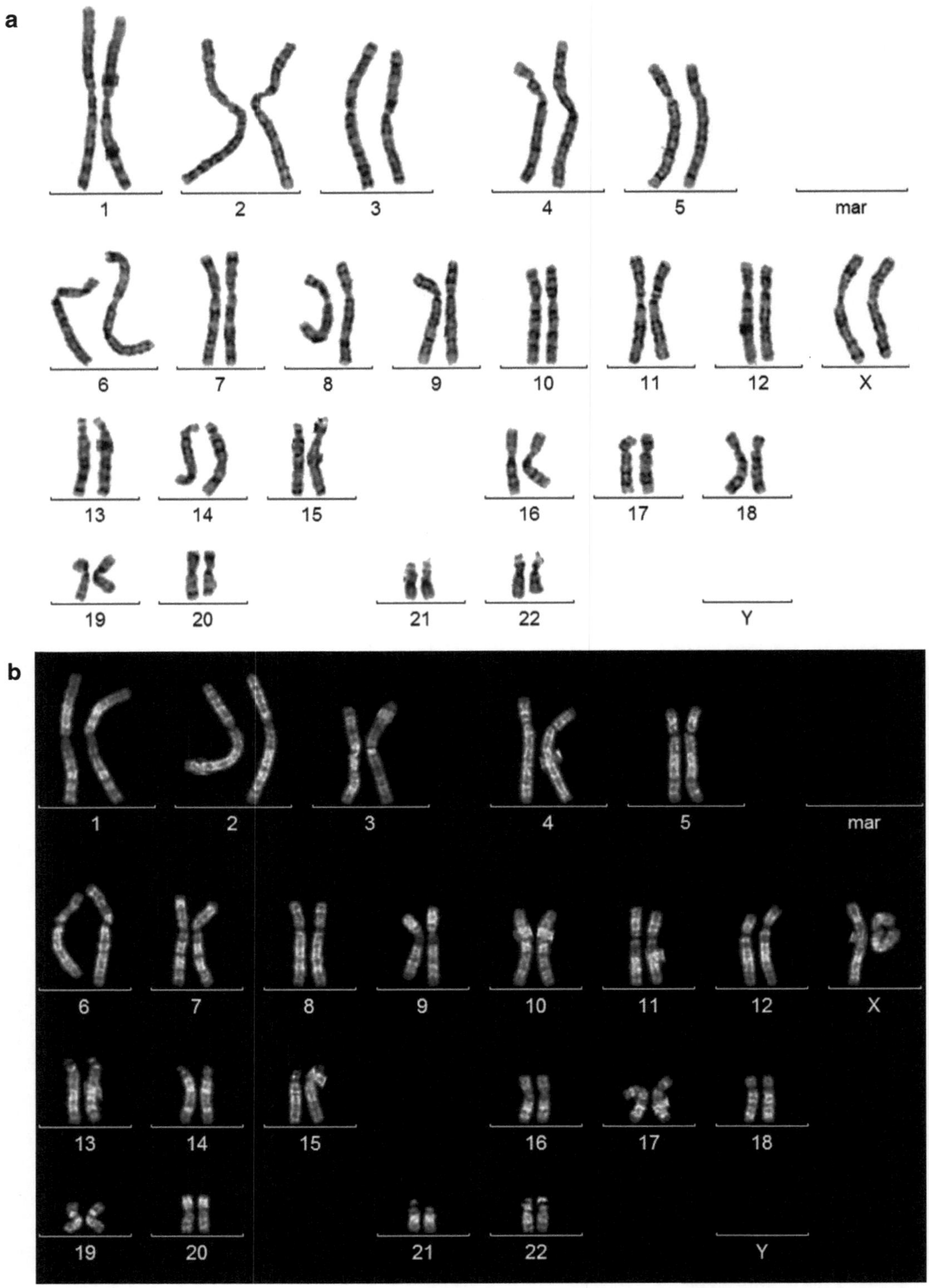

Fig. 11.8 Karyotype 46,X,t(X;6)(q26.1;p25.1). **a** GTG-, **b** QFQ-banding, both 550 bands per genome, **c** QFQ-banding, **d** GTG-banding chromosomes from different mitoses; *left* normal, *right* aberrant chromosomes X and 6, **e** ideograms of normal and aberrant chromosomes X and 6, balanced translocation

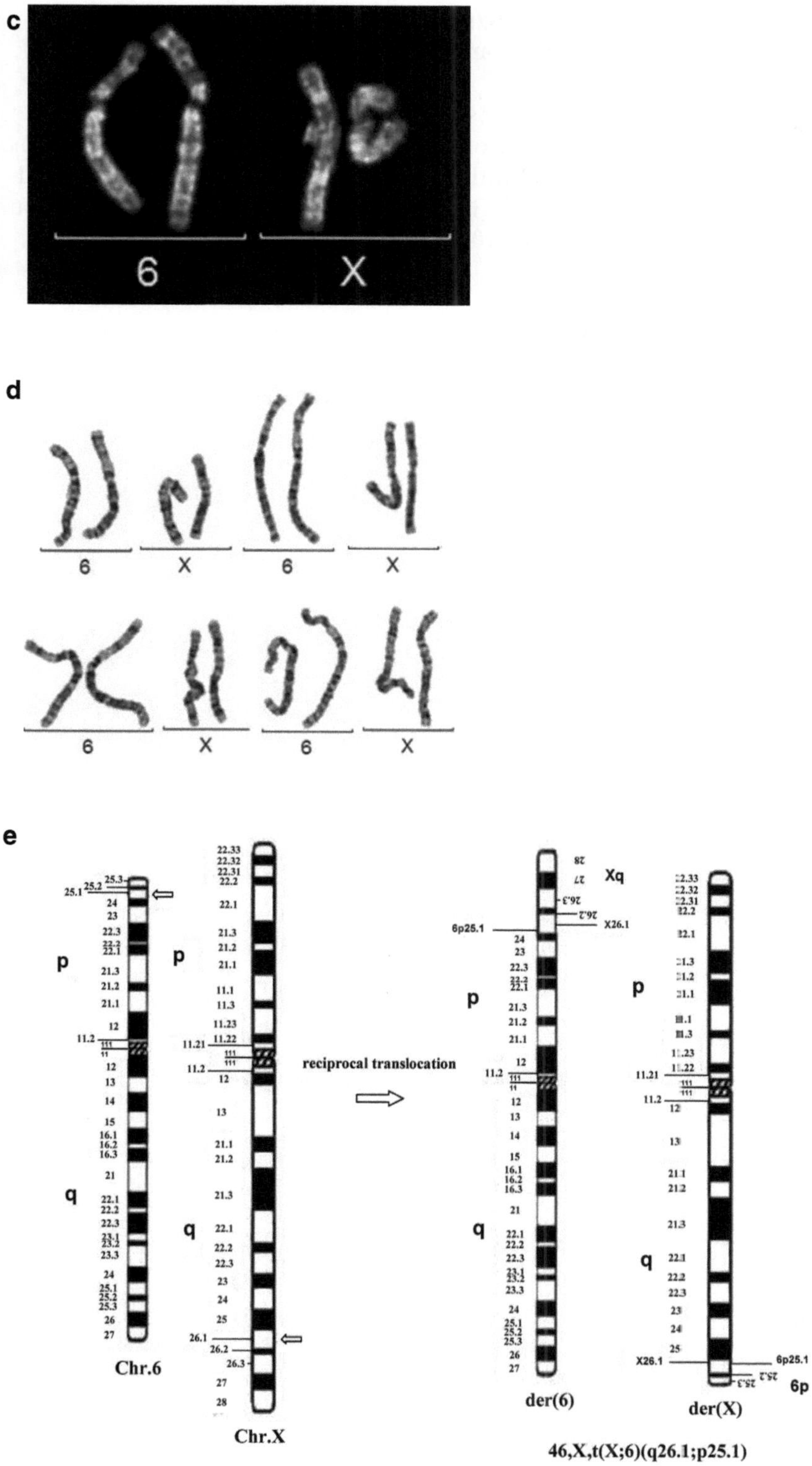

Fig. 11.8 (continued)

▪▪ Case 80: Karyotype: 46,X,t(Y;5)(q11.21;p15.33)

Indication for chromosome analysis: OAT-syndrome.

▪ Morphology

As the distal part of the long arm of the Y-chromosome has been exchanged with the distal band of 5p beside GTG banding, QFQ and CBG must be applied. The banding resolution should be at least 700 bands per genome.

▪ Cave

As to our experience with Y/autosome translocations pairing abnormalities in meiosis are frequently leading to meiotic arrest at the end of the prophase (◘ Fig. 11.9).

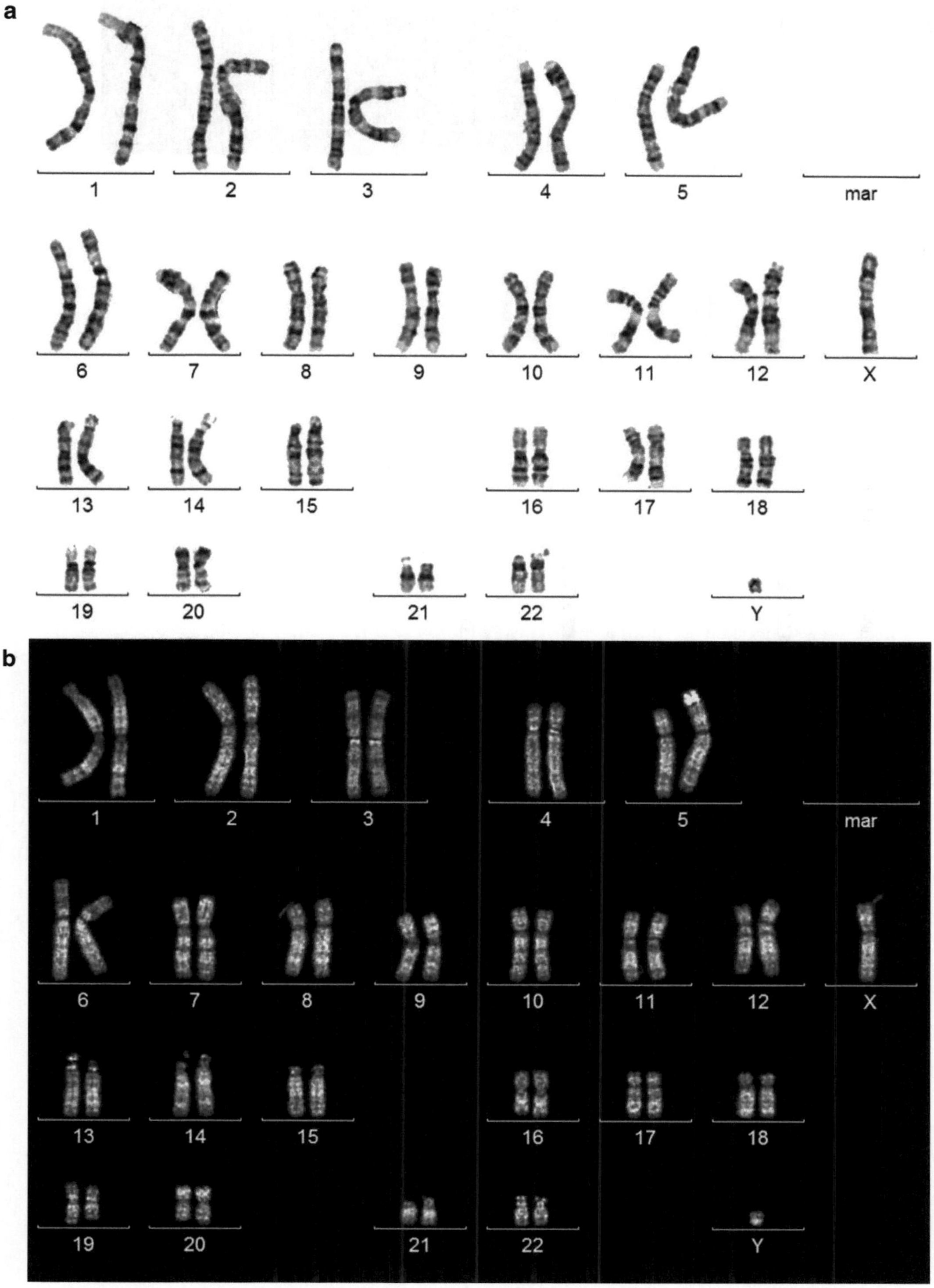

Fig. 11.9 Karyotype 46,X,t(Y;5)(q11.21;p15.33). **a** GTG-, **b** QFQ-banding, both 550 bands per genome, **c** CBG-, **d** QFQ-, **e** GTG-banding, here 1 metaphase with 2 Y-derivates. Chromosomes from different mitoses; **d**, **e** *left* normal chromosomes, *right* the two derivates; **d** on the *right* normal Y-chromosome, **f** ideograms from normal and aberrant chromosomes Y and 5, balanced translocation

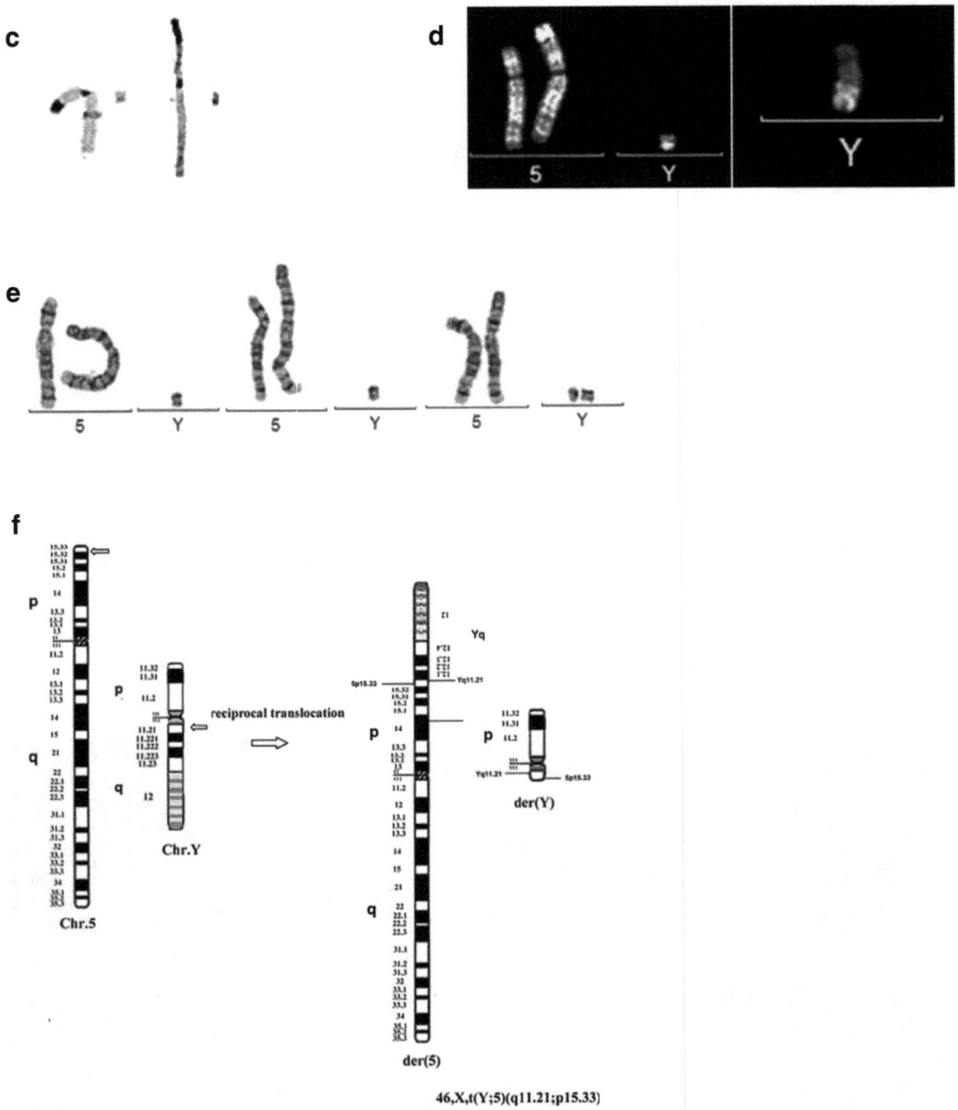

◻ **Fig. 11.9** (continued)

11.2 Summary

A total of nine different reciprocal translocations is presented in detail: seven rearrangements are between two autosomes, one is a X/autosome and one is an Y/autosome translocation.

In the family with reciprocal translocation 2/18, 12 persons from three generations could be investigated. The carriers of an unbalanced translocation showed both possible forms of a partial duplication and partial deletion in equal frequency (see pedigree). Male and female carriers of the translocation were equal often observed. The amount of abortions was—according to information from the family—not increased. Unexpected was the high frequency of unbalanced translocations, inherited paternally (4 out of 8 pregnancies), which according to empiric inquiries is estimated to be about 10%. Thus, this familial case makes it obvious that in reciprocal translocations the genetic risk factor can be estimated more precise after detailed family investigations.

The case of an unbalanced translocation 18/22 by loss of the smaller translocation chromosome is a typical example for an additional risk factor in reciprocal translocations by aneuploidisation. This has to be taken into account when one of the translocation chromosomes is small, so that the common regions for pairing with the homologues in prophase of meiosis I are reduced in size.

The unbalanced translocation 9/12 presented here is a good example for the requirement of additional investigations by microarray to get the precise analysis of the karyotype in cases when the second translocation chromosome cannot be identified by banding techniques in an unbalanced rearrangement.

The X/6 translocation and the Y/15 translocation were both identified in healthy carriers with infertility.

The spectrum of reciprocal translocations presented here illustrates their different mode of ascertainment and risk of estimation.

11.3 Robertsonian Translocations

81. 45,XX,dic rob(13;14)(p11.2;p11.1).

▪▪ Case 81: Karyotype: 45,XX,dic rob(13;14)(p11.2;p11.1)

Indication for chromosome analysis: Multiple abortions, relative with Down syndrome.

▪ Morphology

The total chromosome number is 45. A fusion has occurred between the proximal short arm regions of chromosome 13 (breakpoint p11.2) and 14 (breakpoint p11.1). The fusion product is dicentric, always with inactivation of the chromosome 13 centromere. The region in between the two centromeres can be easily characterised by C-banding or QFQ-banding (brilliant fluorescence). The distal part of 13p and the whole short arm of chromosome 14 got lost, which is of no relevance for the carrier, as they are composed of constitutive heterochromatin.

▪ Cave

When the carrier is asking for genetic counselling beside structural and numerical chromosome aberrations functional abnormalities by partial or complete UPD 14 have to be discussed (◘ Fig. 11.10).

a

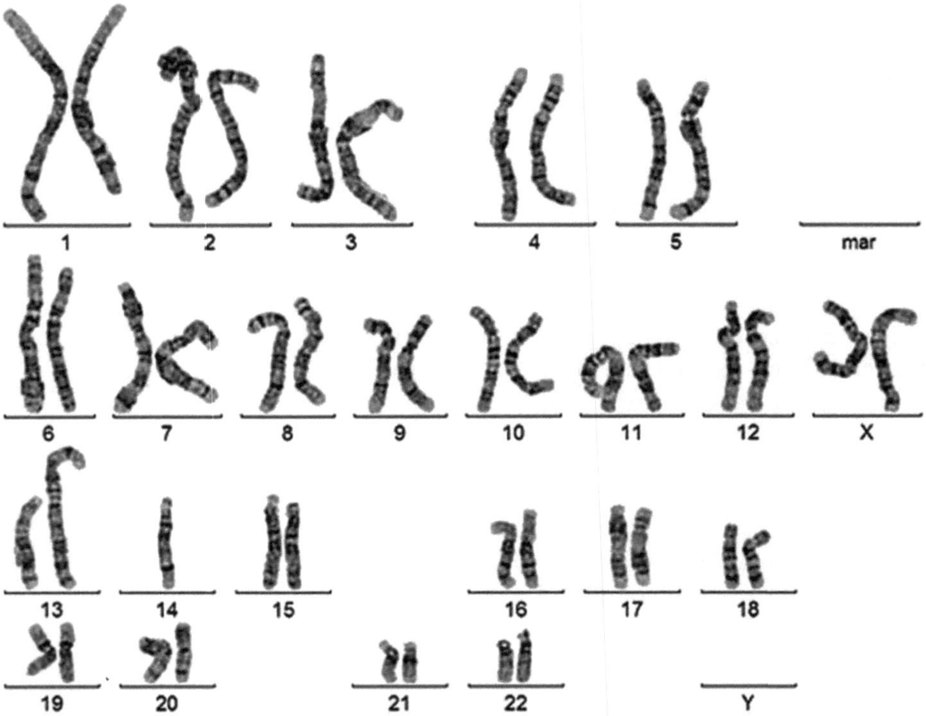

b

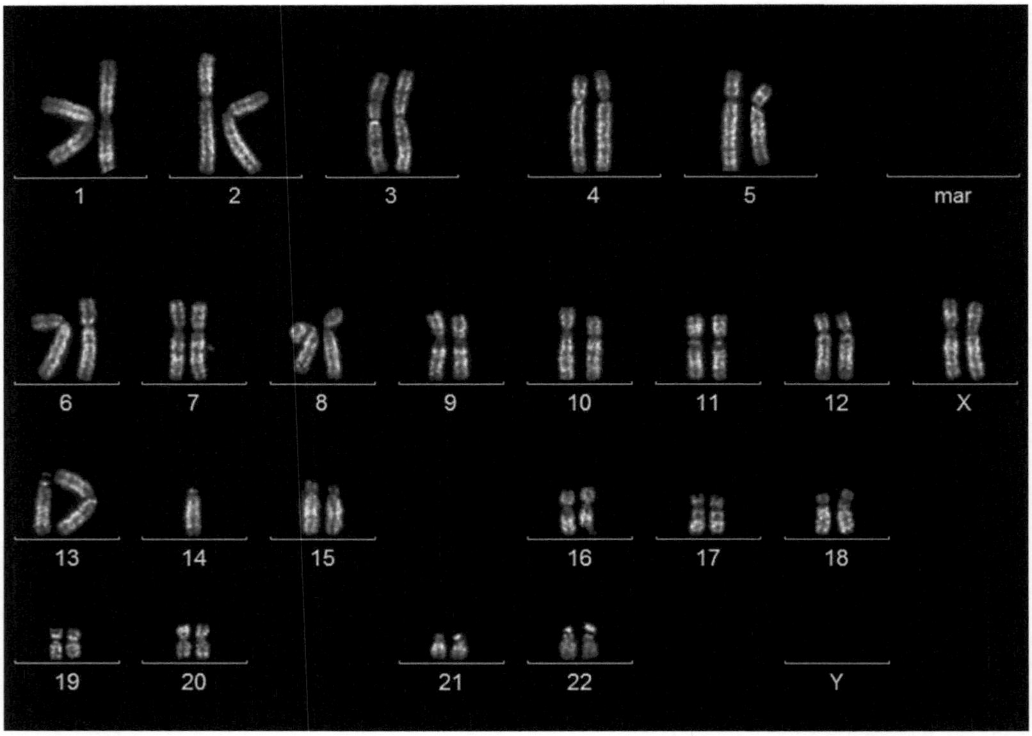

Fig. 11.10 Karyotype 45,XX,dic,rob(13;14)(p11.2;p11.1). **a** GTG-, **b** QFQ-banding, 700 and 400 bands per genome, **c** CBG-banding, dic,rob(13;14), **d** QFQ-banding, **e** GTG-banding, chromosomes from different mitoses; **d**, **e** *left* and *right* normal chromosomes 13 and 14, in between: fusion product, **f** ideograms of normal chromosomes 13 and 14 and fusion product, balanced rearrangement

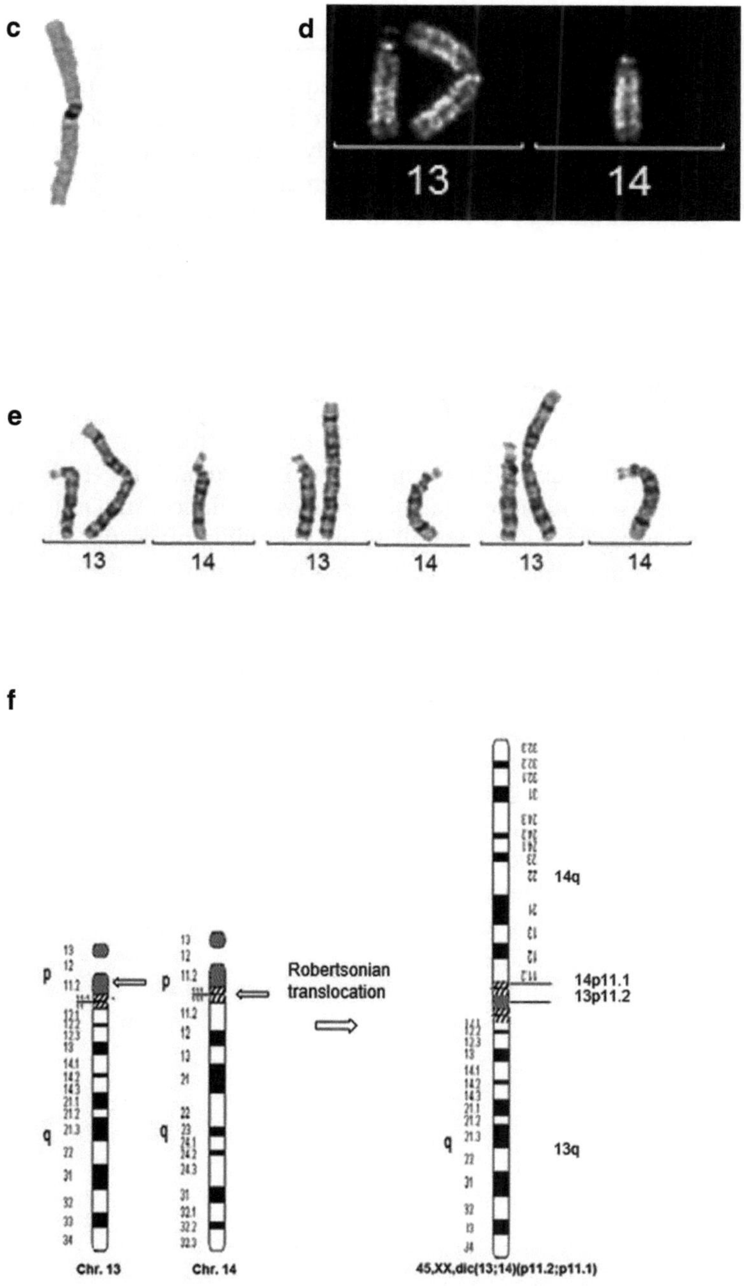

Fig. 11.10 (continued)

11.4 Summary

As an example for a special but frequent type of translocation the centric fusion 13/14 is described in detail. Here, the translocation chromosome is practically always dicentric, caused by homologous but inverted heterochromatic regions in the proximal parts of the short arms in both chromosomes. The balanced carrier has 45 chromosomes as the second translocation chromosome composed of the distal short arm regions is acentric and therefore gets lost in mitosis. The genetic risk for the balanced carrier is complex: The fertility is reduced in both sexes but particularly in the males, the risk of abortion is significantly increased (cases of trisomy 13 and 14), birth of a child with translocation-trisomy 13 is less frequent than 1% and additionally UPD-14 and partial proximal UPD-14 have to be taken into account in offspring with balanced and normal appearing karyotype as well as the possibility of partial duplications and deletions of euchromatin by an unequal crossing over in meiosis.

In the first years of this translocation investigations in familial cases, it was assumed that pairing abnormalities in meiosis can lead to malsegregation of heterologous chromosomes not involved in the translocation especially chromosome 21 and that thus the risk for the birth of a child with trisomy 21 is increased. More detailed investigations made it obvious that the reduced fertility of a carrier with 13/14 translocation is the reason for a higher amount of pregnancies in these couples in an advanced maternal age. This means that the risk for trisomy 21 in the offspring is the well-known age factor and not caused by the translocation.

Insertions

Contents

12.1 Insertions – 310

12.2 Summary – 314

C. Behrend et al., *Human Chromosome Atlas*, https://doi.org/10.1007/978-3-031-10588-3_12

12.1 Insertions

82. 46,XX,ins(6;5)(q13;q31.1q33.1).

■■ Case 82: Karyotype: 46,XX,ins(6;5)(q13;q31.1q33.1)

Indication for chromosome analysis: Identical chromosome rearrangement in the daughter.

■ Morphology

The chromosomes involved in the balanced rearrangement are diagnosed by the altered length of the long arms of chromosome 5 and 6, and by the aberrant banding pattern. The chromosome analysis must be performed at a resolution of 550–700 bands per genome. Additional microarray investigation is indicated.

■ Cave

The family investigation should be extended to all relatives which are possible insertion carriers. As the risk for an unbalanced karyotype in the offspring is high (about 30%), prenatal diagnosis including microarray should be recommended to all carriers (◻ Fig. 12.1).

12

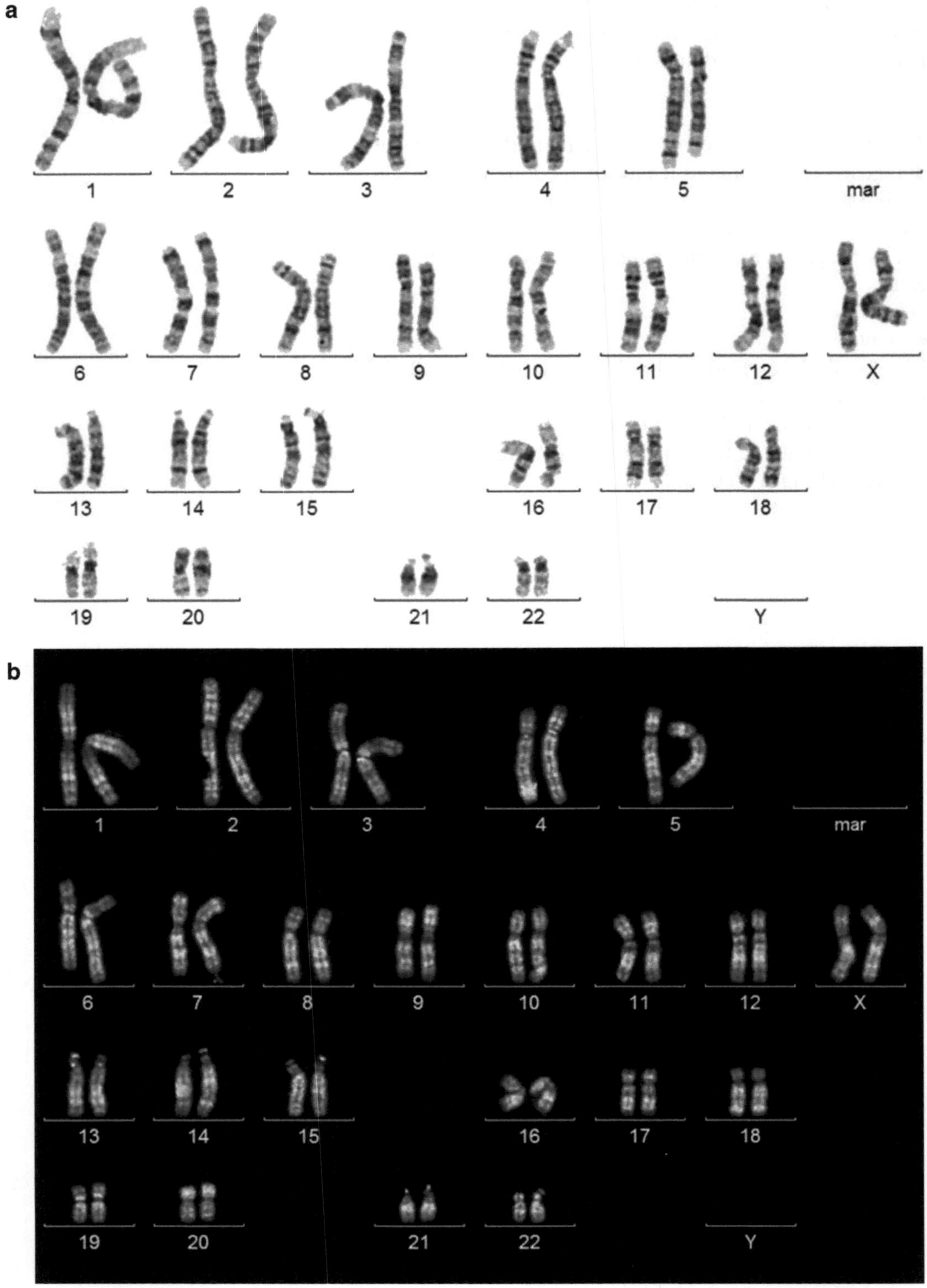

Fig. 12.1 Karyotype 46,XX,ins(6;5)(q13;q31.1q33.1). **a** GTG-, **b** QFQ-banding, 550 and 400 bands per genome, **c** QFQ-, **d** GTG-banding, chromosomes from different mitoses, *left* normal chromosomes 5 and 6, *right* products of rearrangement, **e** ideograms of normal chromosomes 5 and 6 and the products of rearrangement, balanced karyotype

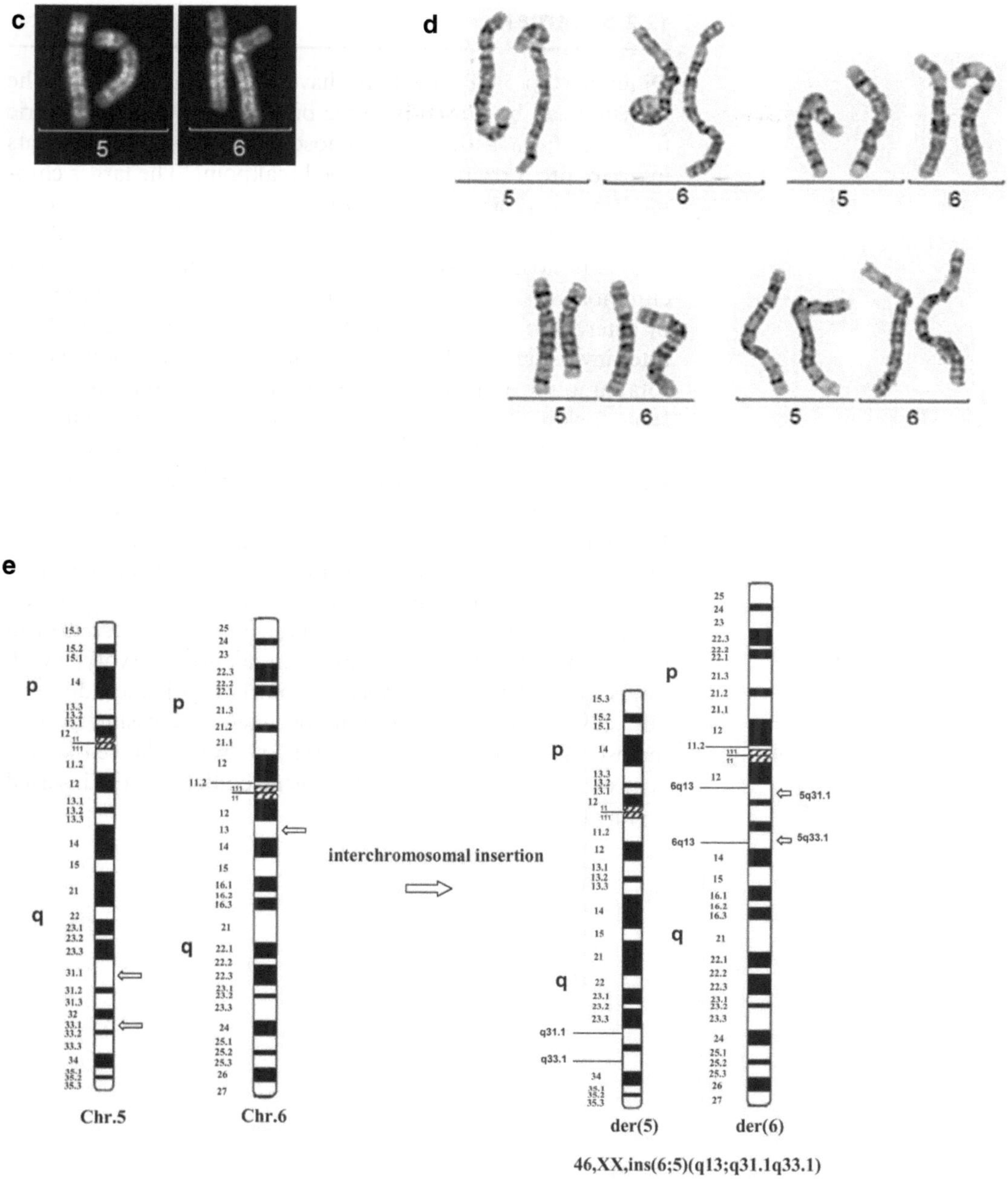

Fig. 12.1 (continued)

12.2 Summary

Different types of insertions have been characterised. The insertion can be caused by three breakpoints and the acentric fragment from a donor chromosome with two breakpoints inserted into a recipient with one breakpoint. The larger chromosomes 1–12 are more often involved than the smaller ones (13-Y). This procedure can occur between the chromatids of one chromosome, between homologous and heterologous chromosomes. The translocated segment can be euchromatic or heterochromatic in origin, and it can be inserted directly or after inversion. The frequency of insertions is very low with a total amount of 1 in 80.000 newborn. The genetic risk is the same for directly inserted and for inverted segments. The risk for a pathologic phenotype in the offspring with unbalanced karyotype is low (<10%) in small and AT-rich inserted segments but high when it is large and/or GC-rich (up to 50%). Lethality is especially high if the unbalanced karyotype has a deletion for the segment in question. Therefore, a detailed analysis including relatives of the carrier is necessary in each case.

The case of a familial direct insertion 5/6 is described in detail. The genetic risk includes an unbalanced karyotype with interstitial deletion or duplication 5q31.1q33.1 and additionally submicroscopic aberrations by unequal crossing over in meiosis. The viability of the carrier of an unbalanced karyotype depends on the size and the gene content of the inserted segment.

Complex Chromosome Rearrangements (CCR)

Contents

13.1 Complex Chromosome Rearrangements (CCR) – 316

13.2 Summary – 326

C. Behrend et al., *Human Chromosome Atlas*, https://doi.org/10.1007/978-3-031-10588-3_13

13

13.1 Complex Chromosome Rearrangements (CCR)

83. 46,XX,inv(2)(p11.2q13),del(18)(q22)
84. 46,XY,del(5)(q14.2q22.1),t(16;20)(q24;q13.11)
85. 47,XY,inv(12)(q15q24.1),+idic(15)(q11.2q11.2).

▪▪ Case 83: Karyotype: 46,XX,inv(2)(p11.2q13),del(18)(q22)
Indication for chromosome analysis: Phenotype of de Grouchy syndrome II.

▪ Morphology

The patient shows a complex structural aberration (CCR), involving two chromosomes with three breakpoints.

1. Pericentric inversion 2

   The rearrangement is balanced. One of the breakpoints (p11.2) is located at the border of pericentromeric heterochromatin and the first euchromatic band, whereas the second breakpoint is in the GC-rich euchromatic band q13.

   (see: Case 2).
2. Deletion chromosome 18

   The partial monosomy 18q can be defined as de Grouchy syndrome II, a well-known and frequent structural aberration.

▪ Cave

Even in cases of de novo structural chromosome aberrations (CCR), it must be excluded by additional microarray if the patient shows additional submicroscopic aberrations.

Chromosome analyses in the parents must be performed before genetic counselling, as they can be carrier of the inversion 2 or a reciprocal translocation involving 18q (▫ Fig. 13.1).

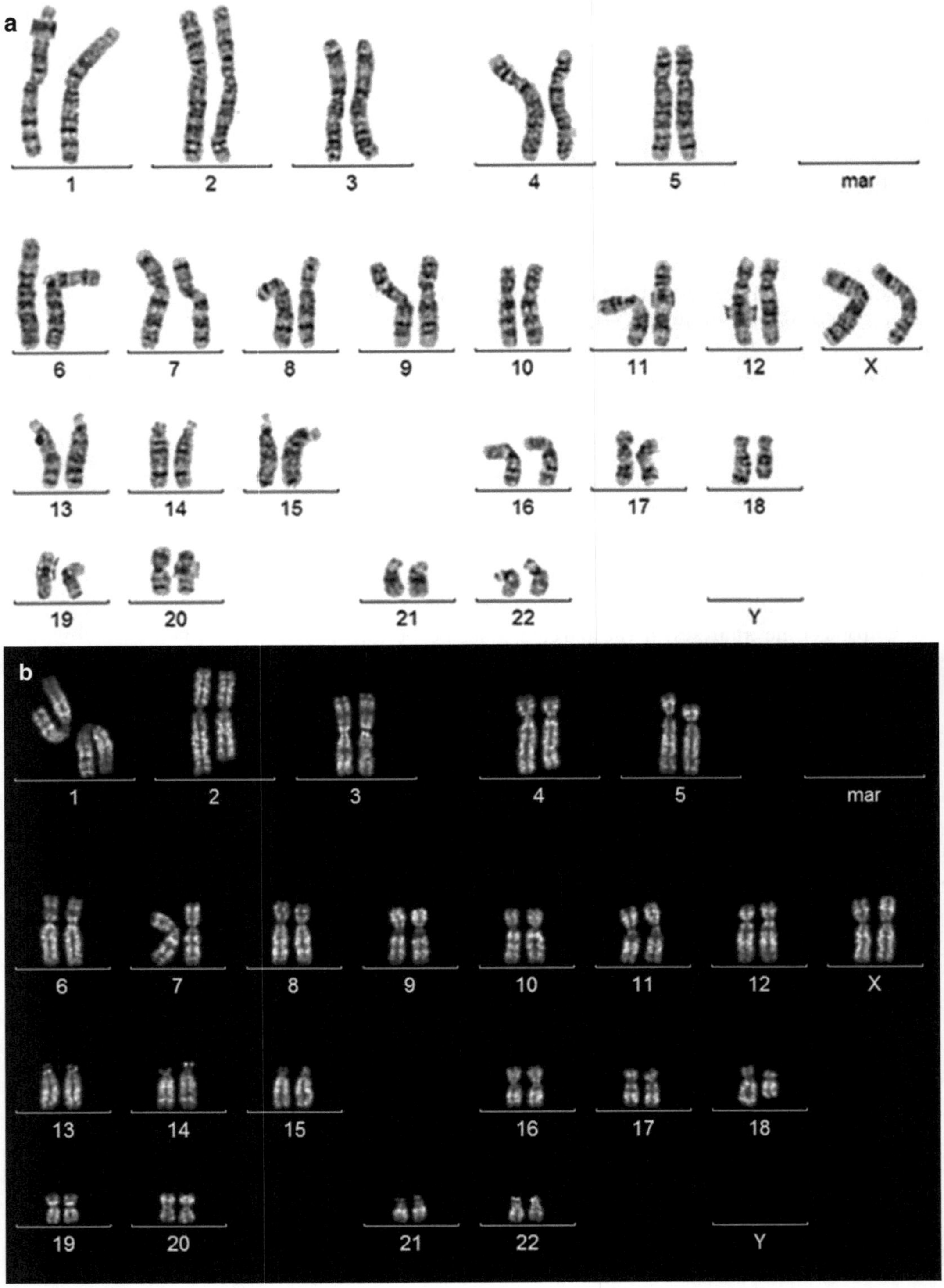

Fig. 13.1 Karyotype 46,XX,inv(2)(p11.2q13),del(18)(q22). **a** GTG- **b** QFQ-banding, 550 and 400 bands per genome, **c** chromosomes from different mitoses (chromosome 2), **d** QFQ-, **e** GTG-banding (chromosome 18), **f** ideograms of normal and aberrant chromosomes 2 and 18; 2: balanced, 18: unbalanced rearrangement

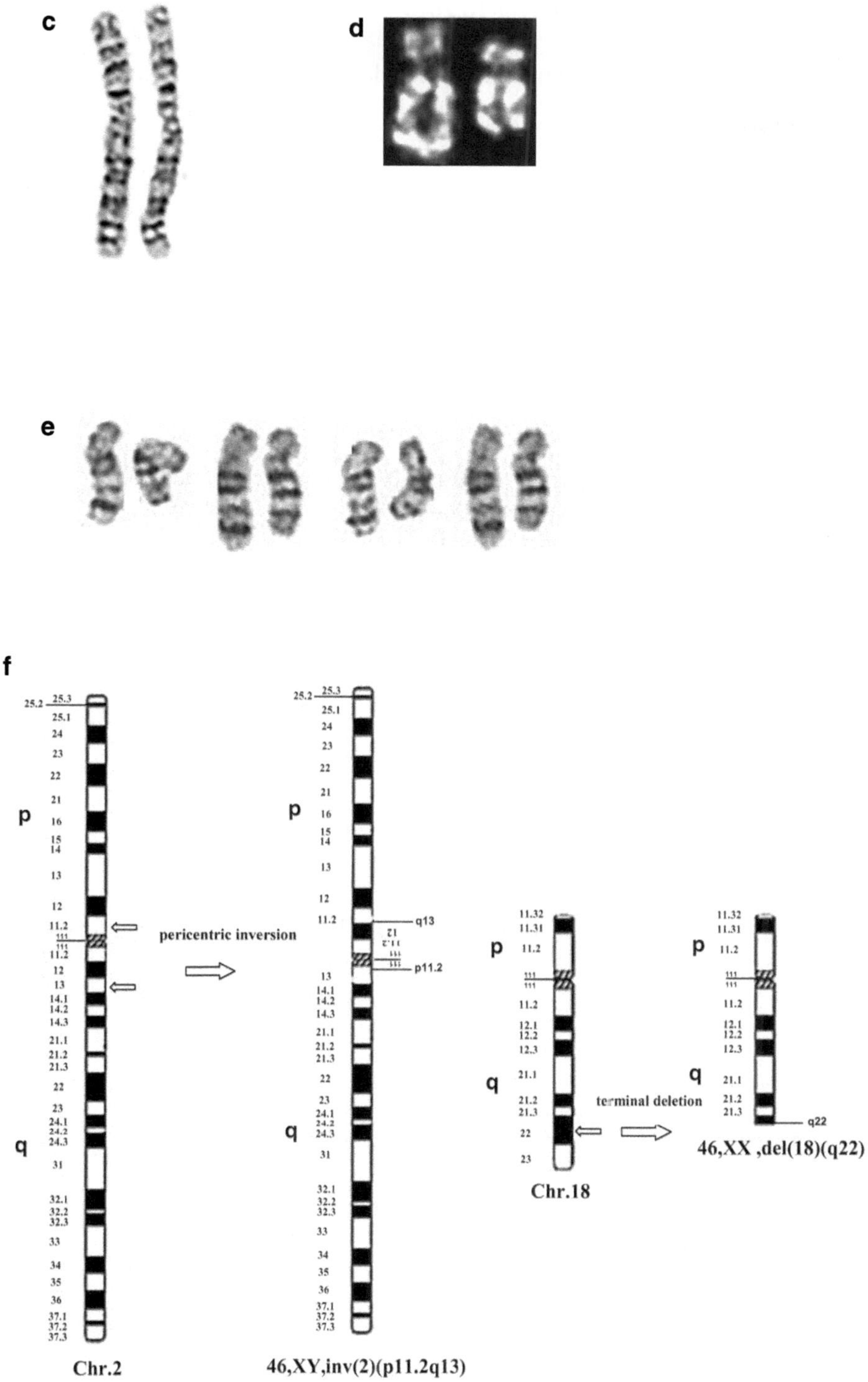

Fig. 13.1 (continued)

Case 84: Karyotype: 46,XY,del(5)(q14.2q22.1),t(16;20)(q24;q13.11)

Indication for chromosome analysis: Developmental retardation, macrosomia, dysmorphic features.

Morphology

The patient shows a complex chromosome rearrangement (CCR) involving three chromosomes and four breakpoints.

1. Deletion 5q

   The phenotype of the child can be delineated from the interstitial deletion 5q.

2. Translocation 16/20

   The translocation 16/20 appears balanced.

Cave

In the patient microarray analysis should be performed to exclude an additional submicroscopic aberration. The karyotype of both parents must be investigated before genetic counselling of the family (◘ Fig. 13.2).

a

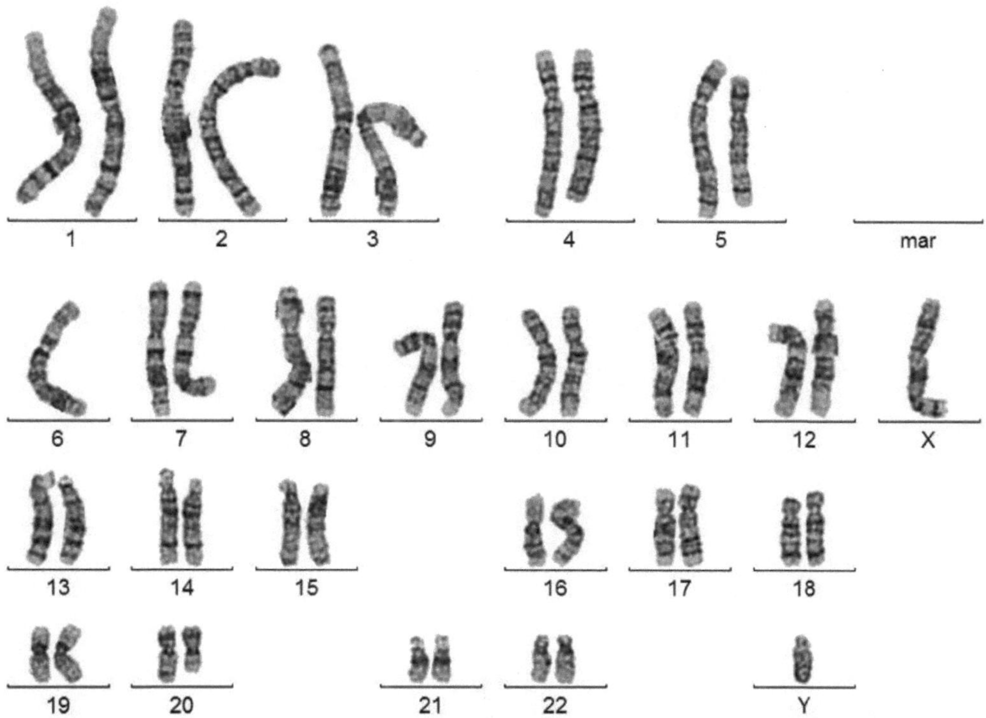

b

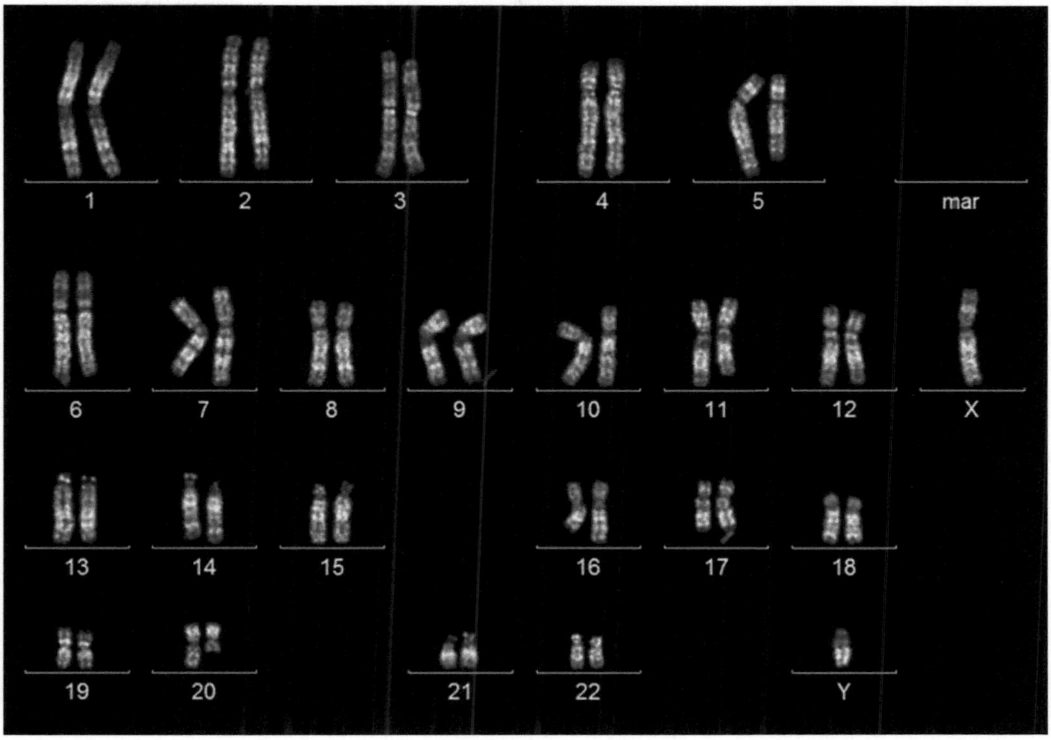

Fig. 13.2 Karyotype 46,XY,del(5)(q14.2q22.1),t(16;20)(q24;q13.11) with 1 chromosome 6 lost by preparation **a** GTG-, **b** QFQ-banding, 700 and 500 bands per genome, **c** GTG-banding, *left* normal, *right* deleted chromosome 5, **d** GTG-banding, *left* normal chromosomes 16 and 20, *right* translocation chromosomes, **e**, **f** ideograms of chromosomes 5, 16 and 20, chromosome 5 unbalanced rearrangement, chromosome 16 and 20 balanced translocation

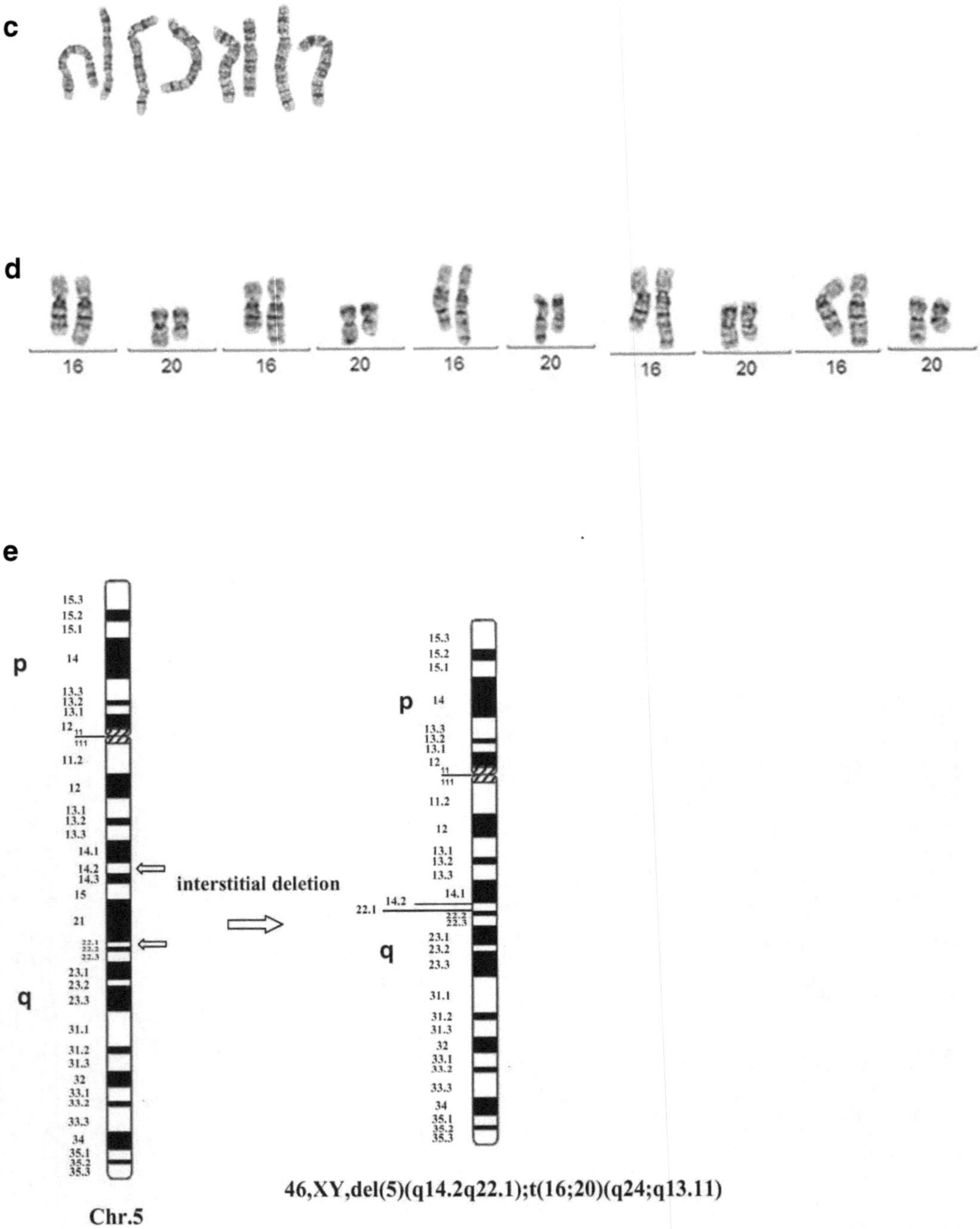

Fig. 13.2 (continued)

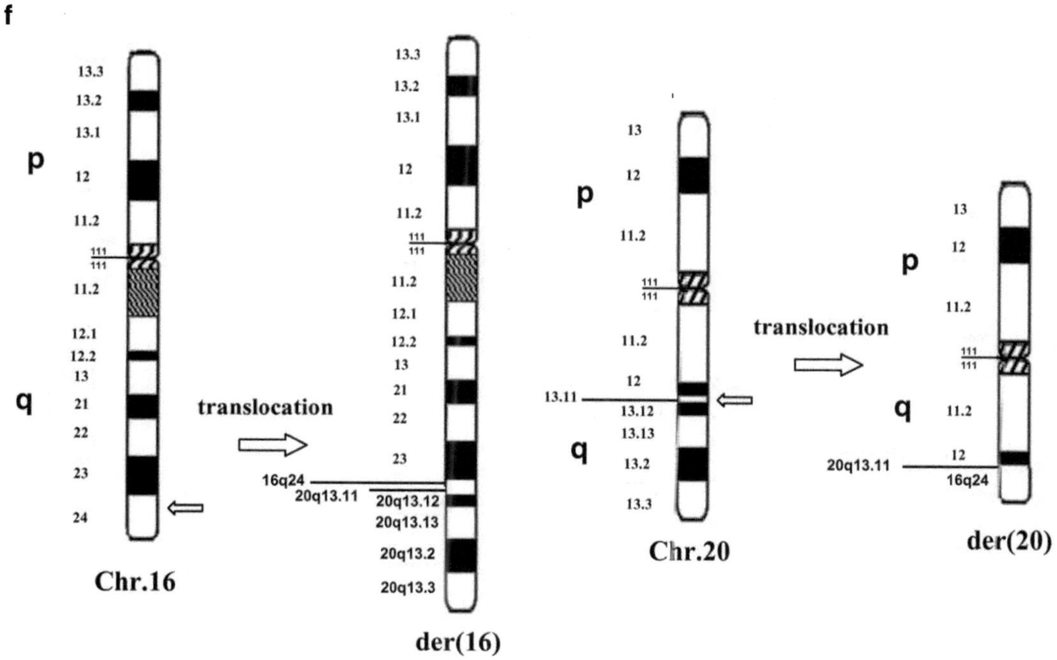

46,XY,del(5)(q14.2q22.1);t(16;20)(q24;q13.11)

▪▪ Case 85: Karyotype: 47,XY,inv(12)(q15q24.1),+idic(15)(pter→q11.2::q11.2→pter)

Indication for chromosome analysis: Increased number of abortions.

▪ Morphology

The complex chromosome abnormality consists of a paracentric inversion and an additional marker chromosome with three breaks in two chromosomes.

1. Paracentric inversion 12

   The inverted region comprises about 30% of the total length of chromosome 12. The genetic risk is low according to the literature.
2. Additional marker chromosome idic(15)

   The derivative chromosome 15 consists of constitutive heterochromatin and is thus without relevance for the carrier himself.

▪ Cave

Prenatal diagnosis has to be offered to the couple, as the derivative chromosome 15 can lead to an unequal crossing over in meiosis and thus to partial UPD 15 and to submicroscopic duplication or deletion (◘ Fig. 13.3).

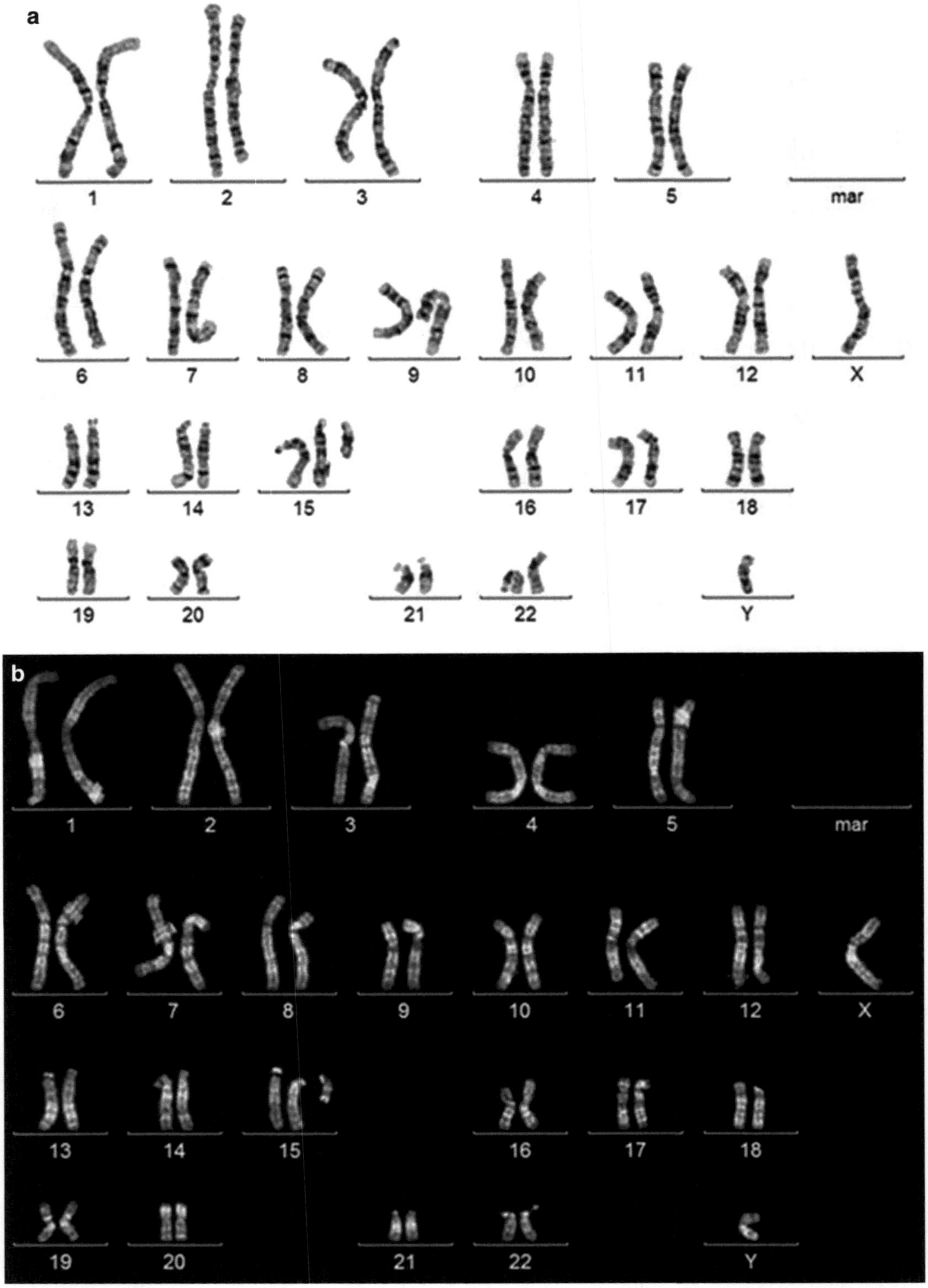

■ **Fig. 13.3** Karyotype 47,XY,inv(12)(q15q24.1),+idic(15)(pter→q11.2::q11.2 → pter). **a** GTG-, **b** QFQ-banding, 550 and 400 bands per genome **c** QFQ-, **d** GTG-banding, chromosomes from different mitoses inv12,+dic(15), *left* normal chromosomes 12 and 15, *right* aberrant chromosomes, **e, f** ideograms from normal and aberrant chromosomes 12 and 15

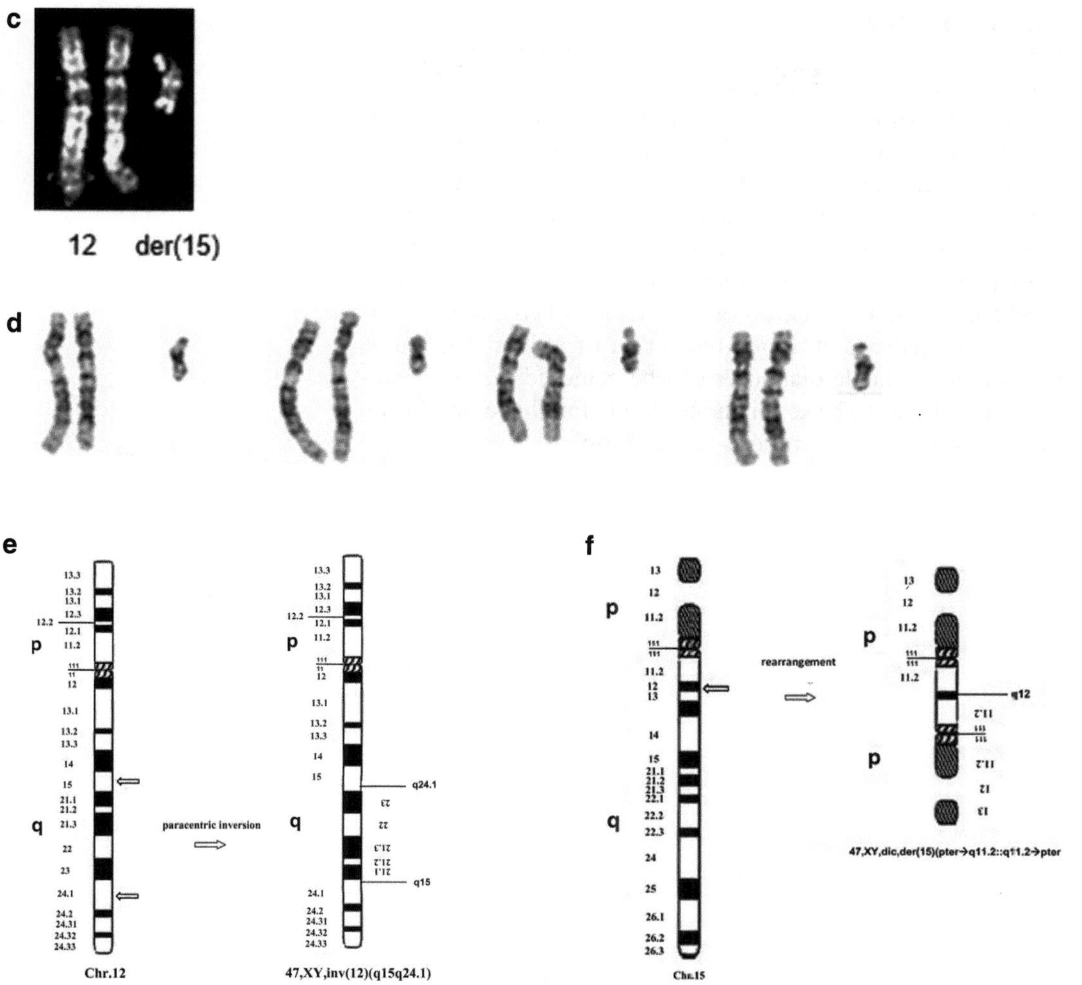

Fig. 13.3 (continued)

13.2 Summary

Three cases are presented in detail. All of them can be assigned to the same subtype of aberration which comprises two different, independent structural abnormalities in the same karyotype. In the first case, this is a deletion combined with a reciprocal translocation, in the second case a pericentric inversion is combined with a deletion, and in the third case a paracentric inversion is combined with a supernumerary heterochromatic marker chromosome. Only, when the chromosomes are analysed at a high resolution of at least 550 bands per genome, reliable diagnoses can be achieved in these complex cases. It has to be kept in mind that in familial cases during meiosis additional aberrations can develop.

Marker Chromosomes

Contents

14.1 Marker Chromosomes – 328

14.2 Summary – 338

C. Behrend et al., *Human Chromosome Atlas*, https://doi.org/10.1007/978-3-031-10588-3_14

14.1 Marker Chromosomes

86. 47,XX,+i(14)(p10)
87. 47,XY,+idic(15)(pter→q11.1::q11.1 → pter)
88. 47,XY,+idic(22)(pter→q11.1::q11.1 → pter)mat.

▪▪ Case 86: Karyotype: 47,XX,+i(14)(p10))

Indication for chromosome analysis: Genetic counselling because of the diagnosis of a marker chromosome in the mother.

▪ Morphology

The derivate is heterochromatic, monocentric and consists of the complete short arms (p11.1–p13) of the acrocentric chromosome, thus bearing satellites on both ends. The marker is not an isochromosome but can be delineated from two different chromosomes 14. The final analysis of the derivate required a FISH investigation.

▪ Cave

The marker is completely heterochromatic and thus of no relevance for the carrier herself. But in pregnancies it leads to an increased genetic risk:

An unequal crossing over in meiosis I can lead to partial duplications or deletions of the proximal euchromatic long arm regions of chromosome 14.

Malsegregation of a chromosome 14 in meiosis can be followed by trisomy or monosomy 14 and can lead after a regulation process to UPD 14.

In some carriers (data from the literature), a reduced fertility was observed (◻ Fig. 14.1).

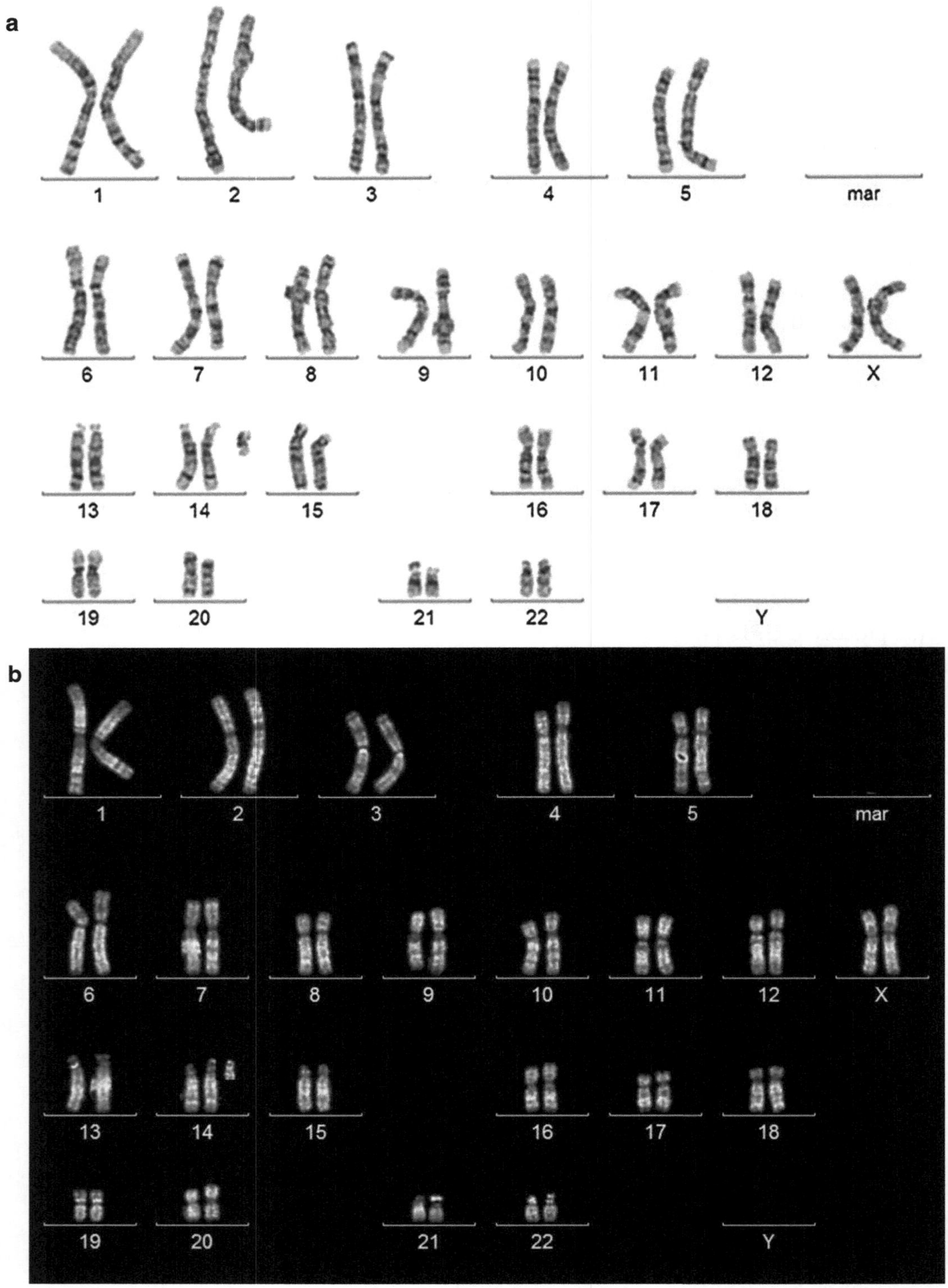

Fig. 14.1 Karyotype 47,XX,+i(14)(p10). **a** GTG-, **b** QFQ-banding, 550 and 400 bands per genome, **c** QFQ-, **d** GTG-banding, chromosomes from different mitoses, *left* 2 normal chromosomes 14, *right* der(14), **e** ideograms of normal chromosome 14 and der(14), the additional marker consists only of heterochromatic segments and therefore is without genetic relevance for the carrier himself

c

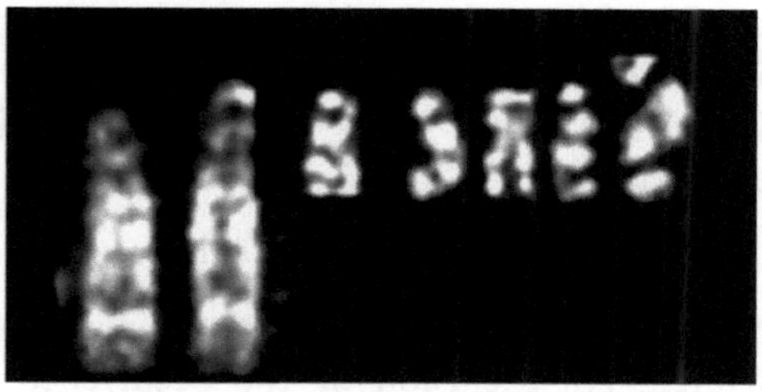

d

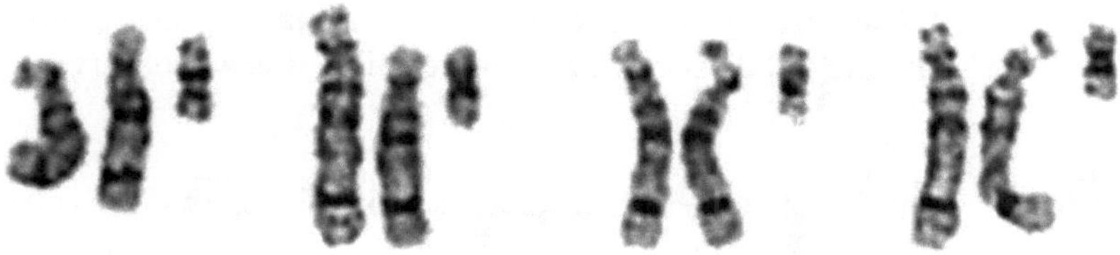

e

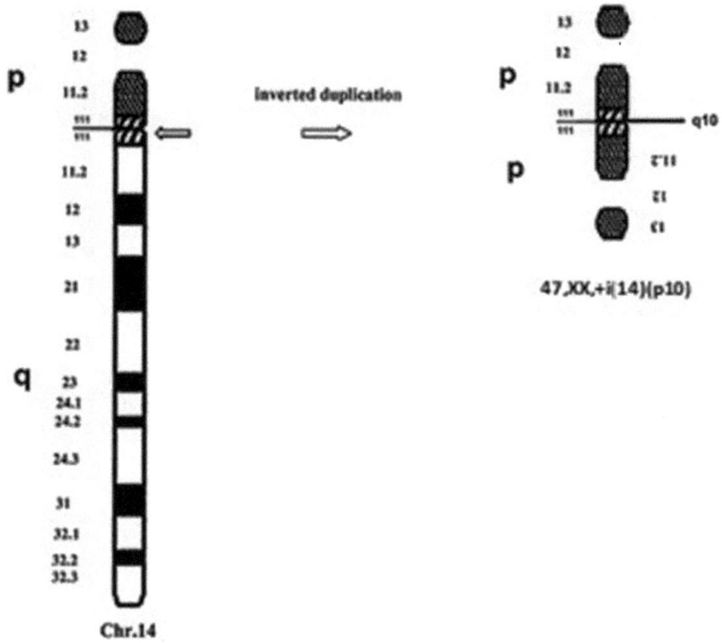

Fig. 14.1 (continued)

■■ Case 87: Karyotype: 47,XY,+idic(15)(pter→q11.1::q11.1→pter)

Indication for chromosome analysis: 4 years old boy with developmental peculiarities.

■ Morphology

The additional marker chromosome is dicentric with one centromere inactivated. An exclusion of the occurrence of euchromatic segments were performed by FISH. The marker originated from two homologous chromosomes 15. In spite of its dicentric structure, the marker was stable passing through somatic cell divisions. A second cell line was not observed.

■ Cave

The chromosomes of the parents should be analysed, as all carriers of a heterochromatic derivate 15 have an increased genetic risk by the possible unequal crossing over in meiosis and by UPD 15 formation. As the developmental peculiarities of the child cannot be caused by the heterochromatic marker, further investigation should be performed (◘ Fig. 14.2).

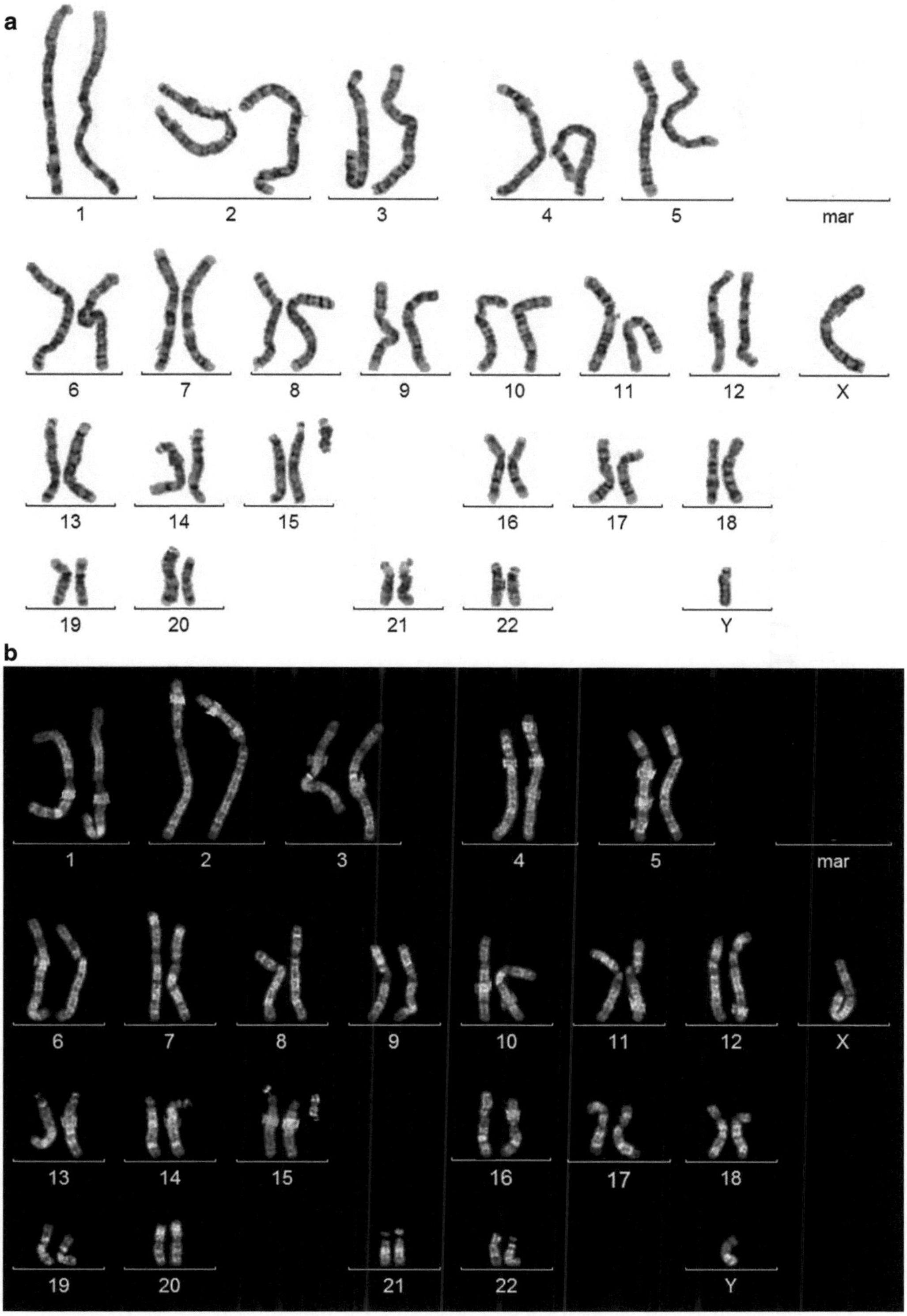

Fig. 14.2 Karyotype 47,XY,+idic(15)(pter→q11.1::q11.1 → pter). **a** GTG-, **b** QFQ-banding, 700 and 500 bands per genome, **c** CBG-, QFQ-, **d** GTG-banding, chromosomes from different mitoses, **d,** *left* normal chromosomes 15, *right* derivate 15, **c** derivate 15, **e** ideograms from chromosome 15 and derivate 15. der(15) is a completely heterochromatic marker, thus without genetic relevance for the carrier

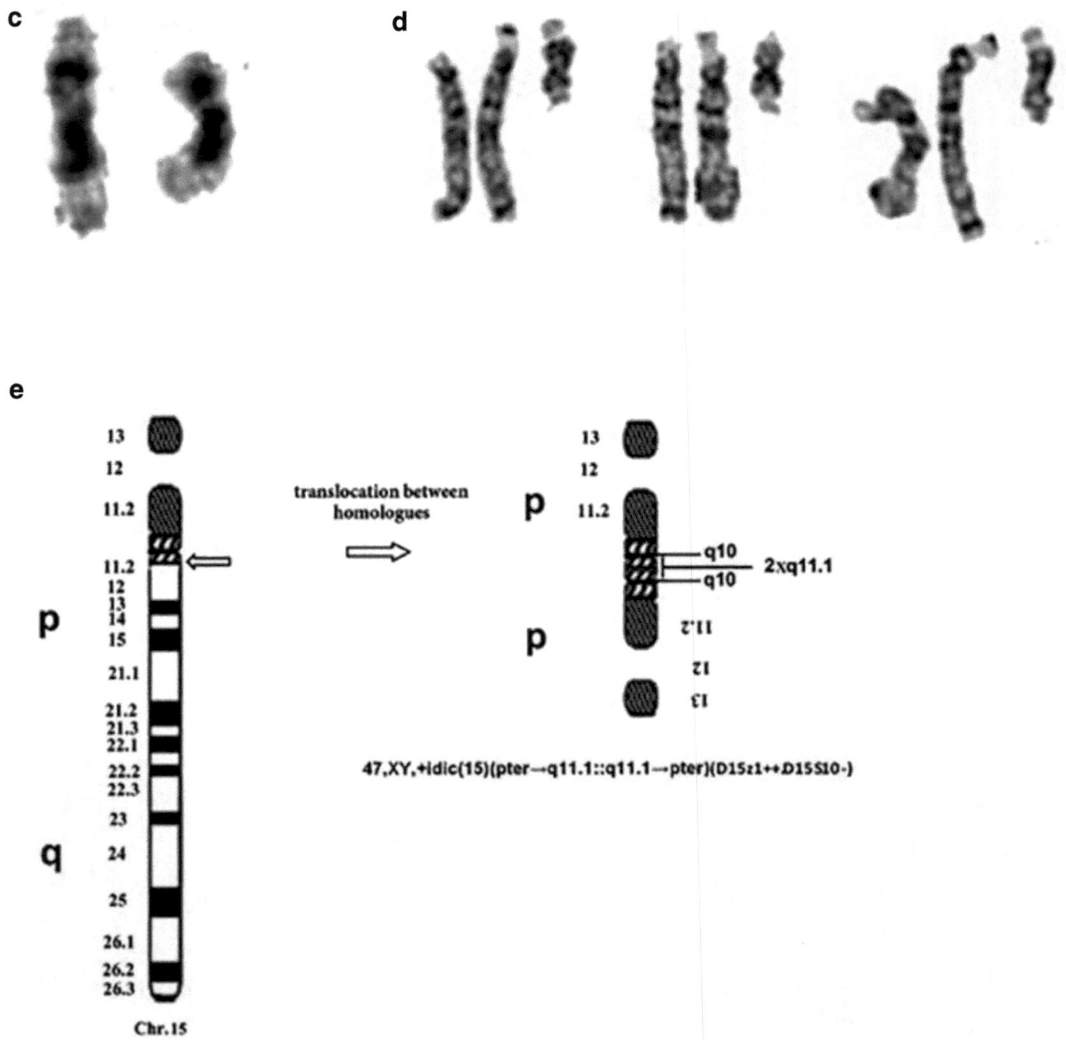

Fig. 14.2 (continued)

▪▪ Case 88: Karyotype: 47,XY,+idic(22)(pter→q11.1::q11.1→pter)mat

Indication for chromosome analysis: Prenatal diagnosis as the mother is carrier of an additional derivate 22.

▪ Morphology

The derivate 22 in mother and foetus are identical. The marker in the foetus is also heterochromatic, dicentric with inactivation of the second centromere. It originated by rearrangement from two homologous chromosomes 22.

▪ Cave

This child has to be informed when it is grown up by genetic counselling that it has an increased genetic risk independent of his age for UPD 15 and microdeletion or duplication 22q11.2. Therefore prenatal diagnosis has to be offered in all pregnancies (◘ Fig. 14.3).

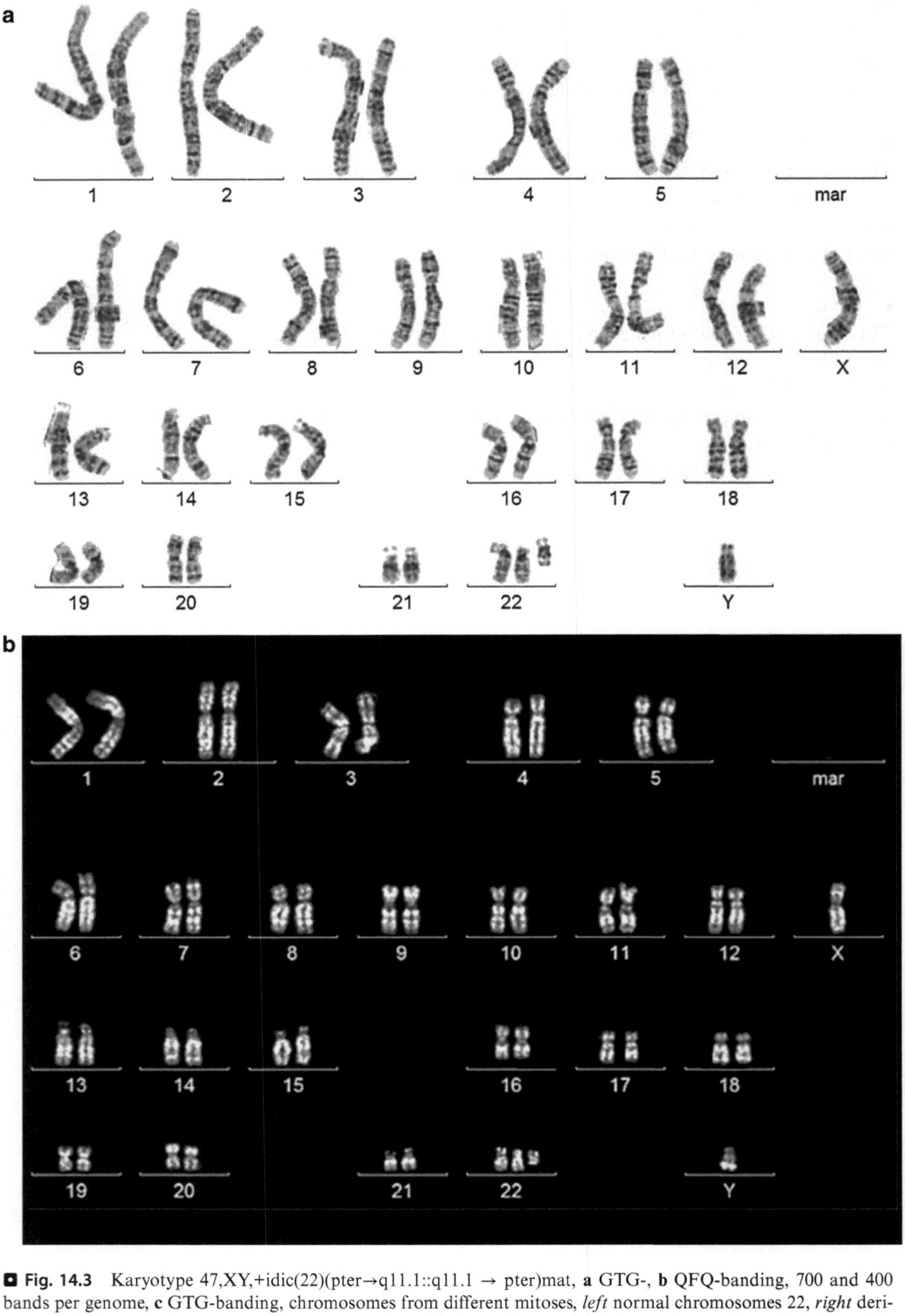

Fig. 14.3 Karyotype 47,XY,+idic(22)(pter→q11.1::q11.1 → pter)mat, **a** GTG-, **b** QFQ-banding, 700 and 400 bands per genome, **c** GTG-banding, chromosomes from different mitoses, *left* normal chromosomes 22, *right* derivate 22, **d** ideograms from normal chromosome 22 and derivate 22, completely heterochromatic marker, thus without genetic relevance for the carrier

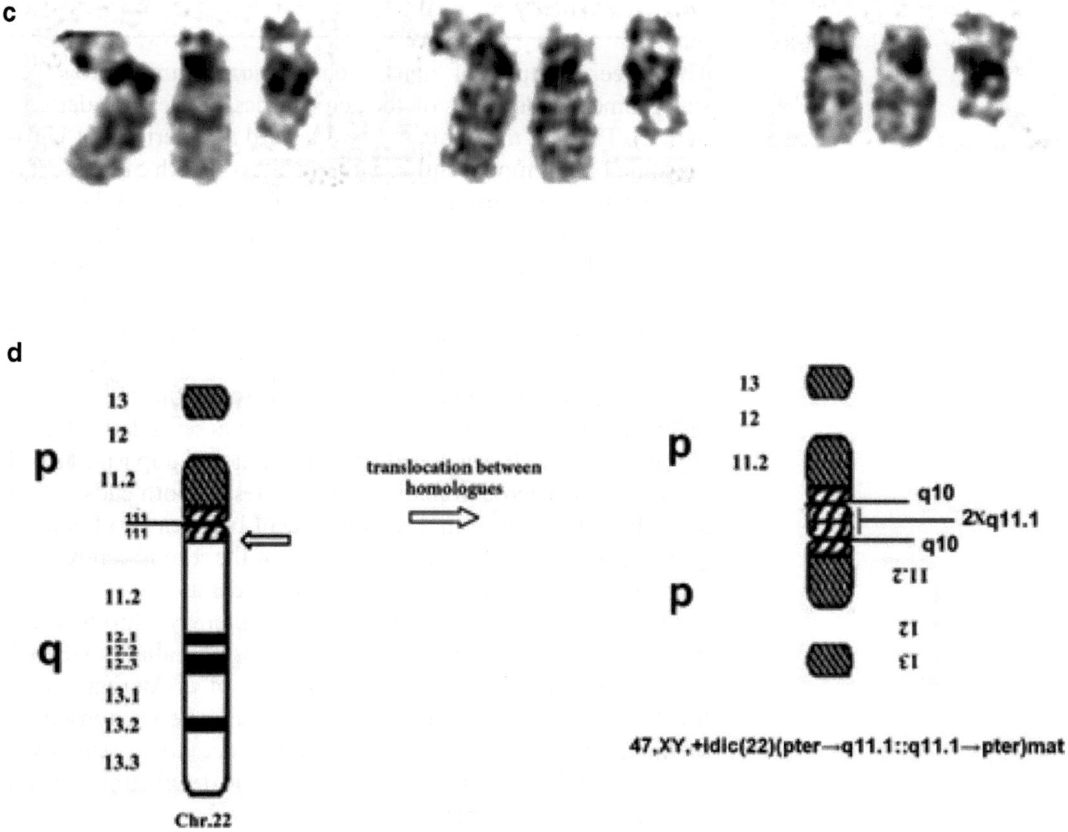

Fig. 14.3 (continued)

14.2 Summary

The three examples of marker chromosomes are all together supernumerary derivates of the acrocentrics, i.e. der(14), der(15), der(22). The first marker could be analysed as a derivate of chromosome 14. It is monocentric, with satellites on both ends, and it is completely heterochromatic. It was familial in origin and derived from an inversion duplication. An isochromosome 14 could be excluded. The second marker is a derivate of chromosome 15, the most frequent origin of a marker. This der(15) is dicentric with one centromere inactivated, again originated from the two homologous chromosomes and with satellites on both ends. The occurrence of euchromatic segments between the two centromeres could be excluded (q11.2). The third marker is derived from chromosome 22. It originated from an inversion duplication and is completely heterochromatic with satellites on both ends. It can again be delineated from a rearrangement between two homologous chromosomes 22. The inactivation of the second centromere is complete; thus, a second cell line is not present.

The diagnosis of the three supernumerary chromosome derivates required in addition to chromosome analyses FISH and microarray for a precise identification of the derivates and for the exclusion of euchromatic segments. Genetic counselling of the families has to include the risk of unequal crossing over in meiosis of the carrier which might lead to an unbalanced karyotype in the offspring.

Heteromorphismes: Mutations in Non-Coding-DNA Regions

Contents

15.1 Euchromatic Variants – 340

15.2 Satellites – 341

15.3 Nucleolus Organising Regions (NOR-Regions) – 343

15.4 Pericentromeric Heterochromatin – 344

15.5 Centromeres – 345

15.6 Heterochromatic Blocks – 346

C. Behrend et al., *Human Chromosome Atlas*, https://doi.org/10.1007/978-3-031-10588-3_15

In the following chapter, frequent heteromorphisms (synonymes: variants, polymorphisms) are presented in an overview. As this Atlas gives examples on structural chromosome aberrations with clinical relevance, the role of variants is only to recognise, define and distinguish them from rearrangements in euchromatic regions.

The different types of variants are described and characterised in the introducing chapter of this Atlas.

15.1 Euchromatic Variants

Euchromatic variant within a heterochromatic block.

▪ Morphology

The proximal heterochromatin blocks 1q12, 9q12, 16q11.2 and the distal long arm region yq12 can be subdivided once or even several times. These interbands are CBG negative and can differ in size. They do not consist of genetic relevant euchromatin, which has been translocated by an inversion or insertion but consist of non-coding repetitive DNA sequences and are defined as genetic non relevant euchromatic heteromorphisms. In chromosomes 1q12, 16q11.2 and yq12, these interbands appear bright after GTG-banding while the interband in 9q12 appears dark.

▪ Example

The heterochromatic region 1q12 is subdivided by euchromatic polymorphic interbands. After CBG staining, they appear light in the otherwise dark stained region. Even after GTG-banding the euchromatic interbands which differ in size are clearly diagnosable (see also case 21, paracentric inversions) (◘ Fig. 15.1).

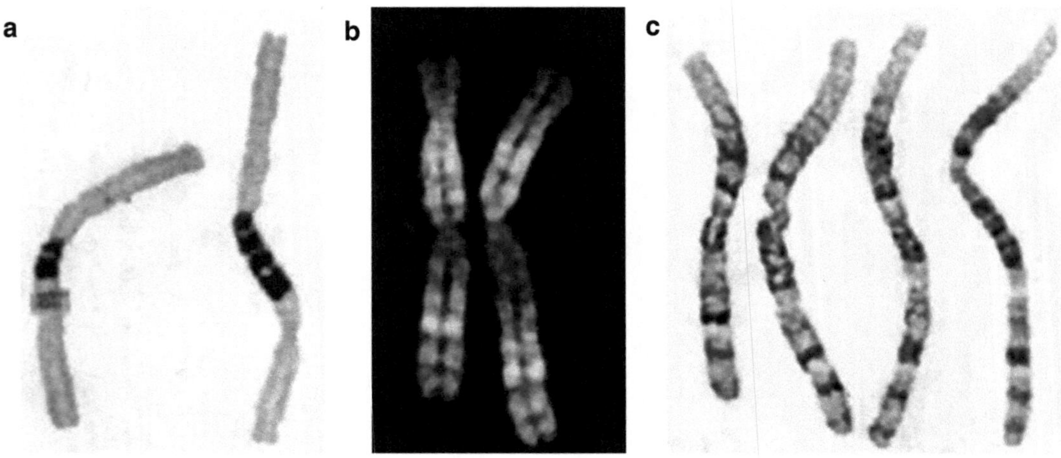

◘ **Fig. 15.1** Euchromatic variants presenting as light interbands in 1q12. **a** CBG-, **b** QFQ-, **c** GTG-banding, left normal chromosome, right amplification of 1q12 with euchromatic polymorphic interbands

15.2 Satellites

The terminal regions (p13) of the acrocentrics can be duplicated, deleted or amplified without relevance for the carrier. Here, we show duplications after QFQ staining combined with a brillant flourescence (i5).

▪ Example

After QFQ-banding, the satellites (p13) of the acrocentric chromosomes present with different fluorescence intensity (i1–i5 according to ISCN). The brighter fluorescence, also defined as brilliant (15) shows different frequency in its interchromosomal distribution (i.e. between 7 and 25% (see also: Case 36, paracentric inversion) (◘ Fig. 15.2).

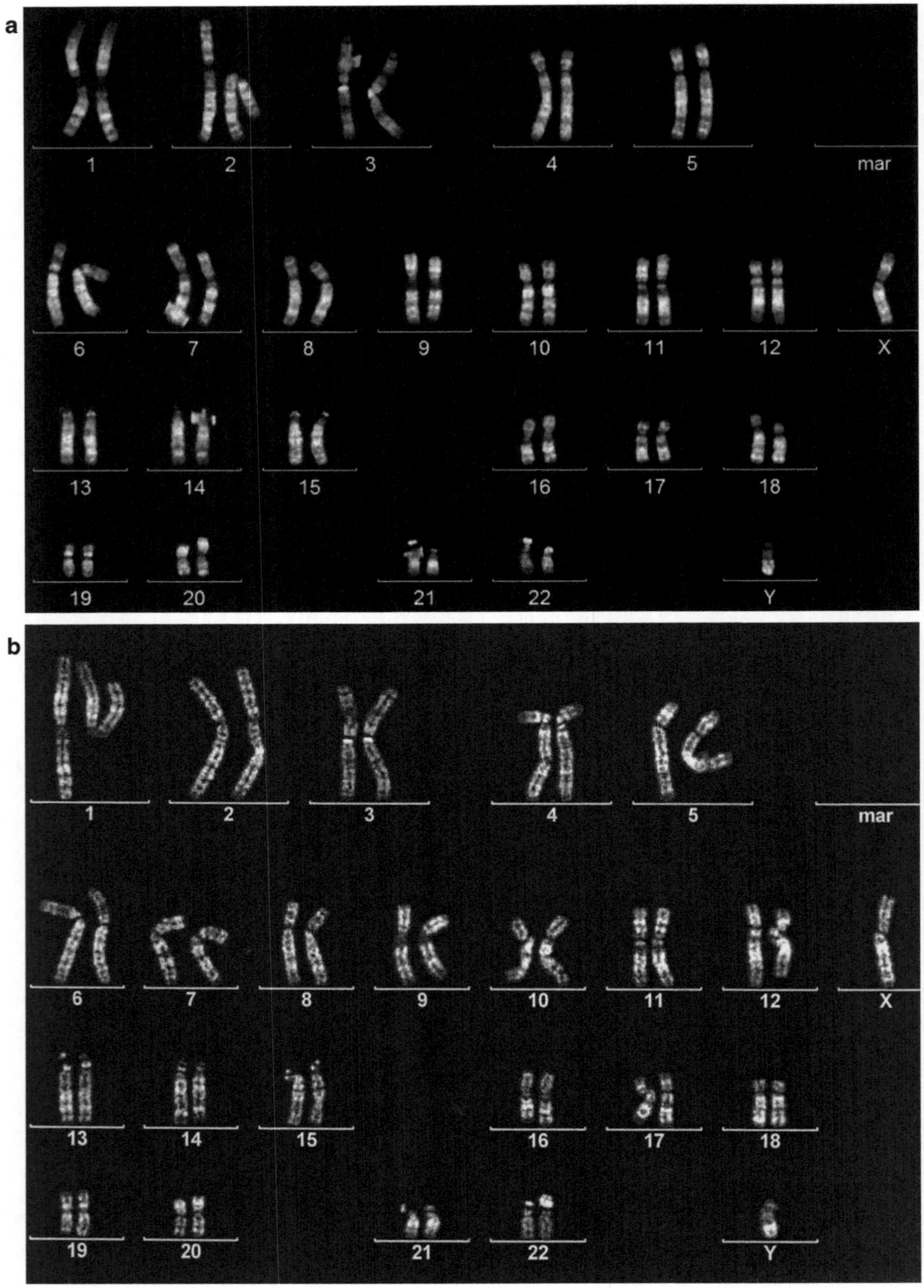

Fig. 15.2 **a** Karyogram after QFQ-banding Brilliant fluorescence in 14p13 (1 homologue), 15p13 (both homologues), 21p13 (both homologues) and 22p13 (both homologues). Besides, the size of p13 differs. **b** Amplification of the band 22p13 showing brilliant fluorescence (a translocation between Yq12 and 22p13 was excluded by FISH)

15.3 Nucleolus Organising Regions (NOR-Regions)

Mutations in p12 of the acrocentrics are frequent. The region can be amplified, deleted or together with p13 duplicated or triplicated (◘ Fig. 15.3).

Examples

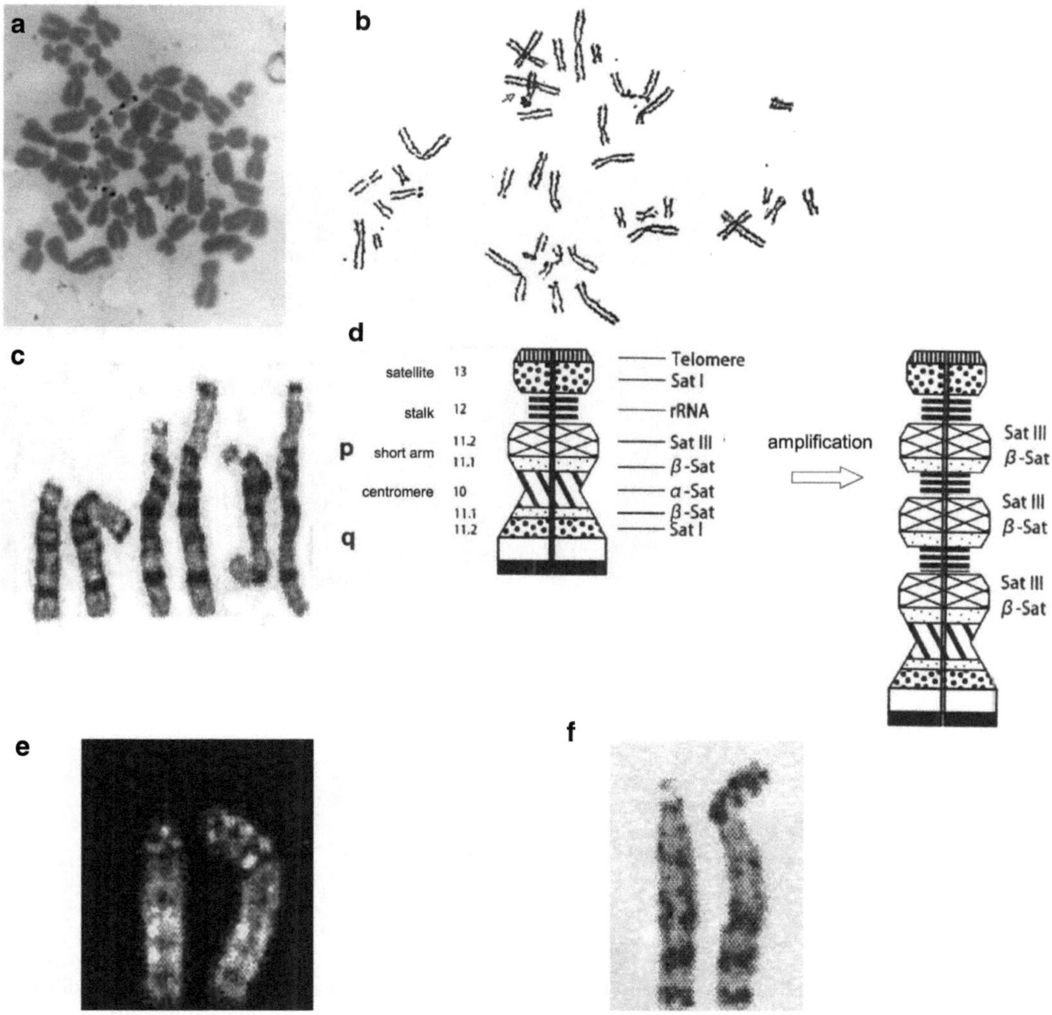

◘ **Fig. 15.3** **a** Metaphase after AgNO3 staining showing NOR-regions (p12) of different size in the acrocentrics, Amplification of the NOR-region 14p12, **b** After NOR-staining, **c** GTG-banding of chromosomes from three different metaphases. The right, aberrant one, additionally shows puffing of p12. **d** Graph of the amplification 13p12, **e** QFQ-, **f** GTG-banding left normal, right aberrant chromosome 13

15.4 Pericentromeric Heterochromatin

■ **Examples**

Neighbouring the centromere on both sides in the position p11 and q11, these bands can be decreased or increased in size. Often, together with the following heterochromatic band q11.2 they are involved in pericentromeric inversions such as 3q11.2, 4q11.2 and 19q11.2 (see also the pericentric inversions: case 3 (chromosome 19), case 4 (chromosome 4) and case 16 (chromosome 3) (◘ Fig. 15.4).

Diagnostic problems were given by deletions and duplications of 15q11.1 in times before the FISH area when the size of the region was misinterpreted and chosen as a diagnostic tool when investigating patients with clinical symptoms of Prader Willi or Angelman syndrome (◘ Fig. 15.5).

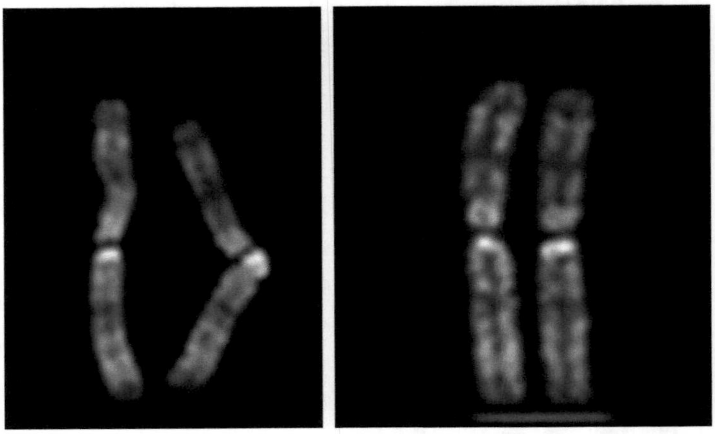

◘ **Fig. 15.4** Chromosome 3, QFQ-banding, *left* normal, *right* duplicated region in the proximal long arm [brilliant fluorescence i(5)]

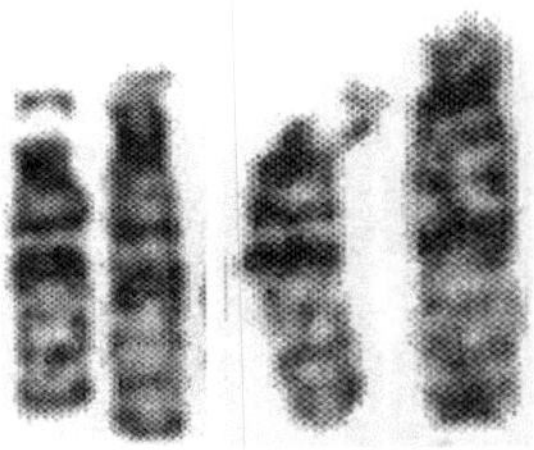

◘ **Fig. 15.5** Chromosome 15, GTG-banding chromosomes from different mitoses, *left* normal, *right* chromosome 15 with duplication 15q11.1

15.5 Centromeres

- **Examples**

A dicentric chromosome with inactivation of one centromere is a well-known rearrangement of the X-chromosome, in derivates of chromosome 15 and in the Robertsonian translocation 13/14. The size of the centromere can be reduced or enlarged, as it is often observed in chromosome 18 (◘ Figs. 15.6 and 15.7).

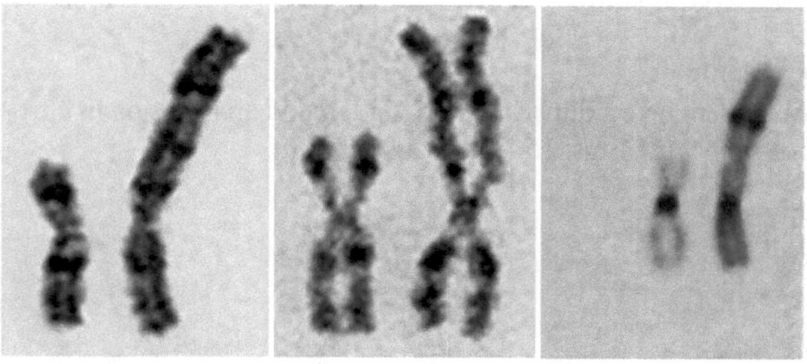

◘ **Fig. 15.6** X-chromosome from different mitoses, GTG-banding, *left* normal, *right* dicentric X-chromosome

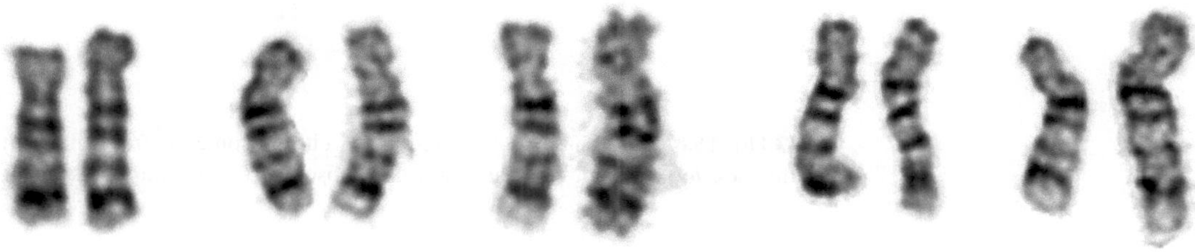

◘ **Fig. 15.7** Chromosome 18 from different mitoses after GTG-banding, *left* normal, *right* aberrant chromosome with increased centromere region

15.6 Heterochromatic Blocks

Mutations in the regions 1q12, 9q12, 16q11.2 and Yq12 show different types of rearrangement, their frequency depends on the population investigated (inversion 9q12 from 1 to 10%). The types of abnormalities are deletions, duplications, amplifications and partial and complete inversions. They are analysed after GTG, QFQ or CBG banding. Some are presented as examples combined with chromosome aberrations even in a homozygous status.

- **Examples**

We present here a duplication of 16q11.2 and an inversion in chromosome 9 (◘ Figs. 15.8 and 15.9).

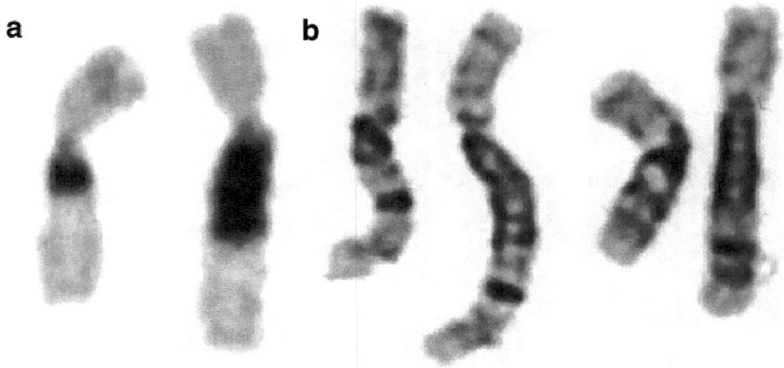

◘ Fig. 15.8 **a** CBG-, **b** GTG-banding of chromosome 16 from different mitoses, *left* normal, *right* aberrant chromosome 16 with amplification of 16q11.2

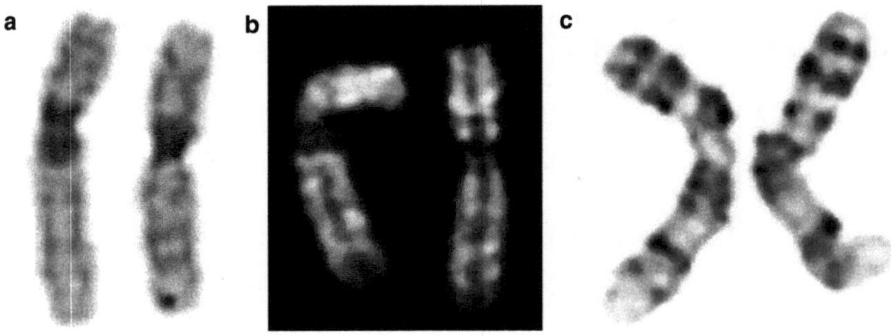

◘ Fig. 15.9 **a** CBG-, **b** QFQ-, **c** GTG-banding of chromosome 9; chromosomes from different mitoses, *left* partial, *right* complete inversion of 9q12

Guidelines

Contents

Chapter 16 **Genetic Counselling Procedures** – 349

Chapter 17 **Life Courses** – 359

Chapter 18 **Support Groups** – 369

Chapter 19 **Diagnostic Procedure** – 375

Chapter 20 **Outlook** – 391

Genetic Counselling Procedures

Contents

16.1 Overview – 350

16.2 Case Reports – 352

16.2.1 Carrier of Unbalanced Translocation 16/21 de Novo – 352

16.2.2 Familial Translocation Y/22 – 353

16.2.3 Familial Translocation 4/22 – 354

16.2.4 Homologeous Translocation 2/2 de Novo – 356

16.2.5 Mosaic Translocation Trisomy 21 – 356

C. Behrend et al., *Human Chromosome Atlas*, https://doi.org/10.1007/978-3-031-10588-3_16

16.1 Overview

Genetic counselling in Germany represents an exchange of information between consult seeking persons and specially qualified medical doctors. In some other countries, counselling may be taken over, to some extent, by qualified personal following additional training. The objective of counselling is the clarification of facts concerning the individually existing chromosomal, monogenetic, or the multifactorial genetic disorder and the developmental disabilities caused by them. The potential risks of a disease for the consulted person, for descendants, and for relatives will be discussed.

On the one hand, this first diagnostic interview functions as an information concerning the medical evidence, on the other hand, it is the basic for the shaping of an independent opinion for the advice seeking person and the forming of a decision based on the assessment of personal values and according to the rating of one's own situation.

In addition, genetic counselling informs about services and provides information to assistance, to support groups (► Chap. 18), and to special therapy centres.

Chromosomal aberrations form a considerable proportion of the causes for genetic counselling.

If a proband, his descendants, or relatives are diagnosed with a chromosomal anomaly, balanced or unbalanced, the first move has to be to inform the primary physician who arranged for the analysis.

This doctor in turn then, after informing the patient and/or his or her relatives or legal guardian, will initiate genetic counselling by a specialist for human genetics.

This counselling is always voluntary, and open and unbiased.

Underlying is the gene diagnostics law (Gendiagnostikgesetz) that defines the requirements and proceedings of counselling (► Chaps. 2 and 3).

In the course of genetic counselling, following the anamnesis of the patient, it is necessary to evaluate data of the family pedigree, including all members and pregnancies of at least the last three generations.

Based on our experiences in counselling following diagnostics of a chromosome aberration, in about 15% of the cases an additional, frequently monogenetic, risk has to be factored in alongside with the chromosomal mutations. In many cases, this risk factor will not be revealed till after the thorough analyses of the family data and then requires additional investigations.

Following the diagnosis of a chromosomal abnormality, this will be explained and demonstrated to the proband by means of a karyogram. In case of a structural aberration, additionally, an ideogram will be used to demonstrate the breakage and exchange points. Under special conditions, it is helpful to discuss FISH images in addition.

It proved to be useful to provide probands with schemata and images of the chromosomal aberration in question so that they can visualise the information after the consultation and talk it over with their family.

This procedure is especially relevant if an aberration is diagnosed that may have affected additional members of the family (► Sects. 16.2.1 and 16.2.2; ► Sects. 17.2.2, and 17.2.3).

In about 20% of our cases of structural aberrations, mostly translocations, the probands asked for a second counselling, and often they were then accompanied by family members.

In the course of genetic counselling, carriers of genetic aberrations will frequently ask specific questions. These are the questions about the possibilities to become pregnant at all; the increased risk of abortion, possibly repeatedly; the potential level of reduced fertility; the delivery and life expectancy of a disabled child.

Frequently, carriers of a balanced chromosomal aberration feel guilty because of their increased genetic risk and experience even difficulties to comprehend the consequences of a prenatally established balanced alteration because they inherit an increased genetic risk factor to the next generation.

While discussing the given genetic risk of the counsel seekers having children with an unbalanced karyotype, it is essential to indicate that there is statistical risk assessment, but individual cases may largely deviate.

Essential factors are the gender of the carrier, the type of structural aberration, a possible involvement of the gonosomes, the position of the breaking points, the size of the involved chromosomes and their gene contents, as well as a potential position effect and pairing anomalies during germ cell formation (pachytene of meiosis I).

A statistically high risk implies a number of pathological pregnancies in the range of 10–50%—with the exception of translocations between homologues being 100% (► Sect. 16.2.4)—an intermediate risk in the range of 1–10%, and a low risk equal to or under 1%.

An additional special group of probands with a balanced structural alteration such as whole arm translocations have no increased risk for pregnancies with an unbalanced karyotype, as all unbalanced karyotypes are lethal, except chromosome 21. But the carriers are sterile.

Certain structural aberrations carry an additional elevated tumour risk even for the carrier of a balanced alteration and this risk factor has to be discussed, if the chromosome breakage point is in the proximity of a tumour suppressor gene—either a protooncogene or its regulatory units.

Besides, based on empiric experience, interchromosomal effects have to be discussed for certain chromosomal aberrations. Separate from the existing structural abnormality, aneuploidies in non-involved chromosomes may occur.

For the subsequent generation of carriers of a balanced structural aberration, genetic counselling should be recommended for them, when they are grown up and want to have children of their own. Among others, the improved diagnostic procedures at that time for the characterisation of the chromosomal rearrangement and the possibilities of prenatal diagnostic must be discussed.

A particular, but not very frequent group of structural chromosomal alterations is represented by heterochromatic mutations. As a rule, they are not relevant for the carrier, but inadequate diagnostics may misinterpret them as euchromatin mutation, and that may even lead to the termination of a pregnancy (► Sect. 16.2.2). This wrong interpretation can especially occur in translocations of Yq12 to the short arm of an acrocentric chromosome. Additional banding techniques are always recommended for the primary investigator if the result of the chromosome analysis is not sufficient according to quality standards.

In summary, genetic counselling is a process of communication including the following contents: decision-making assistance, preventive action, assessment of the personal risk, recommendation for additional clinical counselling, information for relatives, information for further assistance (support groups, ► Chap. 18), and the forwarding of a summary in writing.

In the following, the procedure for counselling of diagnosed structural aberrations shall be presented, on the basis of five case histories of our individual collective.

16.2 Case Reports

16.2.1 Carrier of Unbalanced Translocation 16/21 de Novo

A couple comes for consultation and needs additional assessment about their second child, a son, who was diagnosed with a structural chromosomal aberration. After lymphocyte culture of the child and his parents, the pathologic karyotype of the proband was determined to be de novo:

46,XY,-21,+der(21),t(16;21)(q11.2;p13),GTG,700 B/G.

105 mitoses were analysed in all, one of which showed a normal male karyotype.

Therefore, a skin biopsy as a representative of a different germ layer was analysed in order to clarify whether a chromosomal mosaic might exist. Long-term cultures with the analysis of 34 mitoses consistently revealed the unbalanced translocation, with no indication for a second, normal cell line. An instability of the translocation chromosome caused by a long region 21p12 in the centre of the translocation chromosome was discussed as the probable cause for the formation of the cell line with the normal karyotype.

The proband had been noticeable in prenatal analyses (triple-test, US: week 22), but no further steps were taken.

At birth, the proband showed multiple malformations and dysmorphias, in accordance with literature reference for a duplication 16q.

Development: Constant artificial respiration was necessary. Pneumonia was recurrent, and tube-feeding was required.

The proband could be discharged since his mother was a nurse and was capable of attending to the child's intensive medical care at home.

At the age of 3, the child still needed a tracheal tube. He was capable of turning and pushing up from a prone position. He could make eye contact but did not speak.

The parents wished for an additional child but did not want to take the risk of a possible parental gonad mosaic.

Therefore, a microsatellite analysis was performed for the proband and both parents. 10 STR markers for 16q were analysed, revealing that the translocation was caused by a zygotic failure.

As a result, the risk for the further occurrence of the translocation 16/21 could be excluded for the advice-seeking couple.

In addition, the medical history of the family with a survey of the pedigree was performed, with no indication of any relevant genetic risk factors.

An increased age risk did not have to be factored in. The older daughter of the couple is healthy and develops according to her age.

The advice-seeking woman became pregnant again and carried healthy twins to full term.

The proband passed away at the age of 3.5 years.

When his mother informed us about that, she wrote:

"Our deceased son is leaving behind a large gap. Strangely, I am hardly ever reminded of all the work he caused. Instead, I remember his nature, his smile, his gestures, his positive attitude, indeed, his entire personality. Many a time he made me reflect on what really might count before God".

16.2.2 Familial Translocation Y/22

Before amniocentesis, a couple came for counselling in the 12th week of pregnancy.

Indication for the investigation, besides an elevated maternal age, was the previous pregnancy that was terminated due to the result of the prenatal chromosome analysis. A translocation trisomy 21 based on a fusion 21q/22q was diagnosed; however, the resolution of the chromosomes with about 250 bands per genome was fairly small. The karyograms performed were presented at the present counselling session.

Noticeable was the information that a differentiated pathological analysis of the foetus in the 18th week of pregnancy did not show any symptoms of Down syndrome. The foetus was developed age appropriately and did not show any signs of malformation and/or dysmorphy. Indication for the prenatal chromosomal analysis had been the increased maternal age (38 years).

We performed chromosomal analyses of the parents after lymphocyte culture during the ongoing pregnancy, even before an amniocentesis. The result:

While the husband exhibited a normal male karyotype (46,XY, GTG, ca 700 B/G), the woman presented with an abnormal chromosome 22 which, with additional investigation, could be traced back to a heterochromatin translocation yq12 to 22p, Karyotype:

46,XX,der(22),ish t(Y;22)(q12;p11.2),GTG,QFQ,FISH, 700 B/G).

Additional investigations of the family revealed that the mother of the advice-seeking woman was also a carrier of the translocation.

Unfortunately, the advice-seeking couple had to be informed that there had been no justified indication for the termination of the first pregnancy since the foetus only exhibited the familial heterochromatic translocation. This affected the couple very deeply.

During the ongoing pregnancy, an amniocentesis was performed in the 16th week on the grounds of age. An ultrasound control revealed a timely developed foetus without any indication of a malformation. Cell culture diagnostics yielded a normal male set of chromosomes (46, XY), without the familial heterochromatin translocation.

At date, a healthy boy was delivered.

16.2.3 Familial Translocation 4/22

A couple was seeking counselling after the birth of their third child with the phenotype of Wolf-Hirschhorn syndrome.

The chromosomal analysis, after lymphocyte culture, revealed a derivative chromosome 4: Karyotype: 46,XY,del(4)(p11.2). Further chromosome analyses of the parents showed a normal female set of chromosomes (karyotype: 46,XX,GTG, 550 B/G), but a paternal balanced reciprocal translocation 4/22 (karyotype: 46,XY,t(4;22)(p11.2;p11.2),GTG,ca 550 B/G).

The indication for chromosomal analyses of the parents was given because, according to literature, up to 45% of the short-arm deletions of chromosome 4 can be traced back to a parental translocation.

The child in question died in the neonatal period.

The parents asked for counselling, in spite of the fact, that they did not wish to have any more children. They had already been briefed by their paediatrician about the phenotype and life expectancy of children affected by Wolf-Hirschhorn syndrome.

The couple had two live children, 13 and 14 years of age, both of them exhibiting slight to moderate developmental retardation, as well as growth retardation and facial dysmorphic features.

In a second counselling session, the advice-seeking couple wanted to get informed if those two children might also show an unbalanced karyotype, due to the paternal translocation.

The chromosomal analyses of both siblings revealed the existence of a duplication 4p in each case, which is the other possible transmission of the paternal translocation 4/22.

(karyotypes: 46,XX and 46,XY,der(22),t(4;22)(p11.2;p11.2) pat.

During the following session, primarily, the large variation of clinical symptoms with a duplication 4p was discussed, then the fact that deletions of the same chromosome section versus duplications will lead to a much more severe pathological phenotype of the carrier.

Due to the relatively positive development of their two children, the advice-seeking couple asked a question that was rare in counselling sessions:

Was it possible for their son or daughter respectively to have children of their own later on? How high would be the presumed expectancy for a normal karyotype in a grandchild, and would the grandparents then be legally entitled to raise this grandchild?

We pointed out to them several of the facts to be considered:

- It cannot be implied, without appropriate gynaecological and andrological examinations, that the children are normally fertile.
- Rather, a reduced fertility has to be expected, according to observations on various autosome disorders.
- The theoretical risk for a transfer of the unbalanced translocation amounts to 50%. Empirical surveys of carriers of an unbalanced karyotype, however, revealed a selection against pathological germ cells, leading to a smaller risk.
- In case of a pregnancy and, possibly, the existence of the unbalanced karyotype of the respective parent, the question arises if the indication to a termination of the pregnancy might be given.
- The constellation of the family does not necessarily imply that the grandparents are entitled to take responsibility for the grandchild, in place of their child.
- Further, there should be concern that the choice of a partner for their children might take place in a facility for people with disabilities, and that this partner might also have a genetic mutation.

The couple, after thorough discussion, accepted the situation.

16.2.4 Homologeous Translocation 2/2 de Novo

Before a scheduled sterility treatment, the blood of a couple with the unfulfilled wish to have a child was send to chromosome analyses.

Preliminary investigations had revealed normal gynaecological findings for the wife and aspermia for the husband which led to a discussion about the transfer of spermatids.

The wife had a normal female karyotype (46,XX,GTG,700B/G).

The husband showed a homologous fusion of both chromosomes 2: Karyotype (46,XY,t(2;2)(q10;q12),GTG,700B/G. Both submetacentric chromosomes 2 were replaced by 2 metacentric chromosomes with 2 short and 2 long arms. The mutation had taken place in the centromeric region.

The mode of exchange indicates that the translocation occurred de novo in the carrier.

The most likely phase for the formation of the aberration would be the pachytene of meiosis I with the pairing of the homologues and the crossing-over taking place then.

The presence of a homologous translocation will prevent the formation of germ cells with a normal karyotype, leading to an aspermia of the carrier.

For this reason, it is not necessary to perform any further cytogenetic analyses of family members.

Since no normal haploid gametes (spermatids, sperm cells) can be formed, an ICSI therapy is not possible for the advice-seeking husband.

After demonstrating and explaining the pathological chromosome findings to the couple, they were presented with the two remaining options to fulfil the wish to have a child: either the adoption of a child or a heterologous insemination. They decided to adopt a child.

16.2.5 Mosaic Translocation Trisomy 21

A couple was seeking counselling with their 5½-year-old daughter. The family had switched their paediatrician, and the new one was specialised in genetic disorders and experienced in detection of facial dysmorphias. By reason of the clinical symptoms of the proband, he sent blood for a chromosome analysis. Noticeable were a growth retardation, a delay of the speech development, and a motor retardation, especially in fine motor skills. Organ malformations like a heart defect did not exist, the conspicuous facial features corresponded to moderate symptoms of Down syndrome.

94 mitoses were analysed after lymphocyte culture, 86 of which showed a normal female karyotype, and 8 revealed an isochromosome 21:

Karyotype: mos46,XX,i(21)(q10)[8]/46,XX[86]dn,GTG, 500B/G.

Both parents presented with a normal karyotype.

Motor skills and speech development had been stimulated since early childhood.

After diagnosis, we arranged to have a differentiated development test carried out, in order to facilitate a basis for a specific therapy and to better evaluate the girl's abilities. The test revealed a developmental quotient (DQ) of 80.

The girl was limited mainly in the field of speech and motor skills. This corresponds with studies of probands with Down syndrome, but in this group the symptoms were much more serious when a complete trisomy 21 had been diagnosed.

A special therapy program was being established in order to mitigate the impairments given as much as possible.

This was especially important for school enrolment because adequate speech capability and good command of fine motor skills are necessary for an effective participation in the classroom. Impaired fine motor skills even prevent a child from learning how to write.

The girl's enrolment was placed back for 1 year. After that she had caught up sufficiently and was able to visit school successfully. Only in the subject math there was an agreement with her teachers to not grade her.

The parents were pleased that the cause for the delayed development of their daughter could be resolved that it could be adjusted with the help of a special therapy program and that their daughter now was successfully integrated.

In a final counselling session, the parents were informed that the actual time frame for the formation of the chromosome mosaic could not be determined. It could have taken place before or after the separation of the germ layers. Therefore, the existence of a gonad mosaic cannot be excluded, resulting in the recommendation to have a diagnostic investigation done for each pregnancy, regardless of age.

The parents will make their daughter be aware of the existing mosaic when she is an adult and seek counselling in order to be up to date on the available diagnostic methods.

The presented case underlines the fact that an optimal promotion for a child with a chromosome syndrome can only be assured through cooperation of the experts of the various disciplines.

In this case, two panel discussions were taking place, 1 year apart, with the participation of the human geneticist, the paediatrician, the psychotherapist, and the parents. The prospective elementary school teacher was included in the second session.

Life Courses

Contents

17.1 Overview – 360

17.2 Case Reports – 360

17.2.1 Carrier of Interstitial Deletion 9q de Novo – 360

17.2.2 Carrier of Unbalanced Translocation 14/21 – 362

17.2.3 Carrier of a Duplication 3q Derived from a Familial Translocation – 363

17.2.4 Carrier of an Interstitial Duplication 3q de Novo – 364

17.2.5 Carrier of a Terminal Deletion 18q de Novo – 366

C. Behrend et al., *Human Chromosome Atlas*, https://doi.org/10.1007/978-3-031-10588-3_17

17.1 Overview

As a rule, chromosome analyses of conspicuous probands showing phenotypic peculiarities are performed before their second year of age.

Indications for examination are, most of all, combinations of malformations, dysmorphisms, and development retardation.

To a limited extent, this holds true for prenatal examinations with an appropriately detailed ultrasound result.

On the other hand, documentation of chromosomal disorders and their typical phenotype in older children, adolescents, and adults is available at a lower rate. It is required, though, in order to characterise each specific chromosome syndrome because they are assumed to have age-independent and age-restricted features.

The documentation of Down syndrome constitutes an exception. For this particular chromosome aberration, all age classes are described in detail and disabilities and diseases that might occur with increasing age, as well as their specific therapies are published, and there exist images of the phenotype of Down syndrome probands at different age and of various ethnic groups.

These findings enable the investigator to compare the phenotype of Non-Caucasian and Caucasian probands.

In the context of a consultation, parents of a foetus, a newborn, or an infant with a rare structural chromosomal aberration will inquire about the prognosis of their child that is to say about its life expectancy.

17.2 Case Reports

In the following, the course of investigations and the steps of control over years are given by five case reports.

17.2.1 Carrier of Interstitial Deletion 9q de Novo

Chromosomal analysis of the female newborn was initiated because of multiple morphological abnormalities of the child:

Microcephaly (head circumstance below third percentile), brachycephaly, pronounced frontal humps, heterochromatism (right iris blue, left iris brown), hypertelorism, broad and depressed nasal bridge, small nose, low set and dysplastic ears, shortened frenulum with sublingual attachment, micrognathia, adduction of both thumbs, contracture of the left elbow, clubfeet (subsequently treated conservatively), hip dysplasia on both sides, hypoplastic kidney on the left side, wide spaced nipples, hypotonia, CT: mild diffuse atrophy of the brain. No malformations of any organs.

The paediatric centre in charge indicated the possibility of a generally positive development, supported by findings in the case of an infant in literature. In that particular case, the prognosis was formulated when the proband was 10 months old, though.

Chromosome findings: Karyotype: 46,XX, del(9)(q31.2q33.1) GTG, 850 B/G.

The breaking points were confirmed by FISH with four single copy probes from the region.

The chromosome findings of both parents were normal.

During the first consultation, the chromosome findings were demonstrated and explained. Possibilities for support measures were discussed.

In the course of the next 21 years, the family visited the human genetic institute on a regular basis.

Unfortunately, the comparatively positive prognosis at the newborn stage did not come true, in spite of optimal care.

Psycho-motorical development: Biometrically, length, weight, and head circumference continued to be under the third percentile.

The motoric development indicated: sitting independently at the age of 2.5 years, standing up at the age of 3 years, walking—to a limited extent—at the age of 7.5 years.

Caused by the persistently severe muscular hypotonia and the deformity of the feet, the proband had to depend on a wheelchair.

- *Psychological development*: There was no development of speech, the girl increasingly exhibited autistic symptoms, was hyperactive at times, had problems sleeping through the night, had a pronounced fear of contact, and developed a tendency to stereotype movements. At no time, it was possible to perform a developmental test.
- *Presentation at the age of 21*: Height, weight, and circumference of the head were under the third percentile, the proband presented with a pronounced muscle hypotonia, severe mental retardation, no speech development except for some undifferentiated sounds, hardly establishing contact, and she stereotypically chewed on a rubber band. The patient could no longer stay with her family. She had to be placed into a nursing home.

After publication of the case report in an international journal, a paediatrician contacted our institute because he was treating a boy with comparable clinical symptoms and the same deletion in 9q. Subsequently, the family of this proband got in contact with our institute.

The mother of the boy impressively outlined that in her case it took more than 2 years for her son's chromosome analysis to be done in spite of her repeated requests for this investigation. Only after the diagnosis of the deletion 9q, a qualified examination was performed and a specific therapy was initiated. Up to this time, the developmental abnormalities observed by the mother were not taken seriously by the specialists.

The mother got in contact with our institute asking us to establish contact with "our" family with the 9q deletion, so that she would be able to discuss with them existing or newly arising problems in the course of the children's development with them. This contact proved to be an advantage for both families as it led to psychologic relief for the parents and to the gain of qualified dialogue partners.

17.2.2 Carrier of Unbalanced Translocation 14/21

The female proband was born as the third child of healthy parents. At the time of the pregnancy, the maternal age was not increased (28 years of age). The mother was told early in her pregnancy that a cousin had delivered a child with Down syndrome at that time. The pregnant woman asked her gynaecologist for a prenatal chromosome analysis which was denied because of her age.

At date a girl with the phenotype of Down syndrome was born. The chromosome analysis revealed an unbalanced translocation 14/21. Karyotype: 46,XX,rob(14;21)(q10;q10),+21,GTG. Additional investigations showed that the mother, the maternal grandmother, plus her sister and daughter were carriers of the balanced translocation 14/21. A cousin of the index patient also revealed the unbalanced translocation 14/21.

The index family was fully briefed about possibilities of support and any expected development retardation. The family took advantage of any support possible. The girl was fully accepted in the family, and both older siblings looked after her tenderly.

In the cause of the next 20 years, the parents consulted the human genetic institute regularly in order to discuss any occurring questions or problems.

The girl grew up in a rural environment with lots of snow in the wintertime. She had a pony and learned to ski at a young age. She performed well in cross country skiing. She participated in regional and later on in national championships and earned gold and silver medals on a regular basis.

She participated in the Paralympics three times. In Idaho, USA (2009), she earned 2 gold and 1 silver medal, in Pyeongchang, South Korea, she received a silver and a bronze medal. Subsequently, she was invited by the prime minister of Bavaria to be honoured as an athlete. In the year of 2020, she participated in the winter special Olympics in Obersalzburg, Austria, and earned two silver medals.

The proband is now 41 years old and has been carrying out her training independently for 10 years, including swimming on a regular basis.

She was invited to a swimming competition by an out-of-town booster club 2 years ago and won a bronze medal over 1500 m.

The young woman grew into a confident personality with a wide circle of friends whom she likes to join to go dancing. Her parents rent out vacation homes. She took over necessary tasks and performs them independently.

She is aware of her disability, makes allowances for any given restrictions, and we are all pleased to see her lead a happy life.

The examination of a second family with a familial translocation 14/21 demonstrates how important the social acceptance may be. Chromosomal investigations of 40 persons out of 4 generations were performed. This family showed increasing problems concerning fertility, miscarriages, and children that died at an early age. For two of them, the phenotype of Down syndrome was verified.

In the remote rural region, the family was banned. The minister proclaimed in a sermon that it was a judgement from above. Consequently, the members of this family had no free choice of partners. Some abstained from children of their own, one member married into an isolated family caused by a monogenic mutation.

At the close of every chromosome analysis in this family, one relative embraced us and quoted: "We are not cursed anymore, we just have a chromosome imbalance"!

17.2.3 Carrier of a Duplication 3q Derived from a Familial Translocation

The advice seeking couple had three children, two sons with disabilities and a daughter that developed according to age.

At age 1, detailed clinical investigations were made for the second son, revealing the phenotype of a duplication 3q25-qter. The chromosome analysis showed an unbalanced translocation 3/22. Karyotype: 46,XY,t(3;22)(q25;p11),GTG,700 B/G. Subsequently, this karyotype was verified for the handicapped older brother and, in the balanced version, for the sister, mother, and maternal grandmother, respectively.

A development programme was established for the initial proband. At age 3, he received hearing aids because of moderate hearing difficulties resulting in a distinct increase of development. At age 5, the boy could successfully be enrolled in a kindergarten for developmentally retarded children.

In addition to annual paediatric check-ups, various development tests were performed at 16 months (DQ 50), at 26 months (DQ 37), and at 38 months (DQ 43).

Subsequently, any occurring problems were taken into account, adequately.

The handicapped 10 years older brother lived on a farm with his grandparents.

His development test at age 10 revealed a DQ of 15. The boy could not make himself understood, only uttered inarticulate sounds, could not handle objects, and had meltdowns after failures. His body control was equivalent to that of a 1-year old child.

Conceivably, his prognosis, therefore, is unfavourable.

The clinical findings of both brothers with identical chromosome aberrations emphasised, how fundamentally important the early assessment of a chromosomal disorder can be.

It is the basis for advancement of the child with due regard to any development deficiencies that have to be expected in the case of the present aberration according to literature.

In case of the same aberration of both brothers, the options of interference in targeted support with regular supervision are obvious, but even so on the other hand are the deficits if you forgo a targeted exogene interaction.

17.2.4 Carrier of an Interstitial Duplication 3q de Novo

The proband was the second child of healthy, non-consanguineous parents. The first child, a boy, was phenotypically normal and develops according to his age.

The proband was born after an uneventful course of pregnancy. The indication for chromosome investigation in the newborn period was somehow uncommon: severe perinatal asphyxia, repeated attacks of hypoglycaemia, thrombocytopenia, and feeding difficulties. Biometry at birth was within the normal range.

- *Morphological peculiarities*: High end bossed forehead, deep-seated eyes with long lashes, hypertelorism, laterally elongated eyebrows, broad nasal bridge, low-set and dysplastic ears, hip dysplasia. Malformations of organs could be excluded. The morphological peculiarities persisted in the following years.
- *Chromosome findings*: The conventional chromosome analysis was combined with FISH, microarray, and microsatellite typing. The result was in an interstitial duplication 3q of 35,6 Mb. Karyotype 46,XY, ish dup(3)(q13.11q23), GTG, 850 B/G (wcp3+, pcp3p-, pcp3q+, 939D5+, 904D1+, 949CD++, 967F11++,766D8++, 800G12++, 925B1++ 967H1+).

- The karyotype of both parents was normal, the duplication was of paternal origin.
- *Development*: In the course of his development, the boy suffered repeatedly from severe infections.

  At the age of 4 years, a corneal distortion was diagnosed and treated by special spectacles. The development of the teeth was normal, but the enamel of the incisors was defect and so they had to be removed at the age of 4 years. An apertognathia was treated conservatively, but as the result was insufficient, a surgical correction is planned. From the age of 5.5 the boy suffered from rheumatism (arthritis juvenilis) which was treated by special medication and physiotherapy. During the following year, the proband developed Wegener's granulomatosis syndrome with an increased number of antibodies.
- *Biometry*: At birth, all parameter tested were within the normal range but decreased in the following years to the third percentile and in the length below it.
- *Motor development*: Since the newborn period, the boy had been suffering from severe muscular hypotonia, which led to feeding problems. Until age 2, he received liquid food, after that he started with normal food, but had difficulties with chewing.
- According to the parents, the boy started sitting and crawling at year 1 and he fully achieved it with 2 years. Walking without help began with 2.5 years, but for a long time he was unsteady and tired quickly. Walking became normal at about 4 years but climbing stairs not before 7 years. At that age, he still needed help when dressing, washing, and using the toilet.
- *Speech development*: Hearing was normal, but speech development significantly delayed. Therefore, at age of 2, a logopaedic therapy was started. Caused by this support, he was able to speak single words at the age of 3 years and with 7 short sentences. If he had to concentrate, he produced stereotypical sounds. In contrast, his speech comprehension was developed according to his age.
- *Social abilities*: At age of 2, the boy distinguished between relatives or friends and strangers whom he refused. At the age of 4 years, he had developed his own personality. At the age of 5 years, he was well integrated in a special kindergarten and at 6 years in a special school. Aged 7.5 years, he had friends and according to his teachers he was helpful to handicapped friends, cooperative and sometimes he even dominated his group.
- *Mental development*: Until an age of 6 years, the boy had severe sleeping problems, but in the course of time the intervals of sleep disturbance decreased as well as their intensity. In his special school, he understood given exercises after only a single explanation and then acted appropriately. He learned to read and the basics of writing, but he was handicapped in the latter one by his reduced motor abilities.

To summarise the boy showed a very positive development, in the course of the years in spite of a number of severe handicaps. This was particularly enabled by the regular controls of the paediatrician, by physio- and logotherapy, and the adequate selection of kindergarten and school.

But first of all it was the family, that accepted and integrated the boy with all his handicaps, and who, over the years, supported all the specific therapies, that were performed by specialists and the family at home.

17.2.5 Carrier of a Terminal Deletion 18q de Novo

At age 2, the proband was examined for the first time cytogenetically and additionally again at age 16, in order to localise the chromosomal breaking points.

Two independent structural mutations could be detected.

- *Karyotype*: 46,XY,inv(2)(p13.1q21.3),del(18)(q21.32q23).

  For verification and to prove that both aberrations were independent occurrences, FISH was performed with WCP 2 and WCP 18. The homologous pairs 2 and 18, respectively, were marked thus excluding a chromosomal exchange.

  The deletion 18q occurred de novo, the pericentric inversion was familial and balanced.

  The family is residing in South America, the chromosome analysis was set up by their paediatrician in Germany.

  Indication for a chromosome examination was a delayed development of the boy as well as the existence of morphological abnormalities such as brachycephaly, broad, bushy eyebrows, low-set ears, broad and depressed nasal bridge, pointed tip of the nose, everted lower lip, and a small chin.

  This phenotype exhibits a substantial match with the de Grouchy-II syndrome according to observations from literature.

  The family stayed in touch, the proband was presented five more times up to the age of 39.

  In the forefront were increasing mental and psychological problems that were harder and harder to treat.

  A cardinal symptom of deletion 18q is a narrowing of the ear canal, leading to hearing loss. A hearing test was performed that revealed a 70% reduction of hearing. The boy was supplied with hearing aids and received as much logopaedic therapy as possible.

- *Development*: the proband exhibited contractures of the hands that could only be treated to a limited extent.

  At age 15, a hernia inguinalis and a hernia umbilicalis had to be corrected by surgery.

  The boy presented with hypothyroidism that needed treatment, a mild myopia, incontinence, muscle hypotonia and pes valgus, and a collapse of the mitral valve. Regular treatment with antirheumatics was necessary (Gilbert-Meulengracht syndrome).

- *Mental development*: Due to regular psychotherapy and logopaedics starting at age 2, the boy was able to visit a regular school in South America, benefitting from tutoring. After graduation, he successfully finished college and worked as a computer scientist.
- *Psychological development*: From early on, the proband demonstrated low self-initiative. He developed anxieties, and depressive and compulsive disorders. As a consequence, he was under psychotherapeutic care and medical treatment up to age 39. His teeth were damaged due to permanent grinding.

At age 39, the proband is hyperactive, aggressive, easily frustrated, irritable, and at times demonstrates autistic behaviour.

His vocabulary is good, but he has trouble to express himself. His logical reasoning is reduced. He is non-independent and is unable to monitor his daily routine.

His social integration is successful, though, and the support of his family turns out to be outstanding.

The family worries about the existing behavioural problems, and the physical restrictions that severely inhibit a positive way of life.

Considering his problems, there had not been enough adequate consultants and therapists available in South America.

Therefore, the family decided to move to Germany and take advantage of the existing therapeutic opportunities.

The human geneticists, the family had established contact with, will give advice and counselling.

The prior presentation of case histories depicting carriers of unbalanced chromosome aberrations illustrated how many exogenous and endogenous factors are relevant for the preferably positive development of a proband.

It could be demonstrated that monitoring and support by a human genetic institute can ensure the necessary interaction of the various factors.

The first counselling session will provide the advice-seeking family with the information needed about the proband's existing aberration, its contingency—de novo or familial—and hence an approach to the given consequences in case of any further desire to have another child.

The physician in charge will be informed in detail in writing, and in cooperation with him and the family any additional examinations and development tests of the proband as well as any required therapies will be initiated.

There will be the suggestion to get in touch with a support group.

In addition, there will be the offer to get in touch with their counsellor or staff (assistants, co-workers) at any time. In many cases, this opportunity will be seized, as demonstrated.

Thus, it happened that families encouraged relatives or friends to look for a consultation of their own, when the indication is given.

Furthermore, by means of our own case histories, it could be demonstrated, that a positively responding environment will substantially aid in the curriculum vitae of a proband with a chromosome aberration and be a support for the family.

Therefore, the human genetic institutes aim at classes not only for students of medicine but also of psychology and teaching positions to illustrate outstanding problems of handicapped and their families during their whole lifespan. The goal is to establish a positive attitude towards the handicapped and, if possible, to win the participants of these organisations to a dedicated involvement and sponsorship of people with disabilities in their future professional careers.

For adults with disabilities, for example the Down syndrome, there are nowadays privately and government-sponsored institutions like the "Aktion Mensch" that promote the integration of disabled people so that they are able to lead an active life according to their abilities.

Life careers like the ones depicted here are encouraging developments for the human geneticist, specialist in cytogenetics and genetic counselling and in future they will influence their attitude towards handicapped.

Support Groups

C. Behrend et al., *Human Chromosome Atlas*, https://doi.org/10.1007/978-3-031-10588-3_18

An increasing cooperation between support groups and human geneticists took place over the last 30 years. Upfront in doing so, especially in the beginning, was the "Aktion Mensch" (a charitable non-governmental organisation), which has been founded for the purpose of capacity building. In addition, it provided scholarships and grants for advanced education of physicians in the specialised field of medical genetics.

At the very beginning of the cooperation was the attempt of a dialogue between clinical geneticists and representatives of support groups. Increasingly, this resulted in affirmative action to support the affected people and their families.

In addition, it was the objective of the cooperation to make the public aware of genetically induced disorders and achieve better acceptance of the people concerned. Information was provided via specific brief leaflets, lectures, television programmes, mutual activities of afflicted and non-afflicted people, as well as contacts of organisations for disabled persons with experts of medical associations for specific disorders, such as Down syndrome, muscle disorders, or neurodegenerative syndromes.

Additionally, the cooperation between self-help groups and human genetics focused on the promotion of basic research of the genetic causes of various forms of disabilities. Also, very effective psychological care of the affected and their families took root. Regular events were established, by which teachers of different types of schools, and nursery school and kindergarten supervisors were informed about the opportunities to include handicapped persons.

These events imparted the necessary knowledge and paired them up with support groups and human geneticists. Joint discussion meetings were organised for physicians in different disciplines.

The initial distancing of individual support groups towards human geneticists decreased, due to functions like congresses, seminars, workshops, and discussion panels. Joint goals and joint action were aimed at for the future though not feasible for all areas. Differences of opinion and discrepancies occurred, especially in focus areas like prenatal diagnostics, heterozygotic-testing, predictive analyses, and gene therapy. Increased experience of both groups as far as potential consequences were concerned and led to an objective and constructive cooperation over the years.

For several years, the participation of various support groups at the annual conferences of the German Association for Human Genetics (GfH) was established and turned into an essential part of the congress.

The following will show the goals and practical work of a self-help group by portraying the organisation "Ohrenkuss - da rein - da raus" ("Earkiss - there in - there out". The title was chosen by the disabled members of the journal) for and with people with Down syndrome.

The acceptance of this group's projects in public and in expert committees can be illustrated by listing the numerous awards they received and which are covered below.

The human geneticist Dr. Katja de Bragança came in closer contact with people with Down syndrome and their relatives in the course of the research to her dissertation about the qualitative and quantitative characterisation of specific morphological features of humans with Down syndrome. She experienced their contact problems with people without Down syndrome, the frequent occurrence of poor appreciation of those affected, the lack of understanding about their abilities and talents, and also the very diverse characters of people with Down syndrome.

Based on her professional knowledge and her practical experience including being first chairperson of the European Down Syndrome Association and organiser of the first German symposium "Down Syndrome", Dr. de Bragança was destined to launch an innovative group for people with and without Down syndrome.

The objective was the inclusion of disabled people by promoting their abilities and talents.

The results of the various scientific sub-projects were each presented publicly, thus enabling people with Down syndrome to be heard individually, frequently for the first time, and their capabilities were publicly recognised.

This approach encouraged other groups for caring people with disabilities to take comparable initiatives and activities.

In 1998, Dr. de Bragança established the semi-annual magazine "Ohrenkuss" (earkiss). The articles of the journal were each assigned to a specific subject area. They incorporate a compilation of poems, stories, aphorisms, interviews, graphics, and pictures of the authors with Down syndrome.

Four years later, the initiators established the "downtown-workshop for culture and science". Since then, this workshop has arranged regional and supraregional readings, lectures and seminars, exhibitions, creative workshops and studies.

Dr. de Bragança set up the participatory research institute TOUCHDOWN21 in 2014 (▶ www.touchdown.info). This included the project "TOUCHDOWN—an exhibition with and about people with Down Syndrome". It was established in collaboration with the Kunsthalle (Art Gallery) Bonn. Initially, this exhibition was shown in Bonn, (in winter 2016/17), then in the Kulturambulanz Bremen (in summer 2017), and finally, in the "Zentrum Paul Klee", Bern (the first two quarters of 2018) and reached a high number of visitors each time.

In the years 2018/19, the team organised the exhibition "What is important?" in Kiew (iZone) and in Odessa (Museum of Modern Art), in cooperation with the Goethe-Institut. This resulted in the cooperation and exchange of team members.

The academic projects outreached the standard extent, content, and way of presentation of other groups for and with disabled people.

Dr. de Bragança headed the integral art project "Arche Noah" in cooperation with the "Aktion Sorgenkind" as part of the Aktion Grundgesetz (Campaign Constitution) from 1997 to 1998.

Right after this until 2000, Dr. de Bragança initiated, in cooperation with Professor Schott, head of the Institute of Medical History, University Bonn, the project "How do people with Down Syndrome experience the world—how does the world experience people with Down Syndrome? A contraposition". The project was sponsored by the Volkswagen Foundation and presented the scientific results at the EXPO 2000 in Hannover in the theme park Basic Needs.

The exhibition "Seven saws in color, two balls in blue: Art exhibits and photos of the Arche Noah" took place in 1999.

A further project "People with Down Syndrome and their jobs—an analysis" was supported by the Donors' Association of German Science (2000).

There was one more exhibition in Bonn with the topic "At night", with photos by Michael Bause.

In the year 2005, Dr. de Bragança took a trip to Mongolia, together with the editorial staff of the journal Ohrenkuss (Ear kiss). Since people with Down syndrome still today are referred to as "Mongols", it was the objective of the trip to gain a realistic approach to people in Mongolia. The trip was escorted by the GEO magazine and was a benefit for all participants involved.

On occasion of the tenth anniversary of the magazine "Ohrenkuss" (Ear Kiss), the editors published the book "Ear kiss dictionary". It was awarded Gold in the category Corporate Books BCP: Best of Corporate Publishing in the year 2009.

In the following, the dictionary received the Silver Design Award of the Federal Republic of Germany in the category Communication Design, and after 15 years of publication, the journal received the Matthias Claudius Award for Social Engagement. Besides, TOUCHDOWN21 was honoured as a brilliant site in the land of imagination.

In 2016 the website—► www.TOUCHDOWN21.info—was nominated for the Grimme Online Award in the category Knowledge and Education.

After 20 years of publication, TOUCHDOWN21 achieved the PR-Award in the categories Responsibility and Sustainability Communication, and Digital Public Affairs, the Swiss Diversity Award, and the Golden Chromosome of the German Down Syndrome Center.

In recognition of her merits, Dr. de Bragança received the Order of Merit of the Federal Republic of Germany.

The portrayal of groups with and without Down syndrome demonstrated that Dr. de Bragança and her team succeeded in reshaping the receptiveness of the public at large towards people with Down syndrome and placing it on the objective basis desired. They accomplished this in cooperation with the universities of Bonn and Cologne and by regularly assisting students and interns and involving people with Down syndrome.

With her many years of activities, Dr. de Bragança achieved that partially existing prejudice towards handicapped people is being critically debated and that ethical principles acquire new meaning.

The authors of the Chromosome Atlas want to express their desire that the academic projects depicting the life conditions of people with Down syndrome may be a future basis and encouragement to launch comparable initiatives in Germany and abroad. Dr. de Bragança's future goals for basic research and areas of application fulfil in an outstanding way the goals of the Convention of the United Nations on the rights of persons with disabilities (2006).

The main rights granted are:

- Respect for the inherent dignity of the human being.
- Non-discrimination.
- Full and effective participation and inclusion in society.
- Respect for the diversity of people with disabilities and the acceptance of these people as part of human diversity and humanity.
- Creating equal opportunities.
- Respect for their right to preserve their identity.

Diagnostic Procedure

C. Behrend et al., *Human Chromosome Atlas*, https://doi.org/10.1007/978-3-031-10588-3_19

In the case of clinical suspicion that a chromosome aberration is present, it is necessary, first of all, to consider the somatic cell system suitable for the investigation. The diagnostic limitation has to be taken into account that arises as a result of the restricted availability of most cell types and tissues that may function as a test material. Furthermore, in the presence of a mosaic, it has to be considered that in a cell cultivation in vitro processes of selection may occur against one of the cell lines. Consequently, quantitative predictions may be distorted due to the advantage or disadvantage of growth or not reflect the actual conditions of the tissue examined. Besides, longer periods of cultivation or incubation of sub-vital cells may cause structural and/or numerical aberrations and have to be excluded in the diagnosis. In order to separate those emerging artefacts or pseudo-mosaics from true mosaic constellations, international agreements are in existence with regard to the diagnostic procedures in the way of staged so-called workups.

The presentation of the chromosomes is performed in the prometaphase and metaphase of mitosis. Allowing for the special indication, this ensures an adequate resolution of the bands for the assessment of the chromosomal structure.

The resolution of the chromosomes is defined by the system of international nomenclature (ISCN, 2020) and has to be indicated in the diagnosis. The predominantly applied banding technique is GTG-banding, which might have to be supplemented depending on the cytogenetic findings (QFQ, RBA, CBG, NOR, DA-DAPI). It has to be indicated in the results, if the required number of qualitatively sufficient prometa- and metaphases was not achieved.

In order to eliminate potential mosaics, the *chromosome analysis*, in individual cases, may be complemented by *interphase-FISH* with one or another target chromosome or sequences, respectively.

Furthermore, interphase-FISH will be applied, if all of the investigated cells are in interphase. This is the case in *polar body diagnostics*, the examination of *sperm cells* and *spermatids*, and analyses of *buccal mucosa* to verify or eliminate mosaics in various differentiated somatic tissues, the examination of sub-vital cells, that cannot be stimulated anymore to divide, but clinically present with the phenotype of a defined chromosome syndrome (e.g. trisomy-18) in the proband.

In the case of balanced or unbalanced *complex chromosome rearrangements* (CCR), FISH can identify the chromosomes and chromosomal segments involved with a combination of probes.

Mosaicism: The selection of a specific cell system for chromosomal analysis may result in a false finding caused by a tissue-specific intense selection in vivo and/or in vitro against pathological karyotypes. This is exemplified by the presence of an additional isochromosome 18p, mitotically unstable structural aberrations, triploidies, and trisomies 8 or 15. For these chromosomal anomalies, as a rule, the lymphocyte culture — the most frequently examined somatic cell system — will increasingly yield normal results, while fibroblasts and epithelial cells show a lower selection compared to pathological karyotypes with the result, so that the cell line with an aberrant karyotype is traceable over a longer ontogenetic period of time, in vivo and in vitro.

Unique in this category are cases with a variety of mosaic patterns, such as distributions that occur with and without centreline demarcation. They are associated with a varying expression of clinical symptoms. In case of a restriction of the chromosomal aberration to one side of the body — right or left — or rather only one or several areas collaterally, the clinical symptoms will only be expressed on the side or region with the pathologic karyotype.

In the context of CVS-diagnostics, a secondary reduction of the aberration can be observed. Besides, differences between findings from trophoblast and stroma are possible.

The early partial reduction of trisomy-18, which is a chromosomal disorder preferred prenatally analysed by CVS, occurs, without preferential location, in various sectors of the placenta, according to first-hand observations. Thus, a biopsy can be extracted from a placenta region, which may simulate the presence of a normal karyotype by way of aberration reduction. A second chromosome analysis has to be performed, if the foetus is phenotypically suspicious, for instance by amniocentesis.

Abnormalities of the cell cycle have been accounted especially in the presence of surplus chromosome derivates, for example, i(12p), and are a cause for selection of the pathologic karyotype. They are detectable by interphase-FISH and a specific centromere – or single-copy probe.

If two normal homologous chromosomes are in the G2-phase of the cell cycle and an additional chromosomal derivate is still in the S-phase, this situation may increasingly lead to maldistribution or loss, respectively lagging of the derivate in the ana- and telophase of the next mitosis.

Low gonosome mosaics of a pregnant woman may lead to misinterpretations of a Praena-test. For example, if this examination shows XO-cells, this may be caused by a maternal mosaic, although the foetus is showing a normal karyotype.

In this case with a differentiated normal US-finding, additional analyses have to be initiated.

A category of chromosomal anomalies, that are frequently compiled as special findings of pre- and postnatal chromosome analyses, are the so-called secondary chromosomal aberrations.

They are being evoked endogenously by a DNA-repair defect or exogenously by physical, chemical, or biological mutagens.

These types of aberrations include gaps, breaks, or exchanges on the chromatid or chromosome level. In these cases, the documentation of data includes the number of aberrant mitoses, the number of aberrations per aberrant mitosis, and the type of aberration.

Any evidence of a DNA-repair defect entails molecular genetic analyses in order to account for the underlying genetic defect, by including the clinical symptoms of the proband in the final diagnosis. In case of exogenous exposure by a mutagen, it is necessary to distinguish between weak and strong mutagens and single or permanent exposure.

In the case of single exposure, as to our experience it is advisable to repeat the examination within 4–6 months. As a rule, the result then will show a significantly lower aberration rate compared to the first examination, caused by selection against the pathological cells.

Another specific group of structural chromosomal variants are the fragile sites. As far as the X-chromosome is concerned, an amplification on the DNA-level in the region Xq27.3 — a CGG-trinucleotide expansion in the 5′-untranslated region of the FMR1-gene — was verified to be the reason for the fragile-X or Martin-Bell syndrome.

As far as the autosomes are concerned, so far more than 120 fragile sites are documented — 30 of the rare and 90 of the common type. A common similarity of fragile sites consists in the occurrence of one or more factors that originate in the DNA-architecture and have an inhibitory impact on replication.

Fragile sites may be expressed as a result of specific cultural conditions which may amplify these replication inhibitors as well as other modes of formation. Besides the percentage of cells with a fragile site of a specific chromosome is determined by the analysed cell system.

The vast majority of autosomal fragile sites is without direct genetic significance for the carrier, according to present-day knowledge. At present, a small amount of rare autosomal fragile sites, caused by their direct association with genes, is being linked to the occurrence of certain disorders, particularly in the neurological or neuropsychiatric spectrum. A correlation to tumour genesis, to immunodeficiency, to neurological, and to fertility disorders could be established for various common fragile sites.

A new examination method is the *microarray*, increasingly applied in diagnostics, for initial examinations as well as a supplement to chromosome analyses. Depending on the compiled findings — normal or pathological — varying depictions of the karyotype are being selected (ISCN, 2020). This technique comes into play especially in the localisation of breakpoints, the analysis of subtelomere aberrations and the detection of micro-deletions and duplications due to a pathological course of meiosis.

Finally, the analysis of chromosomes in various phases of *meiosis* has to be addressed briefly, at this point. To some extent, not the chromosomes themselves but the synaptonemal complex is being analysed with the examination of pairing disorders of homologous chromosomes being a leading issue.

These examinations predominantly involve three indication groups:

- Changes in the number of chiasmata,
- Pairing abnormalities of structurally changed chromosomes with their normal homologous chromosomes in the pachytene of meiosis I,
- Abnormal formation of the XY-bivalent.

There is a specific nomenclature available for the documentation of these analysed abnormalities.

Increasingly, chromosomal analyses are being combined with various molecular genetic investigations. These methods will not be outlined here since they would go beyond the scope of the Chromosome Atlas.

Outlook

C. Behrend et al., *Human Chromosome Atlas*, https://doi.org/10.1007/978-3-031-10588-3_20

The information in this atlas of structural chromosome aberrations is depending on methods currently available to analysing the different types of abnormalities. It particularly emphasises the choice of spectrum while combining the different techniques to diagnose any single case, for example, direct preparation of cells and mitoses, short- or long-term cell cultures, FISH, analyses of interphases, microarray, and DNA sequencing. In many cases, family investigations and analyses of the pedigree should be performed additionally.

It is to be expected that new and improved investigation methods will be forthcoming, leading to new fields of investigation and re-analysing of certain groups of patients.

Constitutional aberrations of chromosome structure such as complex rearrangements (CCR) and abnormalities of tumour cells, for example, may be classified as abnormalities originating from *chromothripsis*, enlightening the relevance of the relationship of cytogenetic and molecular genetic techniques.

Of special importance is the increasing number of analyses in the field of *epigenetics*. Functional aberrations caused by *uniparental disomy* (UPD) or by the inactivation of genes caused by changes in neighbourhood (*position effects*) are of increasing relevance especially in prenatal diagnosis.

There is evidence through epigenetic analyses that various somatic tissues of an individual person exhibit differences among them during *imprinting* processes. Further research is needed in order to identify those genes that may activate or deactivate other genes.

Another range of basic research with vital importance for applied cytogenetics in the future is the analysis of the various phases of the *cell cycle*, noticeable entailing specifics and abnormalities and their importance for the potential diagnosis of *aberrations of chromosomes*.

The analysis of the *position of chromosomes* in the nucleus during G0 and G1 phase indicates that the gene function may be altered by the architecture, especially by the arrangement in the interphase nucleus. Future research will have to pursue the question if and in what way the position of the chromosomes in the nucleus is changed in the presence of chromosome aberrations.

After completion of the S-phase of the cell cycle, an *asynchrony* of some chromosomes may be noticed in the various following stages. It is crucial to find out what can cause this asynchrony. Possibly, there are genes that can induce either a positive or a negative selection of normal or aberrant chromosomes.

A *selection* in vivo against pathological karyotypes takes place predominantly during or shortly after the implantation of the conceptus in the early ontogenesis.

This selection is especially noticeable in the presence of monosomies with a frequency of 50% of all the aberrations before implantation of the conceptus and 0% after it.

It is possible, but still not verified, a faulty metabolism comes into play for this selection.

Further has to be resolved if there is an *order in the selection* of normal and monosomal karyotypes depending on the missing chromosome. And the role of monosomy-21 mosaics and their existence beyond the implantation phase could not be stablished so far.

It is known that *exogenic factors* also matter in the formation of aneuploidies and the emergence of structural aberrations of chromosomes during the first mitoses after the formation of the zygote. This is exemplified by the occurrence of *chaotic karyotypes* under certain cultural conditions of the zygote in vitro. These studies were performed, when an ICSI therapy failed. It became apparent that the choice of certain incubation media was responsible for the formation of extremely pathological karyotypes. Further studies of the underlying disturbances in the cell cycle may shed light upon these pathological processes and generate a new approach of applied cytogenetics for the incubation of cells.

The function of *telomeres, centromeres, kinetochores, and the spindle apparatus* for the formation of chromosome aberrations is another unanswered question in cytogenetics.

The reduction of the telomeres with increasing age of the carrier and the parallelly increasing frequency of low-rated chromosome mosaics might be related.

In vitro, as well, the increasing occurrence of aneuploidies, polyploidies, and/or structural aberrations could be verified in extended cultivations of amniotic fluid in ageing cells. Taking into account complementary analyses of the different phases of the cell cycle, further results concerning this problem are to be expected.

Finally, *standards of ethical conduct* are playing an increasing role in the field of applied cytogenetics when analysing inherited chromosome aberrations. It must be clearly defined to whom prenatal findings may be provided (predictive diagnosis), and what analyses may be performed in children and adolescents (not legally able to consent). Legal regulation concerning these topics is rising in various countries, but the requirements are not uniform.

Human geneticists working in the field of applied cytogenetics or in basic research are aware of the importance to develop a network of cooperation between university institutes and private laboratories to enable a regular exchange of experience, to institute regular specific conferences, seminars, and workshops, and to establish joint projects.

Supplementary Information

Appendix – 386

Bibliography – 387

C. Behrend et al., *Human Chromosome Atlas*, https://doi.org/10.1007/978-3-031-10588-3

Appendix

Now in its second edition, this atlas serves as an easy-to-use diagnostic guide for the analysis of the human karyotype.

Split in four parts, it starts with a comprehensive introduction covering the cytogenetic basics, the role of ethic committees, and international quality control in the field of diagnostics. The main parts II and III show the spectrum of different types of chromosomal abnormalities by a combination of karyogram and ideogram. They compare the significance of different banding techniques, give the karyotype formula, and describe morphological peculiarities of each case presented. The final part provides a detailed description of non-coding DNA variants and focuses on potential problems in the detection of aberrations. It also mentions necessary additional investigations and peculiarities to be considered when counselling carriers of a chromosomal aberration or their relatives.

Given its comprehensive scope and practical approach, this atlas is an indispensable resource for researchers, clinicians, and practitioners working in the field of cytogenetics and clinical genetics.

Bibliography[1]

1. Borgaonkar DS (1997) Chromosomal variation in man: a catalog of chromosomal variants and anomalies, 8th edn. Wiley, New York
2. Buechner B (2014) German ethics council on genetic diagnostics: trend setting? Eur J Hum Genet 22:831–832
3. Cremer T et al (2016) Der Zellkern–eine Stadt in der Zelle. Teil 1: Chromosomenterritorien und Chromatindomänen. Biol Unserer Zeit 46:290–299
4. Cremer T et al (2018) Der Zellkern–eine Stadt in der Zelle. Teil 2: Chromatin-Nanoarchitektur und Genregulation. Biol Unserer Zeit 48:45–56
5. Cremer M, Cremer T (2019) Nuclear compartmentalization, dynamics, and function of regulatory DNA sequences. Genes Chromosomes Cancer 58:427–436
6. Cremer T et al (2020) The interchromatin compartment participates in the structural and functional organization of the cell nucleus. Bioessays 42:1–18
7. de Chadarevian S (2020) Heredity under the microscope. Chromosomes and the study of the human genome. University of Chicago Press, Chicago
8. de Grouchy J et al (1984) Clinical atlas of human chromosomes, 2nd edn. Wiley, New York
9. Downing GJ (2009) Key aspects of health system change on the path to personalized medicine. Transl Res 154:272–276
10. Eggermann T, Schwanitz G (2011) Genetics of down syndrome. In: Genetics and etiology of down syndrome. INTECH
11. Eggermann T et al (2015) Fundamental aspects of epigenetic in cancer. In: Mehdipour P (ed) Epigenetics territory and cancer. Springer, Dordrecht
12. Freitag CM et al (2020) Clinical genetic testing and counselling in autism spectrum disorder. Med Genet 32:31–37
13. Fukami M et al (2017) Catastrophic cellular events leading to complex chromosomal rearrangements in the germline. Clin Genet 91:653–660
14. Gardner RJM, Amor DJ (2018) Chromosome abnormalities and genetic counseling, 5th edn. Oxford University Press, Oxford
15. Gesetz über genetische Untersuchungen beim Menschen (Gendiagnostikgesetz Gen DG) (2009) BGBl. I, Bonn and Revised Version 2019
16. Hastings RJ (2012) A common European framework for quality assessment for constitutional, acquired and molecular cytogenetic investigations. Eur CytogenetAssoc Newsl 30:11–19
17. Heidemann S (2020) 10 Jahre Gendiagnostikgesetz–wie kann die Vernichtungspflicht für Ergebnisse genetischer Analysen und Untersuchungen praktisch umgesetzt werden? Med Genet 32:75–78
18. Honnefelder L, Lanzerath D (2008) Bioethik im Kontext von Recht, Moral und Kultur. In: Beiträge zur Ethik in den Biowissenschaften (Bd 2). University Press, Bonn
19. Hunt P (1966) Stigmata. The experience of disability. Geoffrey Chapman, London
20. Kalz L, Schwanitz G (2004) Characterization of constitutive heterochromatin in particular fluorescence polymorphisms in a central European population. Int J Hum Genet 4:1–10

1 References have been selected for publication on basics of Cytogenetics and review articles on techniques which are frequently applied.

Publications on specific problems and on case reports are not cited. They can be ordered, when it seems to be relevant from the authors.

21. Kalz L et al (2005) Polymorphisms of Q-band heterochromatin: qualitative and quantitative analyses of features in 3 ethnic groups (Europeans, Indians, and Turks). Int J Hum Genet 5:153–163
22. Kloosterman WP et al (2012) Constitutional chromothripsis rearrangements involve clustered double-stranded DNA breaks and nonhomologous repair mechanisms. Cell Rep 1:648–655
23. Kloosterman WP, Cuppen E (2013) Chromothripsis in congenital disorders and cancer: similarities and differences. Curr Opin Cell Biol 25:341–348
24. Kunze J, Nippert J (1986) Genetik und Kunst. Angeborene Fehlbildungen in verschiedenen Kulturepochen. Grosse, Berlin
25. McGowan-Jordan J et al (eds) (2020) ISCN: an international system for human cytogenomic nomenclature. Cytogenet Genome Res 160:341–503
26. McLachlan RJ, O'Bryan KO (2010) State of the art for testing of infertile men. J Clin Endocrinol Metab 95:1013–1024
27. Miller OJ et al (2001) Human chromosomes, 4th edn. Springer, Berlin
28. Muss B, Schwanitz G (2007) Characterization of inversions as a type of structural chromosome aberration. Int J Hum Genet 4:1–10
29. Pellestor F et al (2014) Chromothripsis: potential origin in gametogenesis and preimplantation cell divisions. A review. Fertil Steril 102:1785–1796
30. Poot M (2017) Of simple and complex genome rearrangements, chromothripsis, chromoanasynthesis, and chromosome chaos. Mol Syndromol 8:115–117
31. Schallenberg P (2019) Personal dignity in prenatal ethics. Med Genet 31:297–302
32. Schinzel A (2001) Catalogue of unbalanced chromosome aberrations in man, 2nd edn. De Gruyter, Berlin
33. Vogel F (2001) Human genetics: the molecular revolution and its ethical consequences. Int J Hum Genet 1(1):1–9
34. Wegner RD (ed) (1999) Diagnostic Cytogenetics. Springer, Berlin, Heidelberg
35. Weise A, Liehr T (2009) Pre- and postnatal diagnostics and research on peripheral blood, chorion, amniocytes and fibroblasts. In: Liehr T (ed) Fluorescence in situ hybridization (FISH)–application guide. Springer, Berlin, pp 53–60
36. Wyandt HE, Tonk VS (eds) (2017) Atlas of human chromosome heteromorphismes. Kluver Academic Publishers, Dordrecht

MIX
Papier aus verantwortungsvollen Quellen
Paper from responsible sources
FSC® C105338

If you have any concerns about our products,
you can contact us on
ProductSafety@springernature.com

In case Publisher is established outside the EU,
the EU authorized representative is:
Springer Nature Customer Service Center GmbH
Europaplatz 3, 69115 Heidelberg, Germany

Printed by Libri Plureos GmbH
in Hamburg, Germany